The American Environment: Perceptions and Policies

Edited by

J. WREFORD WATSON
Professor of Geography, School of North American Studies
University of Edinburgh

and

TIMOTHY O'RIORDAN
Reader, School of Environmental Sciences
University of East Anglia

JOHN WILEY & SONS

London · New York · Sydney · Toronto

Library of Congress Cataloging in Publication Data:

Watson, James Wreford.
The American environment.

1. United States—Description and travel—Addresses, essays, lectures. 2. Cities and towns—United States—Addresses, essays, lectures. 3. Transportation—United States—Addresses, essays, lectures. 4. Land—United States—Addresses, essays, lectures. I. O'Riordan, Timothy, joint author. II. Title.

E161.5.W37 973.925 74–32224

ISBN 0 471 92221 8 (Cloth)

ISBN 0 471 92222 6 (PbK)

Photosetting by Thomson Press (India) Limited, New Delhi and printed in Great Britain by The Pitman Press, Bath

Contents

III. TRANSPORTATION AND THE AMERICAN ENVIRONMENT

IV. LAND USE: CHANGING PERCEPTIONS AND POLICIES

List of Contributors

DAVID ALLEE	*Professor of Agricultural Economics at the New York State College of Agriculture, Ithaca, New York.*
JOHN DAVIS	*Senior Lecturer in Geography at Birkbeck College, University of London.*
MICHAEL ELIOT HURST	*Associate Professor of Geography at Simon Fraser University, Burnaby, British Columbia.*
RICHARD FRIDAY	*Research Associate in the Department of Agricultural Economics at the New York State College of Agriculture, Ithaca, New York.*
EDWARD HIGBEE	*Professor of Geography at the University of Rhode Island, Kingston, Rhode Island.*
G. MALCOLM LEWIS	*Senior Lecturer in Geography at the University of Sheffield.*
FLORENCE LADD	*Associate Professor in the Graduate School of Design and the Department of Regional Planning at Harvard University, Cambridge, Massachusetts.*
DOUGLAS MCMANIS	*Chairman of the School of Social Studies at Columbia University, New York City.*
TIMOTHY O'RIORDAN	*Reader in the School of Environmental Sciences at the University of East Anglia, Norwich.*
JOHN PATERSON	*Professor of Geography at Leicester University.*
ALFRED RUNTE	*Research Associate in the Department of Environmental Studies at the University of California, Santa Barbara.*
GEORGE STANKEY	*Research Social Scientist at the U.S. Forest Service Intermountain Forest and Range Research Station, Missoula, Montana.*
J. WREFORD WATSON	*Professor of Geography at the University of Edinburgh.*

List of Tables

List of Illustrations

INTRODUCTION
Image and Reality in the American Scene

J. Wreford Watson and Timothy O'Riordan

The purpose of this book is to show how perceptions affect geography: in particular, the geography of America. The views that people have of themselves, and of their country, have a lot to do with how they use (or misuse) their environment. Not all geography derives from the Earth itself; much of it springs from our *idea* of the Earth. There is a geography within the mind to which men may more effectively adjust than to the supposedly real geography of the Earth. Man has the peculiar aptitude of being able to live by notions of reality, which may be more real than the reality itself. Thus mental images are of prime importance to the study of any country.

The first sections of this book look at how settlers perceived America, as they went into it. In a new-found environment it is not what people actually see there so much as what they want to see or think they see that influences their reaction. Actuality exists, of course, but people project what they have heard of it, or what they hope to find in it, and thus see it as something different. As a result a mental image of a country is created—built up of what men have been led to think they would find, or what they want to find, how they set about finding, or how their findings are expressed: and this mental picture then conditions what is, in effect, found. The British and other Europeans who went to America saw it in terms of what they were running away from, and what they were running to, as well as what they actually came across once they were there. This book traces some of their views as they first tested out and settled the Atlantic shore, then pushed up the eastern rivers, next broke out of the Appalachians into the west, again as they climbed up onto the Great Plains, and as finally, cutting across the Rockies and the far western ranges, they came out upon the Pacific coast. Throughout this movement, men discovered what, by European standards, was a comparatively wild world, and their view of nature was to shape what they then did with it. From the first, their colonization sprang from towns or cities. Their views of the city, therefore, greatly influenced the forms their settlement took.

The later sections of this book concern themselves with the way in which images presently affect the use or misuse of the environment. The theme here is centred on landscape change as produced by conflict and conciliation. The opening up of the continent by road, rail and air has led to rapid transformations. Pressures upon America's land and water resources are growing ever more intense as different people expect to use these scarce resources for diverse ends. The forest, the farmland, the metropolis: all must satisfy certain public needs as well as private interests, yet policies which must incorporate this fact are only slowly emerging and show many signs of birth pains. American landscape developments—the rapid spread of urbanization and the rise of the modern metropolis, the need for recreation in the countryside by people cooped up increasingly in the town, and the whole question of the balance between exploitation and conservation in the use of America's agricultural and forest resources—all these were controlled as much by how people thought as by what the land offered. In brief, man shaped his environment by how he saw it (which in many cases can be equated with how he wanted to see it).

The aim of this book, therefore, is to look at the role of collective images, or, if you like, social myths, in the development of the American landscape. The North American scene has been formed out of literally millions of individual acts, each the reflection of individual or collective imaginings.

> 'Look out of the nearest window', advised Nash [1969, p. ix] 'and consider the face of the land. What you see is a human creation ... Its condition, rightly seen, reveals a society's culture and traditions as directly as does a novel or a newspaper or a fourth of July oration, because today's environment, the "natural" part included, is synthetic.'

To leave an area as wilderness is as much an act of cultural preference as to bulldoze a hillside for a superhighway.

In pre-European America, myths concerning nature incorporated sanctions on both individual and collective actions to ensure the long-term stability of the landscape and its life-supporting resource base. Eliade (1968) has shown that such myths were not only very 'real' to pre-industrial peoples, but that they provided an explanation of why certain kinds of human behaviour had to be enforced. The enactment of the myth was both a 'religious' experience in the sense of binding the tribe to rigorously enforced common modes of action, and a convenient mechanism for maintaining the tribe's imagined sense of control over the vagaries of nature and, ultimately, its own destiny. While the modern American can separate without effort the aesthetic from the functional component of the landscape (to an extent which permits him to live in an elaborately timbered home on the one hand yet fight for wilderness preservation on the other), to the native North American peoples the beauty, spirit, usefulness and essence of things were inextricably intertwined.

For example, every significant act of the Kwakiutl Indians of British Columbia had an original ritual and a prayer connected with it. The motives behind the prayers were not always clearly defined, but to tamper with nature was recognized as a transgression of the unity of life and order of things, therefore a plea for forgiveness was necessary to ensure the promise of continued good health. The woman who cut the roots of the young cedar to make a basket sought solace from the tree:

> 'I pray, friend, not to feel angry on account of what I am going to do to you, and I beg you, friend, to tell your friends about what I ask of you . . . keep sickness away from me so that I may not be killed by sickness or in war'. (McLuhan, 1972, p. 40)

The tree is killed but the conscience is salved.

The Eskimo about to spear a polar bear for food and clothing must play out the psychological anguish of killing not simply a bear but the spirit which that bear embodies. To quote a member of the Iglulik tribe:

> 'The greatest peril of life lies in the fact that human food consists entirely of souls. All creatures that we have to kill and eat, all those that we have to strike down and destroy to make food for ourselves, have souls that do not perish with the body, and which must therefore be propitiated lest they should average themselves on us for taking away their bodies.' (Quoted in Tuan, 1971, p. 32)

While the anguish serves constantly to remind the hunter of his dependence upon the continued existence of the polar bear, real-world behaviour may not always be so controlled by such images. The urge to control the earth and to master the forces that shape his existence seems firmly embedded in man: there is evidence that when environmental conditions favoured exploitation, misuse of the earth did occur. For example both Day (1953) and Rostlund (1957) produce evidence that pre-European civilizations were remarkably destructive of their habitat. Ideas, no matter how sensitively expressed, are not always converted into action.

But what of the imagined world of the modern North American? What collective images do North Americans hold of themselves that would have bearings upon the perception of the North American landscape? We have decided to identify four major images which we feel are still influential in determining American thought and action toward the landscape and which pervade the analysis made by the authors of this book. These four are (i) a belief in progress, (ii) the frontier, (iii) the middle ground and (iv) political and social democracy. Lesser images are also important, but will be dealt with in the individual topics, as for example, in the American's view of nature, or of the city.

The Image of Progress

From the very beginning, Americans believed that they were destined to do something new and something better. 'When England began to decline . . .' wrote Edward Johnson in 1654, 'Christ creates a *New England* . . . where He will create a new Common-wealth altogether'. Cotton Mather wrote of 'the Wonders of the Christian religion flying from the Depravations of Europe to the *American Strand*'. There was a divine intent behind the birth of America. The Lord had 'carried some Thousands of Reformers into the Retirements of an American Desert on purpose that . . . He might there, to them first, and then By them, give a Specimen of many Good Things'.

America still believes in these 'good things', now translated into the 'good life', aspired to, indeed regarded as their birthright, by so many. Americans are, in David Potter's words, the 'people of plenty' (Potter, 1954), and for generations that plenty has seemed unending. The American character has always been centred in optimism. The 'New World', symbolizing the rebirth of man, has provided an illusion of hope to the hopeless, freedom to the oppressed, abundance to the starving and, perhaps above all else, the image of a glowing future filled with bountiful promise. The majority of American immigrants were drawn to the new-found land to see this hope fulfilled in their own lifetimes. Many had been persecuted, dispossessed, starved and generally ill-treated in their homelands: emigration held the promise at least of life and at best of prosperity. For most Americans today the future holds neither terror nor fear; its challenge merely inspires further confidence in technocracy and a faith in managerial control.

At an intellectual level optimism is a natural outcome of faith in democratic institutions and an abiding belief in the viability of the American people. Paul Goodman (1966), a noted exponent of the virtues of American culture, feels that the challenges facing the nation today should provide it with a fresh burst of confidence in itself. This image of confidence is most clearly demonstrated by national leaders. Thus it is possible for the President, in his 1973 *State of the Union Report to Congress on Natural Resources and the Environment*, to note that:

> 'there is encouraging evidence that the United States has moved away from the environmental crisis that could have been, toward a new era of restoration and renewal. Today, in 1973, I can report to the Congress that we are well on the way to winning the war against environmental degradation—well on the way toward making our peace with nature'. (Nixon, 1973)

This is precisely what the Congress and the American people wanted to hear: but the crisis was not over. In the latter part of the same year the President had to appeal to the people for restraint in the use of fuel because of 'the energy crisis'. While the factors which contributed to the fuel shortage are complex and controversial, certainly the successful efforts of environmentalists to delay the Trans-Alaska pipeline and block new oil refineries were contributing causes. In the reality of energy scarcity and possible economic recession environmental protection took second place as, in early 1974, the President relaxed clean air standards and dropped some important provisions in hard-won environmental legislation. So optimism had gone ahead of the 1972 Annual Report of the President's Council on Environmental Quality (1972), which had warned (p. 348):

'How little we still know about many aspects of the environment, including its basic physical aspects . . . we need to understand better the environmental impacts which stem from the basic production systems in our society'.

Nowhere is there in reality any demonstrable sign for optimism, but everywhere in the American imagination there is hope.

The Image of the Frontier

Somewhat linked to the image of progress, the image of the frontier has always held a peculiar fascination for North Americans. Beginning with the famous thesis of the American historian Frederick Jackson Turner, the 'frontier' has long been associated with the issue of Americanism.

'American social development' wrote Turner [1920, p. 4], 'has been continually beginning over again on the frontier. This perennial rebirth, this fluidity of American life, this expansion westward with its new opportunities, its continuous touch with the simplicity of American society, furnish the forces dominating American character'.

To Turner and his school, the frontier (which was symbolized by the west), constantly encouraged self-reliance and rugged individualism, a dislike of government controls and a healthy egalitarianism.

While many dispute the validity of the Turner thesis (Zelinsky, 1973, pp. 23–24), the image of the frontier still holds a deep fascination for modern Americans. In an editorial commenting on the rush for jobs in Alaska following the decision to construct the Trans-Alaska oil pipeline (as discussed in Chapter 19), the *Christian Science Monitor* (12 February, 1974, p. 12) comments:

'In these days when there is so much complaint about people not wanting to work, when those with jobs often find them constraining, when families see their financial horizons dimming because of inflation and shortages—the thought of a frontier, of new jobs and a new setting and a will in people to strike out anew, holds distinct appeal . . . we're still glad for evidence that new frontiersmen remain among us, Americans who feel that risk is no excuse for not striking out for better things.'

This quotation illustrates one of the psychological notions associated with the frontier image: the opportunity for self-advancement. To quote Burch (1972):

'If individual freedom was curtailed, one was renewed by "knowing" that it prevailed in the West. If excessive class differences seemed most pronounced, one was renewed by knowing that "true" democratic equality prevailed in the West. It was not so much the objective history of the westward expansion (after all, most societies have had an untamed frontier) as its mythological import that has made it such a significant and distinctive feature of American society'.

Psychologically, too, the frontier meant challenge, the clash of elemental forces of good and evil, of man against nature, and of course it carried with it a romanticism of magnificent vistas and the promise of potentially unlimited individual wealth.

In the 'real' world of the frontier, survival became of paramount importance. Fortitude, hardiness and physical suffering were the ingredients of the real frontier: to 'tame' man's surroundings, to guarantee a livelihood and to enjoy some sort of physical comfort became primary objectives. To frontier people life was no idealized romantic notion of character building or egalitarianism, but a ceaseless struggle and a constant peril.

However, there was one peril the myth did not combat, and that was the peril of the myth itself. The frontier notion did little for the preservation of the American environment: indeed it was on the frontier that the battle for preservation was first lost. Boulding (1966) writes of the 'cowboy economy' which was developed at the frontier: an open economy of exploitation and reckless consumption, antithetical to continued existence in a finite habitat but enormously successful if resources were imagined as abundant. Unfortunately, the cowboy economy helped to perpetuate the illusion that the true American spirit was 'out there', when in fact the real problems were right at home.

The Image of the Middle Landscape

The 'middle landscape' refers to the domesticated wilderness. The American settler has always sought to subdue the wilderness into some sort of ordered tranquillity. Carroll (1969) notes how Michael Wigglesworth assured his fellow Puritans that, as a result of their efforts, the Lord had turned 'a living wilderness . . . into a fruitful paradise'. Intellectually, the agrarian way of life was presented as ideal for man as a whole, an ideal to which the Americans, as Jefferson pointed out, had the best chance of approximating. Edwin Markham puts it in his poem *Earth is Enough*, 'We men of Earth have here the stuff of Paradise—we have enough!' The American rural yeoman became an idealized poetic figure who, according to Smith (1961, p. 138),

> 'defined the promise of American life. The master symbol of the garden embraced a cluster of metaphors expressing fecundity, growth, increase and blissful labour in the earth, all centering about the heroic figure of the idealised frontier farmer armed with that supreme agrarian weapon, the sacred plough'.

The middle landscape of today is the domesticated landscape that is neither wilderness nor city: it is the landscape of the 'middle American; of homogeneity, patriotism and virtue, of men who are Godfearing, law-abiding and hardworking, the 'silent majority' who are appealed to by every politician. Intellectually, the middle landscape provides an idealized image for the anti-city writers who deplore the rootlessness and social fragmentation of American urban life and who praise the virtues of classlessness and simplicity of the agrarian way. It provides a pleasant image of benign social equality which serves to blur the reality of class division and ethnic strife taking place in the cities and the rapid but agonizing acculturalization of native peoples that proceeded at the frontier.

Psychologically, the middle landscape symbolizes the desire for private property and for security, especially of economic investment. Private landholding is an American ideal, the essence of territoriality on a mass scale. Ownership confers the opportunity for transformation and reconstruction of the landscape to make the image of property real and very personal. The desire for privacy and seclusion is so strong that whole communities are now protected by electric fences, security guards and watch dogs. 'Private Property: No Trespassing' notices are posted throughout the American middle landscape, to keep unwelcome visitors out and to preserve cultural identity.

In real life the American middle landscape is not classless, it is entrenched white middle class. As a result, life there has a sameness about it, full of *similar* people who have *similar* values and who wish to do *similar* things. Yet it is still regarded in the imagination as ideal. Lowenthal (1972) reports how first-year female students at Harvard visualized the curving gentle landscape of middle class suburbia as idyllic and by far the most preferred landscape of a choice of six urban and rural settings.

In political terms, the middle landscape became entrenched in the Homestead Act. As is well known, this Act promoted an eastern image of agricultural settlement on a more westerly landscape that was ill-suited for such densities and intensities of agricultural practice. The results were, of course, catastrophic, but the images of the landholder and the domestication of the landscape were obviously tenaciously held since subsequent policy simply enlarged the homestead allotments. Today, the middle landscape is symbolized politically by the suburb administered as a separate municipality, and by the rising second-home market in country districts: both protected by carefully designed zoning codes.

What of the landscape of the future? Two utopian notions spring to mind: the experimental city and the commune. Both centre on the fusion of the frontier and middle landscape images, classlessness and equality yet self-reliance, but in many other respects they symbolize quite contradictory relationships between man and nature. The experimental city proclaims the triumph of technology in the service of social harmony. It is pure order without nature, the totally synthetic city where nature enters only on man's terms. It is free of air and water pollution and escapes the problems of housing and racial inequality. On the other hand, the commune is small and rural, offering a simple way of life that is utterly interwoven with nature. The sterile functional specialization of the experimental city is replaced with a myriad of roles, each demanding co-operation and friendly social interaction.

The commune is not a recent notion although it has become more popular of late. Its roots lie in the very settlement of the north-eastern seaboard and its intellectual expression was well developed by the transcendentalist philosophies of the mid-19th century (Whicker, 1949). The modern commune is an attempt by groups of people to get out of the value systems of middle America (frenetic time schedules, materialism and individualism) to seek a return to the agrarian ideal of a simple life in natural surroundings predicated upon hard work, simple pleasures and self-sustaining companionship. To many environmentalists it is idealized Utopia—'a visionary commonwealth' as Rosak (1972, p. 394) put it, 'shot through with mystical sensibility that determines family patterns, diet, architecture, work ethic, group ritual, relations with land, wildlife, neighbours.' Rosak admits that these experiments may fail, as have so many of their predecessors. 'But I also know it is more humanely beautiful to risk failure than to resign to the futurelessness of the wasteland,' he comments, 'for the springs are there to be found'.

The Image of a Democratic People

America has always been held up as the home of a real, social, popular democracy as well as of a political one. At the intellectual level this was probably most clearly expressed in the famous writings of the French traveller Alexis de Toqueville, who toured America in the 1830s. De Toqueville was by no means in favour of American democratic institutions, his views being coloured by his European aristocratic background, but he did note how democratic ideals gave Americans a sense of rising expectations among the ordinary people and a desire for 'consensus solutions'.

> 'Democracy does not confer the most skillful kind of government upon the people,' he warned [1961, p. 293] 'but it produces that which the most skillful governments are frequently unable to awaken, namely an all pervading and restless activity, a superabundant force, and an energy which is inseparable from it, and which may, under favourable circumstances, beget the most amazing benefits'.

Psychologically, the concept of democracy provides an image of personal efficacy in handling the affairs of the community. The right to enter the political arena, the opportunity to read a free press, and the privilege of holding public office, are presented to all Americans as natural rights. The psychological image of democracy carries with it connotations of equality and equal representation, connotations that are enshrined within the American constitution.

But in real life it is the equal right to compete for resources, rather than an equal sharing of those resources, which has typified the American system. The emphasis on freedom has meant being free with each other to an extent hardly realized by Americans themselves. Freedom, as the hallmark of American democracy, has made men free to exploit the land as perhaps nowhere else. This has stamped itself upon the whole landscape, which is a landscape of individual enterprise and development. In this struggle for the use of the land, political freedom has meant the *effective* vote, which in many cases has meant the vote of power blocs, or power machines.

The quest for power has led to the exploitation of different sub-cultures in America from the earliest days of conflict between white and Indian, to the present. The conflict was inevitable, given the tremendous impact of rapidly expanding American colonization on indigenous cultures that had hardly changed over half-a-dozen centuries. The temptation on the part of colonizing groups to scorn and despise an alien culture is difficult to resist, so it is hardly surprising that derisive images of native Indians and introduced Africans were widely held. Pearse (1965) notes how, in the imagined world of the European settler, the Indian warrior was symbolized as a surviving relic of a savage age, an age which the 'civilized' settler had left with his European ancestry. Therefore to destroy and dispossess the Indian was a necessary expedient to protect immigrant society from possible debasement. Some have even offered a Freudian interpretation of this culture clash. According to Loewenberg (1970) white settlers were projecting their repressed desires for freedom from restrictive social conventions into the Indian and later the Negro.

Whatever the analysis, culture conflict in America has always been widespread and a source of much distress. The Indian and the immigrant Negro have become objects of hate and distrust regardless of their real individual and social characteristics. Thus it has always been possible to exploit such peoples without either stirring a social conscience to any great extent or destroying the image of democracy. Gans (1972) developed this theme by propounding a thesis that poverty in America has persisted because it is convenient to various social groups to maintain it. The existence of a poor stratum in society, he maintains, helps to subsidize the economy by providing a pool of low-cost, unskilled labour to do the dirty jobs, and by consuming low-cost, second-grade commodities which otherwise would go to waste. But the poor also have psychological functions by presenting the élite with a sub-culture both to pity and to despise, and by giving liberal politicians and the intellectuals alike a convenient focus for criticism of the Establishment of which they are a part.

In political terms, democracy certainly exists, but primarily for those who are in a position to make use of it. Citizen actions, and reforms of the courts, provide avenues essentially for the middle class who have the time and the ability to utilize the connections provided. Even the rise of a new breed of advocacy-planners who are providing their expertise and political connections at low cost to needy groups has not proved a panacea. All too often the advocate uses his altruism as a convenient mechanism to push through his own view of desirable social reform at the expense of the very group he is purporting to serve (Peattie, 1968). We live in the days of the 'evangelical bureaucrat' whose well-intentioned efforts fail to recognize the real injustices of the current social system. The growth of 'citizen participation' in American public policy-making is a reflection of a basic mistrust in democratically elected governments to cater to the day-to-day needs of the people. The future of this movement should test the democratic myth quite thoroughly.

Image, Reality and Public Policy

Idealized behaviour can exist separately from actual practice. While myths serve as guidelines for social action, they also serve to blur the discrepancies between what is imagined and what is experienced. It is easy to pay lip service to conventional social rhetoric, indeed even to add to that rhetoric, in the hope that social problems will be solved, while in actuality they linger on and worsen. A tension develops between what one would like to see and what one experiences in day-to-day living. Opinion polls regularly report expressions of concern by upper-income groups over environmental deterioration and a stated willingness to pay to clean up the mess. But it is precisely among this group that conspicuous consumption, leading to environmental damage, is most prevalent: large single-family homes with big energy demands, expensively serviced by urban utilities, two cars, possibly a second home and a demand for an escape to solitude in vast tracts of wilderness. What J. K. Wright (1947) called 'promotional imagining', illusory and deceptive conceptions conforming to what one would like to see rather than necessarily to the truth, constantly inhibits the recognition of the duality between the real and the imagined, and may even encourage certain patterns of behaviour which could result in a worsening of the real problems: for example, when decisions are made to buy a second home as an economic security against inflation, thereby increasing land values, which encourages more decisions to purchase second homes, and so on.

The problem facing community leaders and politicians is to guide some of these images into new directions where conflicts between current practice and desired social outcomes are apparent. This is a difficult task as these images are cherished regardless of the denial of reality. Probably the most fundamental conflict will come over the issue of growth and distribution of both quantity and quality for all Americans (Ridker, 1973). For the American belief in progress has enormously stimulated development, and carried the benefits of growth to an ever wider range of people. As Dollard has said, it is the American tendency to 'middle-classify itself', because through the buoyant effects of its optimism it is, in fact, raising an ever greater proportion of people to the middle class, in the creation of an affluent society. But there are now serious doubts that America has the resources to sustain this growth and maintain a quality environment (Olson, 1973). Therefore the question now is: can it adjust to the need for controlled growth, in which it seeks to relate demand to supply, and maintain the delicate balance between what it hopes from its resources and what those resources will bear? This is the great American dilemma today, to which American illusions about reality will continue to be a major contributor.

References

Boulding, K. E., 1966. The economics of the coming spaceship earth. In *Environmental Quality in a Growing Economy:* (Ed. H. Jarrett). Baltimore: Johns Hopkins Press, pp. 3–14.

Burch, W. R., Jr. 1972. *Daydreams and Nightmares. A Sociological Essay on the American Environment*, New York: Harper and Row.

Carroll, P. N., 1969. *Puritanism and the Wilderness—The Intellectual Significance of the New England Frontier*, New York: Columbia University Press.

Council on Environmental Quality, 1972. *Environmental Quality. The Third Annual Report of the Council on Environmental Quality*. Washington, D.C.: Government Printing Office.

Day, G. M., 1953. The Indian as an ecological factor in the northeastern forest. *Ecology*, **32,** 329–346.

Eliade, M., 1968. *Myth and Reality*, New York: Harper & Row.

Gans, H., 1972. The positive functions of poverty. *American Journa' of Sociology*, **78,** 275–289.

Goodman, P., 1966. *The Moral Ambiguity of America*, Toronto: Canadian Broadcasting Corporation.

Loewenberg, P., 1970. The psychology of racism. In *The Great Fear, Race in the Mind of America* (Eds. Nash, G. B. and Weiss, K.). New York: Holt, Reinhart and Winston.

Lowenthal, D. (Ed.), 1972. *Environmental Assessment: A Comparative Study of Four Cities*, New York: American Geographical Society.

McLuhan, T. C. (Ed.), 1972. *Touch the Earth: A Self Portrait of Indian Existence*, New York: Pocket Books.

Nash, R. (Ed.), 1969. *The American Environment: Readings in the History of Conservation*, Reading, Mass.: Addison-Wesley Publishing Co.

Nixon, R. M., 1973. State of the Union Report to Congress on Natural Resources and the Environment. Washington, D.C.: Government Printing Office.

Olson, M. (Ed.), 1973. The non-growth society. *Daedalus*, **102,** 1–223.

Pearse, R. H., 1965. *The Savages of America, A Study of the Indian and the Idea of Civilization*, Baltimore: Johns Hopkins Press.

Peattie, L. R., 1968. Reflections on advocacy planning. *Journal of the American Institute of Planners*, **34,** 80–88.

Potter, D. M., 1954. *People of Plenty: Economic Abundance and the American Character*, Berkeley: University of California Press.

Ridker, R. G., 1973. To grow or not to grow: that's not the relevant question. *Science*, **182,** 1315–1318.

Rosak, T., 1972. *Where the Wasteland Ends: Politics and Transcendence in a Post-industrial Society*, Garden City, New York: Doubleday.

Rostlund, E., 1957. The myth of a natural prairie belt in Alabama: an interpretation of historical records. *Annals of the Association of American Geographers*, **47,** 392–411.

Smith, H. N., 1961. *The Virgin Land: The American West as Symbol and Myth*, New York: Random House.

de Tocqueville, A., 1961 edn. *Democracy in America* (translated by H. Reeve), New York: Schocken Books, 2 volumes.

Tuan, Y. F., 1971. *Man and Nature*. Washington, D.C.: Association of American Geographers, Resource Paper No. 10.

Turner, F. J., 1920, *The Frontier in American History*, New York: Holt and Co.

Whicker, G. F., 1949. *The Transcendentalist Revolt Against Materialism*, Boston: Heath.

Wright, J. K., 1947. Terrae incognitae: the place of imagination in geography. *Annals of the Association of American Geographers*, **37,** 1–15.

Zelinsky, W., 1973. *The Cultural Geography of the United States*, Englewood Cliffs: Prentice Hall.

Section I. Perceptions of the Unfolding Environment

1
Image Regions

J. Wreford Watson

Although Americans have certain images that are held in common, they also have regional characteristics that set off one group from another. Settlers coming into America moved forward along a very broad front, from Nova Scotia to Florida; they came from many lands in Europe and were drawn from many classes and occupations. Not unnaturally, they had very different views of their bit of America and their part in developing it. Even if an area had its own intrinsic qualities, these could be brought out in various ways according to different perceptions. As Bowman (1934, p. 113) remarked:

> '... the natural environment is always a different thing to different groups. Its potentialities are absolute, but their realization is a relative matter, relative to what the particular man wants and what he can get with the instruments of power and the ideas at his command and the standards of living he demands or strives to attain.'

Note the emphasis on *values* here, as well as objects. Geography is made from subjective thinking, however objective the results may be. The geographer must be concerned with the standards of living men *demand*, with the things they *want*, and with the *ideas* at their command. It is these immaterial things which, as they make a material impact on the landscape, make the geography of a country.

Mens' first notions about America, in the late 15th and early 16th centuries, were to regard it as a passageway to Zipangu or Cathay; thus the through-gates, the great bights, bays, and estuaries, were the important things (Watson, 1969). From these early ideas of the American coast, European powers had to guide their colonization of the New World. The fortuitous landing by Columbus in the Caribbean led to the Spanish exploitation of that sea and of the Gulf of Mexico. The Spaniards then moved north to what became the United States in an extension of their probes about the Gulf. The legend of the cities of Cibola carried them well into what is today the American south-west. The myth of a passage to Cathay likewise led to a deep penetration of the continent by the French, beyond Hochelaga (Montreal) to the Great Lakes and down the Mississippi. Britain was left with the region in between. Here, there seemed to be a break at Cape Hatteras, but the so-called isthmus turned out be no more than a broad and many-fingered estuary. Indeed, nothing but limited sounds and bays were found. This meant that the British could not go far into the continent, a fact which may have made them try to develop the seaboard for its own sake. Thus they evolved comparatively compact colonies which, though they did not embrace the vast territories of the Spanish and of the French, made, for that reason, a much more concentrated impact. In terms of population and production, the Thirteen Colonies came to compare favourably with their rivals.

Once settled, the new immigrants' thoughts turned to the wealth that might be found in the Americas themselves. The search for El Dorado or the Lost Cities of Cibola soon became a search simply for gold and silver mines. By the mid-17th century the search had widened to include sources of iron and copper, and areas where cotton and sugar could be raised. There was no gold in the British sector, but the Southern colonies offered the wealth of a cotton-growing climate. Commerical plantations sprang up. Meantime, in England, social and religious ferment led to colonization by conscience and the rise of communities committed to New World as a new and final home. The New England and mid-Atlantic colonies answered this need.

After the Revolution many factors acted in opening up the West, but none more strongly than the desire for land, especially as the land-hungry of Europe poured in. But the development of minerals, the building of roads, canals and railroads, the rise of trade and industry, and the growth of cities (all mingled with a sense of manifest destiny in opening up the continent), played their individual parts and had a cumulative effect.

These themes will be developed further in this collection of essays; but it might be as well to introduce them here with a general survey of each of the major regions concerned. Sectional differences arose early in colonial America and became stronger with time, culminating in the Republic with the bitter civil war between North and South. Though they are less affective today, as the mass media try to convey to Americans the image of America as a whole, these differences are nonetheless real; they still live in men's minds. As such they live in the American scene, especially as local loyalties are revived to offset common and conformist tendencies.

New England

Regions are formed by group minds sectionalizing the land. The New England mind grew out of 'the expectation of great things' in the belief, as William Stoughton put it, 'that the Lord hath taken us to be a peculiar portion to himself'. From the outset, then, there was a sense of destiny, with the hope of advance.

What people came from was as important as what they came to. Tension was rife in England. King James I threatened to 'make the Puritans conform or harry them out of the land'. Those who would not conform had to fly. This they did, turning flight into a march to freedom, which has characterized New England since. 'As the Lord's free people, they joyned them selves in ... felowship to walke in all his wayes, made known, or to be made known unto them.' They turned into the future. Here, in the words of the great nonconformist Helwys, they hoped for 'a liberty for all men, at all times, and in all places'. This was the driving force that, as Trevelyan put it, led 'many thousand refugees of all classes to abandon good prospects and loved homes in England, to camp out between the shore of a lonely ocean and forests swarming with savage tribes.'

The land was ready for them to make it the basis of a quite new society, formed by free and mutual compact 'on just and equal laws for the general good'. This was to be the rock of American progress. Free people would freely choose to compact with each other for their mutual benefit. Much was made of this dependence of people one upon another. Governor Winthrop argued that families should 'confederate together and ... bind themselves to support each other'. Yet society, agreed on by all, was for the advancement of each, so that every man could, in Bradford's words, 'work for his own particular, and ... in that regard ... trust to themselves'.

There grew up a duality, or perhaps it would be better to say a compromise, between individual interest and group responsibility that was to colour the whole geography of New England. The prevailing landscape was that of the 'compact', i.e. of land arranged into townships, centred in towns governed by town-meetings, that of group movement into the land, the gathering of settlers in group settlements, and of the group organization of duties. Amongst *New England's First Fruits* was the establishment of Harvard College, at Cambridge, Mass., 1636, when 'it pleased God to stir up the heart of one Mr. Harvard to give one half of his estate (it being about £1,700) towards the erecting of the college. After him another gave £300, others after then cast in more ...' The landscape became full of concrete evidences of the stirred heart. Many town services, such as the public library or the fire brigade, were started and run by voluntary co-operative effort.

On the other hand, a sturdy and growing individualism challenged communal control, and led to the rise of separate and often isolated farms, of individual businesses, and of cities stamped with the individual search after status. Though people frequently settled the land in blocks they took up individual lots. As Wright (1969, p. 59) remarks:

> 'Land was easy to come by, and if a settler did not like the particular spot he had chosen, he could go somewhere else. By the labour of his own hands he could hope to attain at least moderate comforts and some degree of prosperity. He had independence and freedom and felt that he was master of his own fate.'

There was an immense feeling of optimism abroad, driven on by a stirring sense of destiny, and the opportunity that was calling. Moving freely into New England, men had the freedom to move on. As a result, the New Englander has spread right across the continent, carrying with him his passion for freedom. It is as well to remember that among the first New Englanders were the first Independents.

This spirit of independence marked their whole landscape. It led, amongst other things, to a sense of opportunism; if a person failed to get on in one job, or even in one section of a concern, he moved to another. Winthrop early complained of the way people had an eye to their own advantage; independent ambitions often threatened communal ideals.

> 'How people can break from [their compact to support each other] without consent is hard to find. Ask thy conscience, if thou wouldst have plucked up thy stakes [in England] and brought thy family 3000 miles, if thou hadst expected that all, or most, would have forsaken thee there.'

Yet rugged individualism frequently struck out to start new farms, new businesses and new settlements, rather than to intensify and refine existing ones. The eyes of freedom, looking out upon a land that was relatively free to take, made free of tradition: Americans saw new opportunities and new ways of doing things. As McManis says, even though the early reputation of New England suggested an 'inhospitable and uninhabitable' coast, yet the vision of men motivated by the thirst for freedom, 'saw the potential to sustain a new England, not just an economic outpost of empire.' Here was a land at last where new things could be done. True, people may have had 'little solace or content in respect of any outward object', yet they had tremendous inward resources, and their rugged spirit proved what could be made out of even a rugged land.

The South

In the South, as in New England, ideals, developed over time, built up an image about the region which in turn shaped its geography. The illusion that the climate was hard on white men was frequently used to explain, if not justify, slavery. Even a person as perspicacious as Jefferson was guilty of blaming the physical rather than the mental climate for the use of slaves—although this very fact was itself part of the mental climate. 'In a warm climate,' he wrote, 'no man will labor for himself who can make another labor for him. This is so true that, of the proprietors of slaves, a very small proportion indeed are ever seen to labor.' A few years later, Senator Hammond made his notorious 'mud-sill speech', in which, claiming that all social systems had their menial class, the mud-sill of society, basic to the build-up of 'progress, civilization, and refinement', he blessed the advent of a race, in the Negroes, 'eminently qualified [to serve the South], in its capacity to stand the climate'. This represented long-standing views throughout the South. Yet the deleterious effects of climate were not really that important: many non-slave-owing whites worked their fields in the heat of the Southern sun. It was the mental belief that climate *was* important which mattered.

This belief in turn led to the prime factor that distinguished the early South: the use of slavery. When the estate owners could not get enough indentured labourers from England (and the supply was often short: who wanted to give up being a virtual serf in the Old World to become practically a bondsman in the New?) the gentry turned to the use of African slaves. Thus the South quickly became marked by slave and by free communities, unlike the North, which early abolished slavery and relied on family or hired labour. Slavery entrenched the manorial system in the South, and increased the power of the élite. Even if the slave-owners were in the minority, they were the minority that made the decision, and their decisions affected the whole landscape.

The non-slave owning sector still supported the system because it gave even the poorest white his sense of superiority. Comparing the poor white in the North and in the South, Debow (1860, p. 169), a famous advocate of slavery, said: 'The poor white laborer in the North is at the bottom of the social ladder, whereas his brother here [in the South] has ascended several steps and can look down on those beneath him [the black slaves] at an infinite remove. Thus Southern whites, both rich and poor, added caste to the class differences prevalent in the North, in their collective fear and hatred of the mounting tide of blacks. The separation of black from white, brought about partly by custom and partly by the infamous 'Jim-Crow' laws, also made a profound impact upon the landscape, separating the South from the other regions of the United States– at least until the rise of the black ghettoes in Northern and Western cities. Thus the Southern outlook wrote its own geography in the land.

The patent signs of segregation put off many immigrants who preferred to settle in the North and West. Consequently, the South did not feel the invigorating injection of North European, Slavic and Italian migrations that added such zest and interest to the rest of the country. Southerners even became proud of this.

> 'The South', [wrote Debow (1860)] 'has maintained a more homogeneous [white] population and shows a less admixture of races than the North. Our people, mainly the descendants of those who fought the battles of the Revolution, partake of the true American character. . . .'

Mental images helped to make the South different from the North. Cash, (1941, p. vii) brings this out in his study of *The Mind of the South* where he says

> 'There have arisen people to tell us that the South . . . is distinguishable from New England or the Middle West only by such matters as the greater heat. . . . Nobody, however, has ever taken them seriously. And rightly!'

The South is not a condition of land but a state of mind. It is a region whose character and uniqueness rest in its outlook, in the character of its aims and the uniqueness of its perceptions.

The South started off with views of the New World that were significantly different from those of the North. The decision-makers in the South were conformists seeking an extension of established ways, not dissenters preaching a new light. The South was never out, like the North, for 'a new course of living', to quote Bradford. Power was centred in the existing (manorial) systems. The South did not claim, like the North, to be 'the place where the Lord will create a new Heaven'. It contented itself with the Church of England! The South did not want, like the North, 'to be a city upon a hill, the eyes of all people upon us'. Contrary to Winthrop's ideal, which was to concentrate settlement in towns, where it would be possible to see that 'the care of the publique must oversway all private respects', the South stressed the land, parcelled out among individual owners. As Captain John Smith said at the outset, 'here every man may be master of his own labor and land'.

If anything, the South was against cities. Its ideal was the soil. One of its greatest sons, Thomas Jefferson, wrote: 'We have no towns of any consequence'. He said this with a certain satisfaction, if not pride, because he disliked industry and the urban way of life. He was the arch proponent of the middle way between the crude living of the trappers in the forest, and the hectic life of the factory-hand in the city. 'Those who labour in the earth are the chosen people of God,' he claimed.

> 'The proportion which the other classes of citizens [i.e. the non-agricultural population] bears in any State to that of its husbandmen, is the proportion of its unsound [urban] to its healthy [rural] parts, and is a good enough barometer whereby to measure its degree of corruption. While we have land to labour then, let us never wish to see our citizens occupied at a work-bench or twirling a distaff!'

John Taylor, of Caroline County, Virginia, went even further. For him, agriculture ensures that 'almost every moral virtue is amply remunerated in this world, whilst it is also the best surety for attaining the blessings of the next'. This agrarian mentality did a lot more than the climate to shape the development of the South, and to make it a firm part of what has been called middle America.

Early views have come down to us from the literature of the region. The mind of the South was made by settlers who, in Beverley's words, 'peopled the Colony . . . to increase their Estates'; or who came, according to Byrd, 'as Adventurers out to make a very profitable Voyage'; or simply on 'the Humor to take a trip to America. This Modish Frenzy being inflamed by the Charming Account given of Virginia made many fond of removeing to such a Paradise.' The Southern mind was also shaped by 'the patronage of certain of the nobility, gentry, and merchants' who, as Captain John Smith pointed out, backed the first developers in their search for wealth. The desire to make a fortune out of the New World spurred many a young cadet from an illustrious Old World family. Soon, an American gentry appeared, and the landscape was marked by manor houses, servants' quarters and tenant farms. As gentry, they did little or no manual work, but hired indentured labour from England:

men and women who agreed to work on estates for a number of years in return for a free passage and a home in the New World. Some of these were city people and were really unsuited as farm workers. Others, who had been estate workers, wanted a farm of their own. In any event few of them remained with their new landlords. As soon as they could, they drifted into towns or broke away from the coastal plantations, went inland, and cleared farms for themselves, where they could live a very independent if limited existence. They were helped in this by the Southern land system which allowed individuals to get warrants for as much land as they wanted where they wanted it. This led to a very irregular lay-out of land holdings, in comparison with the North.

In all these various ways the Southern scene became impregnated with the Southern mind. The South is still dominated by Anglo-Saxon institutions and ways. It still relies to a considerable extent on primary production. It still has its large estates, its tenant farms, its Negroes in a relatively subservient position and its poor Whites. It still remains one of the least citified areas of the United States. Although industrialization and urbanization are now making enormous strides, the relics of traditional characteristics still make the South distinctive. The differences between the South and the North are not due to a summer-rich or winter-straitened land, or unglaciated and ice-scoured rock, to red or grey soils, or anything else geographers normally look on as the 'environment'. They result from the images men had of themselves and the way they expressed those images in the landscape.

The West

The people of the eastern seaboard were not willing to stay cooped up behind the Appalachians and to leave the French and the Spanish with the interior of the continent. The stories from the interior coming through to the coast were of an area much richer and easier to develop than the seaboard. The French war broke the encirclement of the Thirteen Colonies, and although the King's proclamation still stood, making the land west of the Appalachian divide Indian territory, people broke across it, into the waiting West. With the Revolution, the 'proclamation line' was withdrawn to the Ohio and later, with the growth of the Republic, to the Wabash and the Mississippi.

Western America evoked strong images. Titling his novel of the West, *The Big Sky*, Guthrie certainly expresses one of the chief impressions of the region, namely its bigness, its openness, its height. The immensity of the West took away the breath of those who had lived in horizons rimmed by the Blue Mountains or closed by the Appalachians. Here were immense rivers like the Ohio, the Mississippi and the Missouri; here were immense plains never met before; here were immense sweeps of mountain; here above all was the immense sky, that could lift a man up or crush him. It was no wonder that Americans came to 'think big'.

Early impressions of the West were so favourable that they led to a rapid and widespread migration from all parts of the seaboard: both northerners and southerners giving up farms and businesses in their anxiety to improve themselves in the new heartland of their country. All the early explorers and settlers were filled with excitement and often with awe as, breaking out of the Appalachian mountains, they came on the Ohio or the Great Lakes plain. Daniel Boone, who discovered the famous Cumberland gap into the interior, spoke of the 'inconceivable grandeur' of the Ohio basin, and attested to 'an exceedingly fertile country, remarkable for fine springs', as he marched across it up to Lake Erie. As early as 1782, Freneau saw this great interior as 'the future of America'. He wrote that

> 'after passing a ridge of lofty mountains [the Appalachians] extending on the western frontiers of these republics [the old Thirteen Colonies] a new and most enchanting region opens, of inexpressible beauty and fertility. The lands there are of a quality far superior to those situated at the sea coast; the trees of the forest are stately and tall, the meadows pastures spacious, supporting vast herds of the native animals of the country. . . . The climate is moderate and agreeable; there the rivers no longer bend their courses eastward to the Atlantic, but inclining to the west and south . . . fall at length into that grand repository of a thousand streams, the Mississippi, who, collecting his waters, derived from a source remote and unknown, rolls onward through the frozen regions of the north, and, stretching his prodigiously extended arms to the east and west, embraces those savage groves and dreary solitudes, as yet uninvestigated by the traveller . . . till, uniting with the Ohio and turning due south, receiving afterwards the Missouri and a hundred others, this prince of rivers, in comparison with which the Nile is but a small rivulet, and the Danube a ditch, hurries with his immense flood of waters into the Mexican sea. . . .'

The perception of the West as a new unifying agency in America was to create an image of the greatest significance. It was picked up by Turner and Webb, who regarded the West as the welder of the new nation. Here the Americans became a new race. 'In the crucible of the frontier the immigrants were Americanized, liberated, and fuzed into a mixed race, English in neither nationality, nor characteristics', wrote Turner. The eastern seaboard had never been successful as a frontier, because though it had fronted men with the wilderness, men had not been constrained to change: history was stronger than geography and had planted a European seed in American soil which took generations to make itself different. In fact, that difference came when it was transplanted, once again, to the West, where it could flourish as virtually a different species. This species depended as much as anything on hybridization, on a complete mix of what had been before which, somehow, came out a new thing. The West was the great melting pot, melting down the differences between North and South, and between the foreign immigrants who continued to pour across the Appalachians to create a society both more egalitarian and libertarian than ever before. The traits of the European past, such as privilege and position, were eradicated in favour of the American future, centred on equal rights and opportunities. As the frontier advanced it 'carried with it individualism and democracy'. Relics of the feudal system that had still managed to survive in the East were swept away. The West eschewed the tenant system, opposed the holding of land by 'quit rent' and, for its menfolk at least, introduced universal suffrage. It became the home of the common man, thus fulfilling the revolutionary hopes of leaders like Thomas Paine. 'Let the crown,' he cried, 'be scattered among the people whose right it is.' Government did not come from above, but sprang from below. To augment it, *all* were concerned. 'I call not upon a few, but upon all: lay your shoulders to the wheel. . . .' The unifying image of the West certainly tipped the balance during the critical days of the Civil War in favour of maintaining the Union, and has been a major balancing force ever since.

The discovery of the Great Lakes seemed to point the way of destiny for American expansion.

> 'If this new world,' wrote Freneau, 'was not to become at some time or other the receptacle of numerous civilized nations . . . for what visible purpose could Nature have formed the Great Lakes in the bosom of her infant empire, which surprise and astonish the traveller, who, leaving the salt ocean behind him in the east, finds, unexpectedly, new oceans of a prodigious extent in those tracts where Fancy would have surmised nothing but endless hills, inhospitable wilds and dreary forests existed?'

The idea of a waiting West was most attractive to those who had begun to feel that the East was filling up. Adams (1889/1964 edn., p. 11) in his history of the United States in 1800 speaks of how the 'New Englander had begun to abandon his struggle with a barren soil, among granite hills, to learn the comforts of easier existence in the valleys of the Mohawk and Ohio'. In his reconstruction of the South between 1790 and 1860, Eaton (1963) remarks how southerners moved West like a tide, 'led on by the restless hope of discovering in a new country "something we crave and have not".' That this westward drive could help the whole nation forward gave to many Americans a genuine sense of their part, however small, in America's 'manifest destiny to overspread the continent allotted by Providence for the free development of our yearly multiplying millions'. The West, then, revived the early sense of purpose and hope of progress that had brought the first settlers to the Atlantic seaboard, 150 years before.

One sign of this was the freedom with which men transformed the West from a vast area of forest interspersed with wet prairie rising up to the grasslands of the Great Plain, into rich cropland or extensive grazing. The interior forests were at one time among the greatest in the world, marching from the Appalachians virtually to the 'break of the plains'. They impressed every traveller in the early days of the Republic, when immigrants swarmed into the Ohio, Wabash, Kentucky and Tennessee basins. The French visitor, Chevalier (1837/1961) noted how even cities like Cincinnati made a comparatively small dint in the prevailing mantle of trees. 'On all sides the view (from Cincinnati) is terminated by . . . an amphitheatre still covered with the vigorous growth of the primitive forest. This rich verdure is here and there interrupted by country houses surrounded by colonnades provided by the forest'. Gradually the openings widened and coalesced, forest gave way to farmland. Yet on the many glacial ridges and beside the rivers, great forests still remained, and as late as 1861, when Anthony Trollope was journeying across the central lowlands, the general impression remained one of woodiness.

> 'From Detroit,' Trollope [1862/1965] wrote, 'we continued our course westward (to Milwaukee) through a country that was absolutely wild till the railway pierced it. Very much of it is still wild. For miles upon miles the road passes the untouched forest showing that . . . the great work of civilization has hardly more than been commenced.'

However, in addition to forests, there were also wide grassy tracts, like the wet prairie of Illinois or the so-called 'barrens' of Kentucky. Sauer (1967, p. 23) has collected the first impressions of these treeless areas. He notes that Monroe thought 'the absence of timber . . . an evidence of the poverty of the land'. Monroe reported to Jefferson, 'A great part (of the interior) is miserably poor and . . . consists of extensive plains which have not had from appearances . . . a single bush upon them for ages'.

Hoffman is quoted as saying 'The prairie, swept by the fire of summer and the piercing blasts of winter, seemed little better than a desert . . .' Indeed, as Sauer points out 'When the pioneer encountered the grassy plains (of Kentucky) he promptly called them "barrens".' These large open spaces occurred on the flats between the low ridges and above the entrenched streams and were ascribed by early settlers to the repeated use of fire by Indians to stampede buffalo and deer in their hunting expeditions. 'After the Indians were driven away . . . the woods to a considerable extent recovered possession of the areas of open ground.'

The replacement of natural vegetation by cropland and pasture was more extensive in the central lowlands of America than anywhere else. By the time the settlers crossed the Ohio the government was offering cheap land in quarter-section lots of 160 acres, and this attracted scores of thousands of homesteaders to pour in, cut down the forest, plough up the prairie and develop commercial farming as rapidly as possible. The rape of the West had begun. Unfortunately, people did not realize what they were doing at the time: they believed they had come on a land of unlimited wealth. American optimism felt that it would offer unending prosperity. What Eaton (1963, p. 33) quotes as 'the boundless licence of a new region' gripped the people. The myth of the American cornucopia had begun to take shape, and led to virtually unbridled exploitation. Here indeed the image shaped the reality; an illusion of history profoundly influenced the course of geography.

The opening of the West was a new revolution. It went with a new mental attitude and created a new American landscape. With it came the easily plotted and expandable 'section' survey, based on latitude and longitude, a whole countryside laid out on the principle of the regular and equal sharing of the land, where each family had its own quarter-section farm, on which to work out its fortune. Self-reliance became the order of the day. As a result, when industry arrived, the West gave great scope for the laissez-faire expansion of commerce and manufacture which, through the fiercest competition for land, communications, capital and labour, starred the scene with mushrooming towns. These were Western towns, with broad roads, railway lines down their centre, false-fronted stores, big plots and family homes, the centrality of station, hotel and saloon, together with real-estate promotion run wild—the 'Mainstreet' town of Sinclair Lewis, utilitarian perhaps but exciting and not without pretensions. Ugly they may have seemed to easterners like Carol Milford, who came from established cities, but they were all places on the make, about which their inhabitants, like Dr. Kennicott, never failed to boast that they were 'going to have a great future'.

The images of wealth, abundance and bigness were counterpoised by another illusion of an opposite kind which was to grow up, namely that fertility ceased at the break of the plains and beyond this, on the Great Plains, lay the 'Great American Desert'. This was one of the most powerful illusions in American perception, and virtually denied the Great Plains to the plough or the corral for a generation or more. The change from a closed-in forested landscape to an open grassland and seemingly limitless horizons demanded not only major changes in land use, but in mental outlook. Many settlers were mentally unprepared for the change; everything that was familiar and had been taken as God-given swiftly disappeared behind them, and they were left to face the strange and forbidding with fear and trepidation. Hence they grew to exaggerate the differences and difficulties and to see the bareness of the landscape as barrenness. Consequently, as Trewartha (1941) shows, they trekked right across the Great Plains, neglecting what are today America's richest wheatlands and most extensive grazings, for tree-green Oregon. Guthrie, in reconstructing the times of the Great Trek in his book *The Way West*, has a group of people arguing as to whether they should settle in the Great Plains or trek right through to distant but wooded Oregon. The dispute was settled when one of the characters asks, 'Can prairie grow a crop? Land that won't grow a tree won't grow nothing.' Consequently, as Billington (1960, p. 690) points out, 'open country was shunned by the first settlers, whose frontier techniques were adjusted to a woodland country'.

The expedition of Zebulon Pike had done nothing to allay their fears. In his journeys down the Santa Fé trail, 1806–12, Pike had reported that the 'Great Plains were uninhabitable; they were a domain only for the Indian'—not for the agriculturists, who were advised to stay on the east side of the Missouri. Long's Yellowstone expedition of 1819–20 went further, issuing a map that called the Great Plains 'The Great American Desert'. Long claimed that 'In regard to the extensive area between the Missouri and the Rocky mountains, we do not hesitate in giving the opinion that it is almost wholly unfit for cultivation and, of course, uninhabitable by a people depending on agriculture for their subsistence.' From this time on, the myth of the Great American Desert, as Kraenzel (1965) states, 'became a reality'.

An interesting corroboration of the American picture is given by the Canadian experience. British explorers and scientists both in Canada and America served in many ways to strengthen the great American illusion, as Lewis demonstrates in Chapter 3. This illusion in the mind dictated the geography on the land, and the Great Plains remained undeveloped until techniques of dry-farming transformed the situation.

Mechanization and mass-production, both on the farm and in the city, soon dominated the West and spread across America. The so-called 'bonanza' farms of the wheat belt were virtual grain factories: their operation attracted business men as much as farmers. W. G. Moody wrote of 'the new development' in Western agriculture as

> 'a revolution in the economics of the farm. Most persons in reading of fields described by hundreds and thousands of acres can form but little idea of their actual or comparative sizes. Manhattan Island, the site of the City of New York, has an area of about 14,000 acres. The Grandin farm, north of Fargo, comprising 40,000 acres, has nearly space enough for three cities like New York. Whatever else may be said of Western operations, they certainly are not wanting in grandeur.'

It was, however, in Western industry that the penchant for 'doing things big' showed itself most strikingly. Ford, the genius of standardization, uniformity and mass organization came—in Siegfried's (1948, p. 30) view—to typify America, in the way in which Ulysses, the arch extemporizer, represented Europe. As a result of the immense success of that technology, worked out most effectively in the American West, the world was seeing 'the machine supplanting the tool; mass production opposed to craftmanship; the group or the gang replacing personal initiative. Ford seems to be confronting Ulysses, the patron of ingenuity and of *débrouillage*'. Consequently, America was directing civilization down magnificent though constricting avenues. In fact, of course, the American West provided for both individual and mass improvement. The combination of individual opportunity and mass organization, by which people worked together in highly disciplined and standard ways, stimulated both productivity and consumption and made the West an immensely dynamic region which had an impact throughout America. Indeed, it sets its stamp on the whole country.

The North

This far region, as represented in Alaska, also generated its own image although this was perhaps only a modification of the myth of the frontier. But it was the unfinished frontier, the empty North, awaiting development largely because it was harsh and remote. It lay within Fleure's Zone of Difficulty, only to be won by a ceaseless and untiring outlay of effort. Alaska was, of course, bought for strategic reasons, while the tide of 'manifest destiny' was running high, and American optimism could take on anything. At one stroke, America was able to keep imperial Russia out of the continent, and put a giant squeeze on imperial Britain by, as it were, getting a 'half-Nelson' hold on Canada. But there was no marked movement of settlers into the area until the Klondike gold rush of 1898. Even then, the numbers were not great, compared with the vastness of the territory: moreover, the boom was short-lived. Then development had to turn to the less glamorous tasks of fishing, forestry and farming—all remote from the main American market.

The general image of the North was not favourable. To begin with, it was coloured by French and British views, based on Canada. As Cartier

> 'sailed along the Labrador coast in coldness and fog, with patches of ice floating by in grey and bitter waters, watching isolated stony headlands and large empty bays where only stunted trees and mosses were to be found he said "Truly, this must be the land that God gave Cain".'

Britain too had an interest in the North, through the Hudson Bay Company, but did not develop it for more than whaling and the fur trade. When Canada took over from Britain, the North, according to Morton (1964, p. 91) was thought to be 'alien and incomprehensible and was therefore disregarded'. It was in fact 'entered in our inventories as a dead liability'.

Nevertheless, to quote from Joaquin Miller's poem on Alaska, it was part of America's 'unfinished world'. It offered the fascination of the 'untracked and unnamed and unknown', and therefore drew the adventurers of America to it. The adventure of Alaska attracted the entrepreneur who found untapped resources, even if these were difficult and costly to exploit. Success bred a better image. Coyne (1974), in a fascinating thesis on the changing image of Alaska, has collected many views of which the early ones were usually unfavourable, but the 20th century ones more optimistic. He quotes 'The prevailing conception of Alaska as a region wholly given up to glaciers and mountains is strikingly at variance with the facts' and '. . . the agricultural capacities are much in advance of the public sentiment of the country'. Gradually America perceived it had a task to do in Alaska that was well worth doing. The more optimistic views encouraged penetration, and Alaska became a frontier of development, particularly after the second World War. But if it was the last world to exploit, it was also the last world to conserve. The interest of Alaska is that the main thrust for exploitation came during the main drive for conservation. Thus Alaska is a testing ground for the American image of progress. Hitherto, that image had been one of progress through the unbridled exploitation of what were thought to be unlimited resources. Now progress is thought of in terms of conserving from today a heritage for tomorrow. In particular, it centres in conserving parts of 'unfinished America' so that they will never be finished off. America has moved full circle, to come face to face with the wilderness.

The great battle of the Alaska pipeline (discussed by O'Riordan in Chapter 19), although won in 1973 in favour of the exploitation of north-shore oil (and its transportation across wilderness Alaska to south-shore oil ports) nevertheless pointed up the need to conserve as much of that wilderness as possible, and forced the developers to plan for a growth that would not spoil the future. This in itself is progress, and if the image of the North has done nothing else it will have served its purpose in keeping a part of the unfinished world alive: not even Americans should believe, as they did in their westward expansion, that 'every American had an undoubted right to pass into every vacant country'.

In conclusion, it may be seen that each region of America has contributed something distinctive to the image of America, yet each has been subject to American ideals in general in the shaping of their several landscapes. In each some thread of all-American perception has tied them to, and made them representative of, the American myth itself.

References

Adams, J., 1889/1964. *History of the United States of America During the First Administration of Thomas Jefferson*, 1889 edn., New York, Scribner; 1964 edn., Ithaca, N. Y.: Cornell Paperbacks.

Billington, R. A., 1960. *Westward Expansion*, New York: Macmillan.

Bowman, T., 1934. *Geography in Relation to the Social Sciences*, New York; Scribner.

Cash, W. J., 1941. *The Mind of the South*, New York: Knopf.

Chevalier, M., 1837/1961. *Letters on North America*, vols. 1 and 2. 1837 edn., Paris: Gosselin, 1961 edn., New York: Anchor Books.

Coyne, J., 1974. *Alaska: Image of a Resource Frontier Region*, London: unpublished dissertation, Birkbeck College.

Debow, J. D. B. *et al.*, 1860. *The Interest in Slavery of the Southern Non-Slaveholders*, etc. London; Evans and Cogswell.

Eaton, C., 1963. *The Growth of Southern Civilization, 1790–1860*, New York: Harper Torchbooks.

Kraenzel, C. F., 1955. *The Great Plains in Transition*, Tulsa: University of Oklahoma Press.

Morton, W. J., 1964. *The Canadian Identity*, Toronto: University of Toronto Press.

Sauer, C., 1967. *Land and Life*, Berkeley: University of California Press.

Siegfried, A., 1948. *The Mediterranean*, London: Cape.

Trewartha, G. T., 1941. Climate and settlement of the subhumid lands. In U.S. Yearbook of Agriculture, *Climate and Man*. Washington, D.C.: U.S. Department of Agriculture.

Trollope, A., 1862/1965. *North America*, vols. 1 and 2, 1965 edn., London: Colonial History Series.

Watson, J. W., 1969. The role of illusion in North American Geography. *Canadian Geographer*, **13,** 10–28.

Wright, L. B., 1969. *Everyday Life in Colonial America*, London: Putnam.

2
The Perception of the Seaboard Environment

DOUGLAS MCMANIS

The settler's view of the colonial American seaboard included perceptions of its suitability as a whole for permanent European settlement and also of the regional differences within the seaboard. These were influenced by selected beliefs and preferences.

The Seaboard as European Habitation

Perception of the environments of the eastern seaboard as areas suitable for European habitation depended in large measure on the role of the colonies in overseas adventures. Before 1610, assessments of the seaboard environments rarely included their consideration as permanent residences for Europeans, for immigration to those parts of the New World was not then a prominent motive for exploring the coast or claiming part of it. An exception to that generalization was a crudely formulated Elizabethan scheme to settle Norembega (New England) as a replica of English feudalism, but the scheme had little influence on later exploration because the idea of resident overseas settlement was not then important to Englishmen (Morrison, 1971; Sauer, 1971; Cumming, *et al.*, 1972; McManis, 1972).

Explorers, however, sometimes noted environments suitable for overseas posts where men would stay for short periods of time. These might be founded to further the exploitive and territorial aspirations of empire seekers. Although those evaluations were usually made under extremely restricted circumstances (such as unqualified observers, short visitations during non-winter seasons, or only small areas viewed) they nevertheless resulted in conclusions about the habitability of the seaboard environments which were of questionable validity and which became the bases of exaggerated promotional literature that often had elastic territorial application of its laudatory claims. With the notable exception of the writing of John Smith, the pre-1620 promotional literature did not explicity encourage immigration, though that action was frequently implicit (Levermore, 1912). Smith, in his vision of English settlements centred on Massachusetts Bay (first articulated in his *Description of New England* (1616), depicted on his map of New England and restated in several later writings), advocated immigration of Englishmen to an unsettled portion of the seaboard (Arber, 1910; Barbour, 1964).

Perception of the seaboard as an area suitable for permanent European settlement, in contrast to earlier images which were concerned only with short-term European occupance, was the outgrowth of changing factors during the early years of the 17th century. Chiefly, these were the emergence in England of conditions which encouraged emigration to colonies both on the seaboard and in the Caribbean and demonstrated evidence that Englishmen could survive in the varied physical environments of the seaboard. However, the process by which the seaboard was perceived to be habitable differed significantly from the northern and southern portions of the coast, although the fate of the concept of resident European settlement along each portion was intimately connected with two colonizing ventures begun almost simultaneously in 1607.

Jamestown, the southern colony, like its contemporary establishment on the northern coast, was founded in the 16th century mode for colonies—primarily as a centre for exploitive and exploratory activities. Its site had few advantages other than a usable anchorage. It was malarial, lacked a satisfactory water supply and had little tillable land near it; not surprisingly the site was eventually abandoned when the anchorage function was no longer needed and the administrative functions transferred inland. The fledgeling colony of Virginia, however, survived both the liabilities of its first settlement site and the initial years of disorder and inadequate supplies. Survivors of that era were joined by new immigrants amongst whom were women whose presence added family life to the numerous transitions occurring in the colony. As dreams of fabled wealth and hopes of early return to England proved invalid and faded away, settlers turned to activities to sustain themselves, principally farming. Suitable areas for farms were sought out along the James River and other streams which flowed into Chesapeake Bay, and later, on the necks between the rivers. As men and families moved to their farms, population was dispersed far and wide from the colony's original nucleus. With increasing specialization of the agricultural economy in tobacco, the land selection process became more focused on land capable of producing that crop and accessible to tidal streams. After a decade or more of tenuous existence Virginia was well into the process of evolution from a traditional 16th century outpost of empire to a colony of dispersed farms whose settlers considered them to be their homes. This continued existence in environments vastly different from those of England provided dearly won proof that Englishmen could inhabit the southern seaboard.

The Sagadahoc Colony, established on the lower estuary of the Kennebec

River in the late summer of 1607, was the climax of English interest in the northern coast, which had been renewed shortly after the turn of the century. Exploratory and commercial voyages by Gosnold, Pring and Weymouth had been the bases of promotional literature which portrayed that coast in near-Edenic terms as a region of unequalled productive potential. The colony, which was as badly organized and governed as early Jamestown, was abandoned the following year. The cause of its abandonment was generally ascribed to the harsh, rigorous winter, a season which was too quickly assumed to be over severe for the survival of Englishmen, and as a consequence the entire northern coast was branded for many years with the belief that it was uninhabitable by Europeans on a year-round basis. Commercial use of the northern coast by fur traders and fishermen continued much as it had been before the founding of the Sagadahoc Colony, but efforts to re-establish another post comparable to Jamestown were unfruitful before 1620, despite the promotional vigour of John Smith and the activities of Sir Ferdinando Gorges. Thus, during the decade when the Virginia Colony was slowly evolving from an economic outpost into a resident community, permanent settlement of the northern coast was retarded by the belief that the region was uninhabitable (McManis, 1972, pp. 90–115).

The evidence needed to destroy that belief was supplied by the Pilgrim settlement. Why the Pilgrims found themselves in the latitudes of Cape Cod in the fall of 1620 remains a matter of dispute. But their decision to settle at Plymouth had profound effects on the northern coast's future geographic development, for the survival of the group, in spite of a high death rate during the first winter, proved that with proper organization Englishmen could indeed live year-round in those northern latitudes. Soon after, smaller settlements dotted the coast and off-shore islands northward of Plymouth; most of them, in contrast to Plymouth, which was primarily a subsistence agricultural community, were fishing stations or fur-collection posts.

In addition to their role in the ultimate reversal of the negative assessment of the northern coast's habitability, to the Pilgrims there also belongs the honour of being the first group of Englishmen to assess a portion of the seaboard solely on the basis of its ability to sustain an immigrant population. Unlike their predecessors on the seaboard the Pilgrims came to the New World for one purpose: to make homes for themselves. Exile, not the lure of adventure, the quest for treasure hordes, or the search for a passage to the Indies, brought them to the shores of Cape Cod.

Burdened by late arrival and low supplies, the Pilgrims surveyed only a small portion of the Cape and adjoining mainland before the selection of their settlement site at Plymouth. The record of their survey and selection process is found in *Mourt's Relation or a Journal of the Plantation at Plymouth* (Dexter, 1865), a volume published in London as promotion for the fledgeling colony within two years of its founding. In contrast to earlier promotional efforts, *Mourt's Relation* was documented by the success of the Pilgrim colony. During their first wanderings on the Cape the newcomers noted local fauna and flora, terrain characteristics, surface water bodies, and such scant evidence of native occupance as fields, caches and trails, for the natives themselves proved elusive. Each potential settlement site was viewed as a composite of features and judged against the optimum conditions which the Pilgrims wanted for their settlement: a sheltered anchorage, two or more sources of surface water supply, cleared areas for fields and village, timber distant from the village but accessible to it, defence from attack by sea or land, productive soil and natural forage. Of the several sites visited by the Pilgrims, Plymouth was chosen as the one best meeting the conditions which were wanted. In time the Pilgrims recognized that their choice was not as satisfactory as originally thought, and within a decade dispersal of people in search of more and better land had begun (McManis, 1974).

Most Europeans who came to the New World knew from their prior existence that variations of physical and cultural phenomena existed on the Earth's surface. When settlers arrived in North America, enough was known about the New World to assume that such variations existed there as well. The problem for European emigrants was to resolve the physical and cultural diversity encountered in their new environments into meaningful patterns of regionalization and management. Thus the first settlers to the seaboard or those first into an area behind the coast had a three-fold task; to make an inventory of the phenomena which existed in an area, incorporation of exotic items into their system of knowledge, and ordering of areal characteristics into associations and patterns.

Perceptions of Regional Differences

Environmental perception functioned at two levels of scale. One was localized and dealt with the problems of site location, field allocation or road layouts in much the same manner as the Pilgrims had described; the other focused on regionalization of the seaboard. As information accumulated it became apparent that gross physical differences existed: to give two examples, that New England lacked any appreciable amount of coastal plain which south of Chesapeake Bay was extensive, and that a section of gently rolling hilly land lay between the plain and the Appalachians. In the perception of those regions values about their uses were included, some being considered highly worthy of settlement, others less so. South-eastern Pennsylvania acquired one of the best reputations as a farming area. The outer coastal plain was considered less satisfactory for settlement than the inner portion of it. Certain valleys of the eastern Appalachians were held in high repute for agriculture, while some in northern New Jersey were valued chiefly because of the ores in the surrounding hills. Regional reputations occasionally changed with the passage of time. New England, which initially enjoyed esteem as a place for farming, had that assessment severely altered by the mid-18th century (McManis, 1974).

Of the physical factors on which environmental perception was based, vegetation types and soil productivity or fertility were among the most important. In New England the desire for satisfactory pasturage in both old and new settlements resulted in a premium being placed on two types of land which ordinarily would not rank high as agricultural preferences. Off-shore islands and small peninsulas connected to the mainland by narrow necks were valued as pastures because the surrounding waters reduced the need for fences or herdsmen, although forage on such lands was often minimal. In contrast, the numerous salt or tidal marshes which dotted the coast were considered rich in forage and provided hay for winter fodder as well, although cattle were apt to wander into adjoining fields. As settlement pushed into the interior the high value placed on the coastal marshes was extended to freshwater marshes and riverine meadows, the latter being thought of not only as pasturage but also as fertile, tillable land. The role of availability of meadows in attracting settlers to an area is illustrated by the example of the Connecticut Valley.

Knowledge of the vast expanse of meadows and numbers of fur-bearing animals to be found in the lower Connecticut Valley had been brought to the coastal Puritan towns around Massachusetts Bay by Indians and fur traders. In response colonists soon formed groups which left those towns, crossed the unsettled country between the coast and the river, and established the four river towns of Hartford, Windsor, Wethersfield and Springfield. Later, when the larger valleys with wide expanses of meadows were taken up by settlers, founders of new towns or men seeking farms in the older towns sought out tiny patches of meadow to be had in smaller valleys and even came to value the bits of grassland bordering glacial ponds and bogs, for no farmer wanted to be without natural meadow, even after the relative lack of nutrients in native American grasses became known.

New Englanders were not the only colonial group to value the eastern grasslands, for in newly settled frontier regions where cattle were expected to survive with a minimum of care and supplemental feeding, the presence of grasslands was a critical factor not only in the general attractiveness of the area but also in determination of early settlement sites. In the Middle Colonies meadows, whether or not they were the result of Indian burning or regrowth on former Indian fields, were considered excellent pastures and usually were among the first lands chosen by the earliest settlers. In the South open grasslands were likewise used as pasture, although they were, at least in the outer coastal plain, not considered to be tillable. Southern settlers, like their counterparts in other colonies, often burned over the grasslands in order to encourage new plant growth and to control ticks (Merrins, 1964, p. 37ff).

The New England preference for coastal wetlands because of their forage potential, was not always duplicated elsewhere along the seaboard, particularly south of Virginia where a larger proportion of the coast was either swamp or marsh, or tidally subjected to inundation and where livestock was not as important to the agricultural economy. Many of those wetlands were avoided by settlers because of the difficulties and costs of draining before they could be used for cropping. In a study of colonial North Carolina Merrins (1964, p. 44) concluded that the large amount of wetland in the outer coastal plain accounted for the belief that it was less satisfactory for settlement than the inner coastal plain.

For many southern agriculturists the question of soil productivity was more essential to success of a settlement than the presence of meadows or grasslands. While the earliest settlers in a region often followed the example of the Pilgrims and chose sites where there were former Indian fields, standards other than prior use were available to gauge soil fertility. It must be noted, however, that no exact determination of soil fertility or quality was then possible, and most settlers in fact were satisfied with vague, general estimates of productivity in the hopes that time would prove the assessment accurate. An example of that sort of assessment was de Brahm's statement of the soils of the valleys of the southern Appalachian Mountains, 'Their Vallies are of the richest soil equal to Manure itself, almost impossible in Appearance ever to wear out ...' (de Vorsey, 1971, p. 105). Two standards were used to determine soil fertility: vegetation and colour, factors which modern soil scientists have long since put into proper perspective as fertility indicators.

Vegetation was by far the more widely used criteria. During the colonial era and well into the federal period settlers associated the productive quality of land with the species of trees which grew there. Hickory and walnut trees usually indicated a rich productive soil, while fir, pine, spruce or birch was a sign of unproductive land or at best the least productive soil in the neighbourhood. By the end of the colonial period certain regional variations in the application of generalizations had appeared, a circumstance which should have undermined the whole vegetative concept but failed to do so. Thus a speculator at the beginning of the 19th century could claim that chestnut trees in eastern Pennsylvania meant poor, thin soil, while in western New York the same specie marked good wheat land.

During the 18th century there was general agreement among settlers, speculators and travellers that the most fertile soils of the seaboard were found over limestone. That generalization was particularly associated with the Middle Colonies. A recent study of south-eastern Pennsylvania noted certain correlations between the belief of limestone soils being the best and the area's settlement process. Population densities tended to be higher in townships underlayed by limestone; the Lancaster Plain, one of the largest sections of limestone soils in that part of the colony, attracted colonists sooner than non-limestone areas which surrounded it. But the study also recorded the strong influence of other factors which had been critical in site selection in New England and Virginia decades before Pennsylvania was founded: access to water supply and preference for riverine bank or terrace locations, again suggestions that perception and utilization rested on an *association* of physical features rather than on one characteristic.

In some instances, perception of an area's agricultural potential was revealed more by suggestions for management than by direct assessment of soil quality.

Jared Eliot's (1934) *Essays upon Field Husbandry in New England* contained many proposals for farm management and improvement which were influenced by the English Agricultural Revolution then under way. By comparison with earlier writings about the soil quality of New England and with evidence of productivity in other colonies one may see, as Eliot did not, that a major part of the problem was the unsatisfactory physical base in New England, compared, for example, with that of the Middle Colonies. De Brahm recognized that many parts of the Carolinas and Georgia should not be cultivated but rather used for timber and livestock; settlers had learned the problems of cropping on the coastal sandy soils long before he argued that they were usable only under 'proper tillage'. Since he wrote before much of the southern Appalachians was invaded by settlers, if they had paid heed to his notice of soil erosion some of the terrible scarring which those highlands have undergone might have been prevented (de Vorsey, 1971, p. 105).

Conclusion

At the end of the colonial era Europeanization had to some degree touched all portions of the eastern seaboard. Its environments had been perceived as satisfactory for an agrarian society, and a scaling of regional and economic variations had emerged as the indigenous landscapes were transformed into areas dominated by European cultural patterns. For all those variations, however, the basic premise of the perception which had occurred over the previous decades was valid—the seaboard was a land which not only sustained but also rewarded those who worked its soils. With the benefits of hindsight, it may be noted that in the new political context and in the new century the geography of the seaboard changed radically. Urbanism, industrialization, technological innovations and agricultural adjustments were among the most important factors which differentially transformed the seaboard during the 19th century. The dynamics of geographic change were associated with changing perceptual patterns. Perception ceased to be dominated by the needs of an agrarian society and economy. Indeed, it was the non-agricultural elements to which the future belonged, and the seaboard was viewed as the best place for many new activities.

References

Arber, E. (Ed.), 1910. *Travels and Works of Captain John Smith*, Edinburgh: Grant.
Barbour, P. L., 1964. *The Three Worlds of Captain John Smith*, London: Macmillan.
Cooper, W., 1810. *Guide to the Wilderness*, Dublin: Gilbert and Hodges.
Cumming, W. P., Skelton, R. A. and Quinn, D. B., 1972. *The Discovery of North America*, New York: American Heritage Press.
Dexter, H. M. (Ed.), 1865. *Mourt's Relation or a Journal of the Plantation at Plymouth*, Boston: Wiggin.
Levermore, C. H. (Ed.), 1912. *Forerunners and Competitors of the Pilgrims and Puritans*, Brooklyn: New England Society of Brooklyn.

McManis, D. R., 1972. *European Impressions of the New England Coast 1497–1620*, Chicago: University of Chicago Department of Geography Research Paper.

McManis, D. R., 1974. *A Geography of Formative Years: Colonial New England*, New York: Oxford University Press.

Merrins, H. R., 1964. *Colonial North Carolina in the Eighteenth Century*, Chapel Hill: University of North Carolina Press.

Morrison, S. F., 1971. *The European Discovery of America: Northern Voyages*, New York: Oxford University Press.

Sauer, C. O., 1971. *Sixteenth Century North America*, Berkeley: University of California Press.

de Vorsey, L (Ed.), 1971. *De Brahm's Report of the General Survey in the Southern District of North America, Columbia:* University of South Carolina Press.

3
First Impressions of the Great Plains and Prairies

G. MALCOLM LEWIS

Early explorers penetrating deep into the heartland of the North American continent saw the landscape in quite different terms from the settlers who were to follow in their wake. Here the role of motivation and expectation colours perceptions and enriches the writings and descriptions of these hardy men.

The Hudson's Bay Expeditions

In 1690 Henry Kelsey became the first British citizen to approach any part of the cis-Rocky Mountain West. A native of London and an employee of the Hudson's Bay Company, he had probably been educated at one of the foundation schools from which the Company liked to draw its apprentices. He may therefore have known something of the geographical myths about the interior of North America before he left England at the age of fourteen. At this time, however, only the Spaniards had any first-hand experience of the cis-Rocky Mountain West and that was limited to parts of the area south of the Arkansas river. The French did not begin to approach the area from the east until a decade or so after Kelsey did so from the north-east. Kelsey may have heard something about conditions in the interior during the six years in which he had contacts with Indians on Hudson Bay. He certainly had no first-hand experience of comparable terrains and climates in other parts of the world. His journal (Kelsey: Doughty and Martin edn., 1929), part of which is in verse, is unstructured and hard to follow, so much so that it is difficult to decide how far west he went. His directive was not to explore or describe the areas through which he travelled but to increase the inland Indian's awareness of the Hudson's Bay Company and to persuade them to make peace with each other, in the hope that they would concentrate on hunting beaver with which to trade at the Company's forts. He almost certainly got within sight of the Touchwood Hills in what is now south-central Saskatchewan and, in moving south-westward from the lakes and forests around the northern edge of Lake Winnipeg he left 'ye woods' for a plain, which it took him three days to cross. This plain was one of several to which he referred and together they indicate that he was in the region of aspen groves (he called them poplars), scattered spruce and grassy openings, which forms the 100-or-so-mile wide zone of transition between the unbroken spruce and fir forest through which he had passed and the wheatgrass prairies beyond. He referred to these openings as plains, champion lands, barren lands and heathy lands, the four terms apparently being interchangeable. They were described as devoid of water and supporting nothing but short, round, sticky grass, buffalo and a great sort of bear (probably the grizzly), in neither of which animals the fur trade had much interest. The desirable lands were the aspen groves which, in addition to being associated with water, were the habitat of the beaver, the ultimate reason for his journey.

Observations such as these were doubtless communicated in conversations between fur traders but they had little impact beyond this group. Within the Hudson's Bay Company the vague idea of increasing amounts of grassland and decreasing wood, water and beaver towards the south-west probably inhibited further exploration in that direction, deflecting further activity westward and north-westward from the Bay, deeper and deeper into the spruce/fir forests. Few accounts were written down and none were published at the time. Kelsey's journal would seem to have had little impact, having for a time been buried in the archives of the Hudson's Bay Company, from which it would appear to have been removed in the 18th century, thereafter lost and not rediscovered until the 1920s.

More than 60 years after Kelsey returned from the fringes of the grasslands of what is now the western interior of Canada, the Hudson's Bay Company sent another of its English employees to encourage the Indians of the interior to bring beaver pelts to the Bay. Anthony Henday's journey in the winter of 1751/5 is, like that of Kelsey, difficult to reconstruct but he certainly went several hundred miles further west, into the area of mixed grasses (Henday: Burpee edn., 1907). However, he did not apparently travel south of the South Saskatchewan into the short-grass country. In contrast to Kelsey's writings, Henday's journal anticipates the nature of the country which lay ahead. When still in the forests to the east of the site of the present town of The Pas, he was informed by an Indian chief that he was 'on the confines of the dry inland country, called by the natives the Muscuty Tuskee'. When some 50 miles or so to the east of the site of the present city of Saskatoon he wrote that 'we are now entering the Muscuty Plains' and 78 days later he wrote 'Left Muscuty Plains, which I have been on since 13th August'. During the interval he had been on the grasslands between the North Saskatchewan and Red Deer Valleys in what is now south-central Alberta. Like Kelsey he was struck by the absence of wood and water but his perception differed from that of his predecessor in that it was structured, albeit crudely, according to a regional concept, that of the Muscuty Tuskee. On reaching the region he immediately recognized it as something different and was struck by its vastness, repeatedly observing in his journal 'We are still on the Muscuty plains'. However, his descriptions of these plains were brief and imprecise, the recurring elements therein being level land, short grass, dry land, salt water and the buffalo. He had no means of knowing the extent of these plains or of recognizing that the plains of which he had experience were but the northern tip of a much more extensive region. About 12 years before, in 1742, Vérendrye (Vérendrye: Burpee edn., 1927) had passed through grasslands 'which were bare and dry' some 600 miles to the south, in what is now western South Dakota. Three years before that the Mallet brothers (Folmer, 1939) crossed plains where they could not even find enough wood to light a fire, some 400 miles still further south, in what is now southern Kansas, and reported that similar conditions extended all the way to the mountains near Santa Fé. Henday almost certainly knew nothing of these French experiences or of the 16th-century Spanish manuscript accounts of the grasslands still further to the south in what is now western Texas.

New England and Other Explorations

Eleven years after Henday returned from the South Saskatchewan, Jonathan Carver, the New England-born son of an emigrant from Wigan, approached (but did not reach) the interior grasslands via the gallery forests of the St. Pierre (now Minnesota) river. In describing the various tribes of Indians (Carver, 1778) he referred to three bands who dwelt to the west (in Dakota) 'on plains that according to their account, are uninhabited, and probably terminate on the coast of the Pacific Ocean'. The impact of this brief published comment must have been considerably reinforced by a legend placed on the accompanying map at the western headwaters of the St. Pierre river, which read 'From this place the Plains are unbounded suppos'd to extend to the South Sea'. Though this information as to the latitudinal extent of the Plains was false and though it added nothing to what little was already known of their character, it did herald a new scale of thinking, in that for the first time the Plains were implied to occupy a large portion of the continent.

In a letter to Sir Joseph Banks, Alexander Henry (1781), another New England-born fur trader, referred to the 'Great plains to the east of the Stony Mountains'. This gave a truer indication of the latitudinal extent of the grasslands and was probably the first occasion on which that name was used by which we now refer to the region. However, Henry made no attempt to delimit the region. Four years later, the Frenchman St. John de Crèvecoeur made a copy in New York of a map, by Peter Pond, yet another Yankee fur trader, who had travelled extensively in the western interior during the previous ten years. The map is remarkable in that it delimits by a pecked line 'ye Eastern Boundaries of those immense Plains which reaches to the greats Mountains' (Pond, 1785). The boundary extends from just south of the confluence of the

Ohio and Mississippi Rivers, north-north-westward on a course parallel with and just to the west of the Mississippi River, Red River and Lake Winnipeg, westwards along the line of the low Saskatchewan and finally north-westwards to the west of Lake Athabasca and Great Slave Lake. This early attempt to delimit cartographically the region as conceptualized by the British was quite remarkable in that it was not to be repeated until the late 1850s, when several American scientists independently delimited cis-Rocky Mountain regions.

Though far ahead of its time, Pond's map delimited only the north-eastern boundary of what was later to be conceptualized as the Great Plains region. At this time the group image of the interior of North America by British-born fur traders from the United Kingdom and New England was based almost entirely on the accumulated first-hand observations of relatively uneducated but intelligent and perceptive men, who entered the region either from the forts and factories on Hudson's Bay or via the St. Lawrence valley and Great Lakes. They gradually recognized that beyond the forest habitat of the beaver and other fur-bearing animals lay an extensive region which was different. As they approached it, forests were replaced by grasslands and the fur-bearing animals were restricted to the valley floors and a few other wooded areas. To the west this region probably terminated at the Rocky Mountains but there was no knowledge of its southern limit. Since the region contained so few fur-bearing animals it repelled rather than attracted the British and there was little incentive to explore, enquire or even speculate about conditions to the south of the Missouri–Saskatchewan watershed. Even if there had been, the lands in that direction had been French before 1762 and Spanish thereafter.

The French and Spanish perceptions of the area to the south must have been unknown and unknowable to the British and Yankee fur traders. They were for the most part in the minds of administrators and entrepreneurs based either in St. Louis or New Orleans, with whom communication was almost impossible. In so far as they had been recorded they were mainly in manuscript form and hence unavailable to the British. They were not to be reflected in published maps until the beginning of the 19th century.

Thus, by 1785 (one year after the ratification of the treaty of peace between the United States and Britain which marked the end of the American Revolution) the northern grasslands had been approached mainly by the British, conceptualized by a few of them as a region which differed from the forest region to the east, delimited, albeit inadequately, on a map and appraised as being of relatively little value as a source of furs and therefore of no significance. Descriptions of the area's essential characteristics were vague, based on the limited observations of a few unscientific observers and badly written in the vernacular of either England or New England. Words such as plain, barren ground, heathy land and dry land were repeatedly used but never defined. Conversely, prairie, plateau and desert were used rarely or not at all.

A Synthesis of Experiences

During the quarter of a century which followed the American Revolution the meridianal as well as the latitudinal extent of the grasslands of the cis-Rocky Mountain West emerged more clearly. This was partly the result of further exploration and partly due to the beginnings of a synthesis of the experiences of four regimes: the British from the north-east; the French and French Canadians from the east; the Americans from the east; the Spanish from the south. Writing of his travels in the western interior in the 1760s and '70s, Alexander Henry (1809) was doubtless incorporating subsequent advances in conceptualization when he referred to the Plains which 'run southward to the Gulf of Mexico'.

One of the earliest Anglo-Spanish contacts involved the Scotsman James Mackay and the Welshman John Evans. Mackay had been a fur trader in the western interior of what is now Canada from approximately 1776 to 1793 and, during this time, had first-hand experience of the grasslands around the site of the present city of Regina. Unlike most fur traders he had received a good formal education and is reported to have been a good linguist. His background made it possible for him to change his allegiance to Spain and to assume the managership of the 'Company of Explorers of the Upper Missouri'. In 1795 he was joined by Evans, who had arrived in the United States some three years before to look for a tribe of Indians, believed to have been descendants of Madoc, a Prince from North Wales who was supposed to have discovered the New World in 1170.

Evans spent two years on the upper Missouri, getting as far as the Mandan villages near the site of Bismarck, the present capital of North Dakota. During his eight months in the villages he talked with the Indians and with fur traders from the Hudson's Bay and North West Companies. On the basis of what he saw of the lands adjacent to the Missouri and of what he was able to learn in conversation he wrote that

> 'The Land on both sides of the River is at one time Mountainous and barren and at other times even and fertile, but in the Back part a tree can hardly be found. The best Quality of Land is found in the Mandaine Country, this quality of Land Extends itself on the West as far as the East chain of the Rocky Mountains—it is at these Mountains where the great Meadows and Prairies terminate. The Country then begins to be Absolutely Covered with trees. . . .' (Evans: Nasatir edn., 1952, pp. 495–499).

Mackay (1807), who had been responsible for sending Evans up the Missouri, clearly paraphrased this description in stating that

> 'The best land is that possessed by the Mandans. This quality reaches west as far as the eastern chain of the Stony Mountains. . . . It is at these mountains that the great plains and prairies terminate.'

The term 'best land' must be interpreted from the point of view of their employers, whose primary objective was to develop the fur trade. The upper Missouri and Yellowstone valleys were potentially the best source of furs within the Spanish sphere of influence and 'Best Land' was almost certainly an appraisal from that point of view. It was not an appraisal of agricultural potential. Indeed, whilst Evans was at the Mandan villages Mackay was exploring the area between the Loup and Niobrara Rivers about 300 miles further south. He indicated on his map in what is now Brown County, Nebraska, a '*Great Desert of drifting sand without trees*, soil, rocks, water or animals of any kind. . . .' Nearby, in what is now Loup County, he indicated 'Sandy rolling country' and in what is now south-western South Dakota 'Sandy hilly country'. These legends referred to small areas but together they tended to confirm a description of a few years before of the area between the Missouri River and the foothills of the Rocky Mountains made by Truteau (Abel edn., 1921), Mackay's Spanish predecessor as manager of the Company's affairs. Truteau's description of the cis-Rocky Mountain West in the latitude of what is now northern Nebraska and southern South Dakota was of 'large prairies, or great waste lands . . . completely sterile, scarcely grass grows there'. These ideas were very close to those of the Americans who soon followed the British. At least one British explorer anticipated Pike and Long by showing how arid the American plains were. This was Mackay, whose description of the cis-Rocky Mountain West between latitudes 42°N and 43°N was far harsher than that of the region to the north of 49°N, where he remembered 'immense plains and meadows beautiful and fertile'. Hitherto, this implicit contrast between the northern or Canadian and central or American interior grasslands would seem to have been overlooked. Mackay was probably unique in having experience of both.

Another notable addition to the recorded perceptions of the cis-Rocky West was made between 1786 and 1788 by David Thomson, originally another of the Hudson's Bay Company's poor boys from London who, while still not 20 years old, crossed the grasslands between the North Saskatchewan and BOW Rivers. Thomson's (Tyrrell edn., 1916) narrative was not written until he was almost 70 years old and remained unknown for almost another 80 years. In it he explicitly distinguished between the meadows of long grass (the Americans and French Canadians called them 'Prairies') which could be cut for hay and which characterized the eastern fringes of the grasslands, and the plains to the west, where the grass was too short to be cut with the scythe. He predicted that the meadows to the east of the plains proper would prove suitable for the raising of cattle and sheep once the wolves had been destroyed and markets were established. He was less precise concerning the characteristics of the short grass plains, except that he expressed the opinion that the soil deteriorated southwards until, south of latitude 44°N the land was

> 'barren for great spaces, even of coarse grass, but the cactus grows in abundance on a soil of sand and rolled gravel; even the several rivers that flow through these plains do not seem to fertilise the grounds adjacent to them.'

Although the editor of the narratives found that they agreed closely with the notebooks which Thompson had compiled some 30 to 50 years before, the account of the central and southern grasslands clearly incorporates opinions obtained from the reports of 19th-century American explorations.

Scientific Expeditions

A new and more scientific interest in the northern grasslands began in 1857, when the British Government sent an expedition under the leadership of Captain John Palliser with the Scottish Naturalist Dr James Hector as an assistant. The accounts of the three years' work were published as Parliamentary papers between 1859 and 1863. They rapidly became known for a new and emotive sub-division of the grasslands which in certain respects was a more precise statement of Thompson's earlier but unknown attempt to differentiate. The General report (Palliser, 1863) described a semi-arid zone 'forming a triangle, having for a base the 49th parallel from longitude 100° to 114°W, with its apex reaching to the 52nd parallel of latitude'. This was surrounded to the north-east and north-west by a second sub-region called the Fertile Belt and the two were clearly outlined on an accompanying map. Although the detailed reports emphasized that there were exceptions to the general aridity of the arid zone, it quickly became widely known as Palliser's Triangle. The concept was reinforced by the report of a Provincial Government scientific expedition to the same general area in 1858 under the leadership of the Nottingham-born naturalist Henry Youle Hind. The official report of 1859 was a confusing document but the popular narrative of the following year was both more readable and widely known (Hind, 1860). According to this, the Great American Desert extended northward into Rupert's Land, where its northern limit coincided almost exactly with that of Palliser's Triangle.

The concept of a desert, or near desert, surrounded to the north-east and north-west by a broad fertile zone, would seem to have been an essential component of the perceptions of the northern grasslands during the 1860s. However, the Deed of Surrender of 1869, by which the Hudson's Bay Company gave up its monopoly and proprietary rights to the area, ignored the distinction and used the term 'Fertile Belt' to refer to the whole of the area between the Red River and the Rocky Mountains. This was probably the fault of Earl Granville, the British Colonial Secretary, who on 9 March 1869, had hastily drafted a proposed settlement, which was presented to the Company and the Canadian Government as a *fait accompli* after protracted triangular negotiations involving the British Government. In this proposal he defined the Fertile Belt as all the land lying between the American Boundary, the Rockies, the North Saskatchewan, Lake Winnipeg and Lake of the Woods. Within this context, failure to recognize an arid zone was of little immediate consequence but just over a decade later the existence of the arid triangle was challenged by the Irish-born botanist, John Macoun. In a widely circulated book Macoun (1882) admitted that

> 'Cactus flats, hills of pure sand, and large areas of excellent agricultural land will attract the attention of different observers (i.e. give rise to different perceptions) so that varied and conflicting accounts are being and will be given of it'.

His own opinion was that, apart from the areas of sand and gravel, the region would prove to contain first-class wheat land. This optimistic appraisal conflicted with observations which he had made in the field in the summer of 1880 and it is considered by some to have been influenced by his patrons, the newly formed syndicate responsible for building the Canadian Pacific Railway. The syndicate had reasons for wishing to route the railway through the heart of Palliser's Triangle, some 200 miles south of the route through the Fertile Belt which had been agreed to in 1880 in the contract with the Canadian Government. Macoun was certainly summoned to a meeting with members of the syndicate in St. Paul, Minnesota, in the spring of 1881 and would appear to have been used to promote a more favourable image of the area to be crossed by the rerouted railway. (See also Eliot Hurst's account in Chapter 14.)

Macoun's book had the effect of polarizing opinion about the environment at the heart of the Canadian grasslands. In 1883, Hind, the cautious explorer-naturalist of almost a quarter of a century before, privately published *Manitoba and the North-West Frauds. Correspondence with the Department of Agriculture, Etc., Etc., Respecting the Impostures of Professor John Macoun and Others.* Among other things he accused the Canadian Department of Agriculture of the publication and official distribution of palpably false and misleading data in over-optimistic emigration pamphlets. In many respects the debate was similar to that between Generals Hazen and Custer almost a decade earlier, concerning the value of the Northern Pacific Railroad Company's lands some 200 miles further south.

Conclusion

The British and Canadian views of the Great Plains, by showing up a marked contrast between 'the fertile belt' that ran right across the Canadian Prairies (from the Red River to the Rockies) and the 'Great American Desert' (of supposedly arid land stretching from the Missouri to the mountains), influenced those who knew about Palliser's and Hind's reports to think things worse than they were on the American Great Plains. Palliser, who was struck by the dryness of the area bordering the American plains, wrote of 'arid plains where . . . sage and cactus abound, and the whole scanty vegetation bespeaks an arid climate'. It should be pointed out, as Watson (1967) has done, that sage and cactus do not abound to anything like the extent of grass: this was grassland not sagebush country, *but to the eye of a man who already had a mental image of the Great American Desert*, sage and cactus doubtless sprang to view.

In other words, people look for what they think should be typical and, finding some examples of this, may 'type' an area, all but regardless of what in fact are its truly typical forms. It is in this way that mental images help to shape the geography of the land. In 1863 Palliser went on to report of the American Great Plains that

> 'The fertile savannahs and valuable woodlands of the Atlantic United States are succeeded on the West by a more or less arid desert, occupying a region on both sides of the Rockies, which creates a barrier to the continuous growth of settlements between the Mississippi valley and the states on the Pacific coast.'

This immense illusion, shared in by the Americans, and which as Webb (1931) points out was at its height between 1850 and 1860, just as the tide of immigration rolled up to the Great Plains, enormously affected the peopling of America, and thus the geography of the country, by keeping the plains relatively inert while the Pacific coast became the active frontier of the United States.

References

Carver, J., 1778. *Travels Through the Interior Parts of North America in the Years 1776, 1777 and 1778*, London: published for author.

Evans, J. T., 1952. Mr Evans' Journal. *Before Lewis and Clark* (Ed. Nasatir, A. P.), St Louis: Historical Documents Foundation.

Folmer, H., 1939. The Maller expedition of 1739 through Nebraska, Kansas and Colorado State to Sante Fe. *Colorado Magazine*, **16,** 161–173.

Henday, A., 1907. The journal of Anthony Henday (1754–5) copied in 1792 by Andrew Graham. In *Proceedings and Transactions of the Royal Society of Canada* (Ed. Burpee, L. J.), Toronto.

Henry, A., 1809. *Travels and Adventures in Canada and the Indian Territories Between 1760 and 1776*, New York: Riley.

Hind, H. Y., 1860. *Narrative of the Canadian Red River Exploring Expedition of 1857 and the Assiniboine and Saskatchewan Exploring Expedition of 1858*, London: Longman, Green, Longman and Roberts.

Hind, H. Y., 1883. *Manitoba and the North West Lands* ... Windsor. Nova Scotia: published for author.

Kelsey, H., 1929. A journal of a voyage and journey. ... In *The Kelsey Papers* (Eds. Doughty A. G. and Martin, C. B.), Ottawa: Public Archives of Canada.

Mackay, J., 1807. Extracts from the manuscript journal of James Mackay. *Medical Repository*, Hexade 2, 4, 27–37.

Macoun, J., 1882. *Manitoba and the Great North West*, Guelph: World Publishing Company.

Palliser, J., 1863. *Journals, Detailed Reports and observations* ... London: HMSO

Pond, P., 1785. Map presented to congress, copied by St. John de Crèvecoeur. London: British Museum Additional Manuscript 15, 332c.

Thompson, D., 1916. Narrative. In *Publication No 12 of the Champlain Society* (Ed. Tyrrell, J. B.), Toronto.

Truteau, J. B., 1921. Description of the upper Missouri in 1796. In *Mississippi Valley Historical Review* (Ed. Abel, A. H.), **8,** 157–179.

la Vérendrye, P. G., 1927. Journal of the expedition ... to reach the western sea. In *Publication No. 16 of the Champlain Society* (Ed. Burpee, L. J.), Toronto.

Watson, J. W., 1967. Mental images and geographical reality in the settlement of North America, Nottingham: Cust. Memorial Lecture.

Webb, W. P., 1931. *The Great Plains*, New York: Ginn.

4
The West: Wealth, Wonderland and Wilderness

ALFRED RUNTE

Samuel Bowles (1869), editor and publisher of the Springfield Massachusetts *Republican*, had no reason to exaggerate unnecessarily. The facts about to be written down in the preface of his latest book were far more fascinating than any fictional embellishments. As he picked up his pen, vivid recollections of his two recent journeys across the United States, undertaken in 1865 and 1868, flashed and reflashed through his mind. Again the great disparity between the trans-Mississippi West and the East held his fascination. 'The two sides of the Continent', he began, 'are sharp in contrasts of climate, of soil, of mountains, of resources, of production, of everything.' Nowhere but in the 'New West', he exclaimed, were 'broader and higher mountains; nowhere richer valleys, nowhere more bountiful deposits of gold and silver, quicksilver and copper, lead and iron; nowhere denser forests, larger trees, nowhere so wide plains, nowhere such majestic rivers.' Throughout the Rocky Mountains and along the Pacific Coast, he continued, was 'a nature to pique the curiosity and challenge the admiration of the world', all in the midst of 'a wealth of minerals and a wealth of agriculture' that awed one by their 'boundlessness'. There could be no doubt, he concluded, that the West was 'destined to develop a society and a civilization, a commerce and an industry, a wealth and a power' that eventually would contribute to American world supremacy.

The West as Climax

However nationalistic, Bowles' impressions about the West typified contemporary thoughts about that vast territory. The striking contrasts of the region, in particular its magnificent, overpowering topography, amazed 19th-century Americans, especially those native to the more subdued physiography of the eastern United States. As Bowles put it, in the West, nature had 'created originally, freshly, uniquely, majestically', and many thoughtful people were in strong agreement. Somehow the scope and splendour of the region symbolized every quality attributable to the nation: strength, greatness and freedom. Beyond the Mississippi, Americans confidently exclaimed, were limitless resources and spectacular scenery admirably suited for a country on the rise to the peak of prestige in the world community of nations.

This claim had even more impact because of the pattern of national development. The United States expanded from east to west, thus it was the Far West, not the East, that finally stood out as the most impressive portion of the country. Eastern regions were predominantly characterized by hilly provinces, rolling valleys and mountains of moderate height. By the 19th-century, however, in spite of their picturesque beauty, these physical features seemed tame and common when compared to the spectacular topography of the West. The point is that the forces of nature and the fortunes of history combined to make the West the climax of national development. Just as Americans opened their final act of continental expansion, the boldest and most magnificent setting in their history unfolded before them. The growth of the United States was not a drama in reverse, one that closed on an anticlimactic note in a tranquil, ordinary environment. Instead, as Pomeroy (1957) has noted, the conquest of the West symbolized the zenith of national expansion, as Americans rounded out their country on what they believed was a grand crescendo, struck in the midst of natural wealth and scenic resources unparalleled in their collective experience.

This popular conviction was strengthened by the relationship between the Far West and the nation as a whole. Added to the unique qualities of the section was its physical separation from the emerging urban centres in the eastern half of the United States. Significantly, no other time in American history found so many people merely an audience to what was taking place on the frontier. For the great majority of 19th-century Americans the Rocky Mountains and the Pacific Coast lived only in books and periodicals, not in actual experiences. Moreover, the separation of city dwellers from nature reinforced nostalgic attachments to the land. These factors do much to explain why the West became and remained an enduring symbol of American culture. Distance lent even greater mystery and romance to an already fascinating region, one that most Americans could encounter only from the confinement of their new urban-based reality. Under these circumstances the serenity and beauty of the West became cherished spiritual resources, especially in the 20th-century.

The West as Wealth

Of course, it was the promise of new wealth and opportunity, not scenic beauty, that first attracted American attention to the Far West. The earliest major economic lure was the beaver fur trade, which flourished throughout the Rocky Mountains between 1810 and 1835. By 1840, however, the enterprise was in rapid decline, victimized by over-exploitation and style changes in the East and Europe. Although a few hardy mountain men were still able to exist by trapping, the thrust of economic development in the West shifted to land settlement and mineral exploitation. In 1846 the United States peacefully gained exclusive control of the Pacific North-west from Great Britain and, just two years later, won California and most of the South-west from Mexico, as spoils of war. These major acquisitions assured the United States of exclusive dominion over the rich valleys of the Pacific Coast, which since the late 1830s had beckoned to an ever increasing number of prospective American farmers and ranchers. Of even greater immediate significance, it was the United States, not Mexico, that controlled California when the Gold Rush opened in 1848–49. For approximately five years thereafter, the western slope of the Sierra Nevada hummed with the frenzied activity of 100,000 prospectors. Then, as the accessible strikes played out, the mining frontier shifted into other territories, including Colorado, Nevada and Montana. Indeed, as Billington (1965) indicated, long before Samuel Bowles first crossed the continent in 1865, the presence of extensive mineral deposits in the West, not to mention timber, grazing and agricultural lands, had assured the economic importance of the region.

Exploitation Period

Even as the tempo of exploitation in the West quickened, a few Americans began to consider the territory as something more than a vast storehouse of wealth. As early as 1832, for example, George Catlin (1851), a Pennsylvania-born artist dedicated to the study and painting of Indian life, proposed that 'A Nation's Park' be established on the Great Plains, 'containing man and beast, in all the wild and freshness of their nature's beauty'. In support of his suggestion he argued that civilization would soon overwhelm the West and push its untamed inhabitants aside. Of course, conquest was precisely what most Americans of the period had in mind. Yet the passing decades of the 19th century, characterized by increasing urbanization and industrialization in the United States, gradually influenced others to adopt a preservationist point of view. Close behind Catlin in concern for the American landscape were such noted scholars as the transcendentalist Henry David Thoreau, who in 1851 proclaimed that 'in wildness is the preservation of the world'. Additional Americans would soon say as much, especially in conjunction with the enlargement of towns and cities. Fostered by the pace of urban life, the idea of unspoiled open spaces to refresh the human spirit acquired greater meaning and value, as Nash (1967) has indicated.

By the 1870s this change of heart had become more apparent. To be sure, the overwhelming number of Americans continued to praise economic exploitation, yet interspersed with these accolades were some signs of doubt. Destructive industrial enterprises, especially mining, no longer struck a totally positive chord in the minds of travellers and writers.

Recalling his 1868 visit to the gold and silver mining districts west of Denver, Colorado, for example, Samuel Bowles exclaimed: 'The clang of mills, the debris of mines, the waste of floods leave nothing that is inviting, except money-making.' Similarly, Verplank Colvin (1872), a noted eastern surveyor and scientist, called the same region of the Colorado Rockies a 'barren and desolate . . . wilderness of stumps', swept clean by the miner's axe. Like Bowles, who preferred to 'seek a more pleasing neighboring valley', Colvin searched for unspoiled beauty that 'compensates for the vandalism'. Nor was this a difficult task in 19th-century Colorado. Merely a short stagecoach ride enabled Colvin to escape the blight of the mining region and stand on a 'lofty ridge' where the view opened out onto 'loftier, haughtier summits, dazzling in their spotless robes of white'. Perhaps mining was a necessary and lucrative enterprise, but neither Bowles nor Colvin could justify more than a backward glance at it, especially when they were in the midst of such 'grand surroundings'.

Similar feelings were strongly evident wherever mining took place. Benjamin P. Avery (1874), editor of *Overland Monthly Magazine*, described the Sierra Nevada goldfields of California as 'desolate'. In place of 'foaming cascades which used to gleam like snow in the primeval woods', he lamented, were 'cataracts of mud' and 'chocolate-colored rivers, choked for a hundred feet deep with mining debris'. As for the woods: 'alas!' he exclaimed; in most instances they had been 'obliterated by the insatiate miner' (Figure 4.1).

Figure 4.1. Hydraulic mining in the Northern California river valleys. Reproduced through the courtesy of the Bancroft Library, University of California, Berkeley, California

However, like Samuel Bowles and Verplank Colvin, Avery encountered no difficulty in his search for quiet solitude. He urged Sierra travellers to follow him 'into the unbroken forests' and 'deep canyons' where they could 'be alone awhile, and free'. Indeed, he wrote, it was 'a relief to get out of sight of the crater-like chasms' left by the miners, and to 'leave the line of travel where the ravaging axe converts the solitude into noisy blanks'. Yet the seemingly endless abundance of unspoiled scenery led Avery to ignore the future implications of his own message. Not only did he consider destructive exploitation part of the normal course of events, he firmly believed that nature would 'heal the wounds inflicted by man'. Moreover, a 'new era of permanent settlement and culture', he maintained, eventually would restore other 'rude places of old'.

The similar manner in which Avery, Bowles and Colvin found the destructive effects of mining objectionable, yet shrugged them off, is indicative of the deeply ingrained American attitudes they still shared, in spite of their embryonic concern. Almost three centuries in the presence of wilderness had convinced most people in the United States of the inexhaustibility of natural resources. This belief was accompanied by a strong commitment to material progress. Americans wanted to settle the wilderness, to build towns, cities and businesses where previously none had existed. Understandably, during the 19th-century the true extent of the West strongly reinforced these ideas. The point is that the region *did* seem limitless, especially during the 1860s and 1870s when its overall population was relatively small and concentrated. Even if one abhorred the ruinous waste of mining or lumbering operations, it was still far too easy to turn away and look off in some unspoiled direction.

Exploitation and Conservation

This sidestep was a typical reaction of Americans in every part of the West, whether native-born or newly arrived in the region. Abundance led to complacency and the inability to comprehend the true rate and force of change. In 1872, for example, Joaquin Miller, the noted western poet and author, described the forests of the Columbia River basin in Oregon and Washington as 'splendid! green! black! boundless!'. Although he observed many lumber mills in operation along the edge of the river, their activity failed in the least to upset him. Instead, he considered the woodlands limitless and maintained that lumbering had 'hardly made a dimple in the inexhaustless sea of timber'. Frances Fuller Victor (1872), the renowned Oregon writer and historian of the Pacific North-west, shared a similar feeling of optimism. 'The panorama of grandeur and beauty' along the Columbia River 'seems endless', she wrote. Nor were her enraptured impressions dimmed by what she described as 'unsightly' and 'unlovely' land clearance practices at the river's mouth near the town of Astoria. Stand 'facing the sea of the river with your back to the half-cleared lots', she suggested, then 'the view is one of unsurpassed beauty'. Yet when even this simple turnabout failed, the destruction was still easy to justify. 'In compensation for the ugliness of burnt forests', she counselled, 'the shape of the country is partially revealed, and one discovers fine level benches of land fit for farming'. However unpleasant ravaged woodlands might seem, in actuality their loss promised a new age of 'wealth, both mineral and agricultural'.

Although this view did not go unchallenged, prior to 1890 the American myth of inexhaustible resources was under only limited attack, especially in the West. The apparent unending abundance of natural wealth in the region lulled most of its inhabitants and visitors into complacency about the future. Yet there were exceptions. Even as Joaquin Miller described the forests of the Pacific North-west as 'exhaustless', Evans (1871) termed this assessment an 'erroneous and unjustified presumption'. Eight hundred lumber mills were in operation along the Pacific Coast, he maintained, while every year improvements in the railroad network made previously inaccessible forests penetrable. Soon a 'war of extermination' against these woodlands could begin, he sadly concluded, and the 'magnificent forests' of California, Oregon and Washington Territory would 'be numbered with the things of the past'.

Such dismal projections, however, were understandably discounted or ignored. As in all frontier cultures, 19th-century westerners depended for a living on the use of natural resources. Consequently, they believed that the economic health of their region depended on greater exploitation, not less, and viewed with suspicion early proposals to regulate lumbering, mining and grazing. Not until the turn of the century did western businessmen, settlers and politicians begin to appreciate the conservation needs of their region, and then only reluctantly. Conservation thought in the West would not gain added force and credibility until the destructive results of resources-devastation became more visible, permanent and widespread. As long as the section seemed vast enough to absorb the ruinous impacts of exploitation, only a few people perceived the true limits of the region's wealth.

The West as Wonderland

The concept of protecting the natural wonders of the West was far easier to adopt. Perhaps forests seemed endless and mineral deposits bottomless, but the distinctive features of western natural marvels required little effort to prove their uniqueness. The reason is that such topographical phenomena were rare. There was only one Yosemite Valley with its sheer cliffs and plunging cataracts, only one Yellowstone wilderness filled with steaming geysers, boiling pools and other thermal attractions. From the moment of discovery these features fascinated the entire nation. Of even greater significance, an appreciation of their spectacular qualities led to successful efforts for their preservation.

The first natural wonder in the West to receive nationwide attention and acclaim was Yosemite Valley, a spectacular gorge located high in the central Sierra of California. A party of white adventurers undoubtedly saw this

Figure 4.2. Yosemite Falls in Yosemite Park, California. The Falls drop a total of 1900 ft in three separate cascades: a sepectacular sight at any time but especially during the spring. Reproduced by courtesy of the United States National Park Service

valley in 1833, but not until 1851, when a battalion of miners pursued a band of marauding Yosemite Indians into the chasm, was its existence finally confirmed and publicized. Within a few years the soaring cliffs and majestic waterfalls of Yosemite had given it the distinction of being the most famous 'wonder' and 'curiosity' in the West (Figure 4.2).

Shortly after the discovery of the valley, however, the Sierra yielded up another of its great natural secrets, the Sequoia Gigantea, or Sierra redwoods. In 1852 a hunter stumbled upon a grove of these giant trees in the Stanislaus River high country, some miles north of the Yosemite region. Soon afterward, as Farquhar (1965) points out, news of their whereabouts spread rapidly across the country, and once their existence had been confirmed, the Sequoias joined Yosemite Valley in the category of great western wonders.

Yet the significance of Yosemite and the Sierra redwoods transcended their fame as unusual natural attractions. Of equal importance was the indisputable fact that only in the United States did such wonders exist. For decades the nation had suffered the embarrassment of a dearth of cultural achievements. Unlike the established countries of Europe, the United States lacked great art, literary attainments and time-honoured traditions. Whenever Europeans pointed out these deficiencies (which was often) it was always at the expense of the national pride of their young trans-Atlantic rival. As early as 1800, Americans sought refuge from these barbs in the scope and magnificence of the North American continent. Surely, the argument went, a nation with such grand surroundings was destined for a great future. The only problem with this claim was that Europe also had impressive landscapes, either comparable with or sometimes superior to America's most famous natural attractions. Not until the 1850s, after the conquest of the Far West and the discovery of Yosemite Valley and the Sierra redwoods, did Americans feel truly confident about the scenic superiority of the United States.

Strengthened in their conviction that no country could match the spectacular scenery of the United States, Americans proudly challenged the landscapes of even the most magnificent trans-Atlantic nations. An early victim of this renewed surge of scenic nationalism was Switzerland, long recognized as the gem of Europe. The claims of Lt. Col. A. V. Kautz (1874), a noted Civil War veteran, were typical. Recalling his adventures on the slopes of Mount Rainier in Washington Territory, he asserted that the region possessed 'mountain scenery in quantity and quality sufficient to make half a dozen Switzerlands'. Other writers were even more specific. 'When we come to the Yosemite Falls proper', declared a California minister, 'we behold an object which has no parallel anywhere in the Alps'. Nor could any Swiss valley, he maintained, match the symmetry and magnificence of Yosemite. William H. Brewer (1930), a member of the California Geological Survey, shared identical views. The 'crowning glory' of Yosemite Valley, he wrote, 'is the Yosemite Fall. . . . It comes over the wall on the far side of the valley, and drops 1,542 feet the first leap, then falls 1,100 more in two or three more cascades, the entire height being over 2,600 feet! I question if the world furnishes a parallel', he concluded, 'certainly there is none known'. But even Yosemite Valley's delicate Bridal Veil Fall only a third as high as Yosemite Fall, seemed 'vastly finer than any waterfall in Switzerland', he maintained, 'in fact finer than any in Europe'.

By the 1860s such statements about Yosemite were commonplace. Even Horace Greeley (1860), the sceptical editor of the New York *Tribune*, declared the valley 'the most unique and majestic of nature's marvels'. Unfortunately, he saw Yosemite in August 1859 when the great cataracts were only mere trickles, a disappointment that led him scathingly to denounce Yosemite Fall itself as nothing but 'a humbug'. Yet such derisive comments were unusual, and statements like that by Samuel Bowles (1869) remained the overwhelming view. 'THE YOSEMITE!' he exclaimed, 'As well interpret God in 39 articles as portray it to you by word of mouth or pen.' Indeed, he continued, 'only the whole of Switzerland' surpassed the valley, while 'no one scene in all the Alps' could match its 'majestic and impressive beauty'. Although he too saw Yosemite Fall in August, when the streams were at 'their feeblest power', he nevertheless believed that 'earlier in season, when ten times the volume of water pours down, it must, indeed, be a feature of fascinating, wonderful beauty.'

Assured that the finest scenery in Europe was inferior to landscapes in the United States, Americans confidently compared the natural marvels of their own country. What emerged from this self-examination was an even greater appreciation of the tremendous topographical differences between the West and the East. Awed by the magnificence and grandeur of the Columbia River, for example, Frances Fuller Victor was moved to claim that the Hudson River in New York State, 'which has so long been the pride of America, is but the younger brother of the Columbia'. Niagara Falls was another eastern wonder frequently brought up for comparison. Although most writers acknowledged its inspiring beauty, to many observers the size of western cataracts seemed far more impressive and overwhelming. Yosemite Falls 'is fifteen times as high as Niagara Falls!' exclaimed an astonished Samuel Bowles, but for Albert D. Richardson (1867) of the New York *Tribune*, the height of the California wonder was even greater, 'sixteen times higher than Niagara'. Yet, regardless of what imperfect statistics were called upon for measurement, the startling effects were the same. 'Think of a cataract of half a mile with only a single break!' Richardson commanded. As for Yosemite gorge, 'Niagara itself', he predicted, 'would dwarf beside the rocks in this valley'.

The importance of these and similar impressions on the general public cannot be minimized; wonders like Yosemite Valley filled a vital national need. For the first time in almost a century Americans were confident that the United States had something valuable to contribute to world wonder. Let Europe have its castles and old ruins, they exclaimed, the United States had 'earth monuments' and giant redwood trees that were standing long before the birth of Christ, let alone the origins of European civilization. The point is that the agelessness of nature lent a feeling of continuity and stability to the young nation. Natural marvels in the West compensated for America's lack of old cities, aristocratic traditions and other reminders of European human and cultural achievements. As Richardson put it: 'In grand natural curiosities and wonders, all other countries combined fall far below' the United States. Indeed, the validity of such claims was great comfort to people still living under the shadow of Milton, Shakespeare and the Sistine Chapel.

The Need for Protection

It is therefore not surprising that Americans eventually supported measures to protect the natural wonders of the United States. Moreover, it was only logical that the first successful efforts for scenic preservation occurred in the West, for it was this section of the country which contained the rarest landscape phenomena. Concern first led to action in 1864 when a small group of Californians, anxious to preserve the Yosemite Valley and a grove of Sierra redwoods from growing private abuse, persuaded their junior United States senator, John Conness, to propose legislation to set aside these wonderlands for public recreation. The timing of the preservationists could not have been more appropriate. By 1864 Yosemite and the Sequoias had been the subjects of scores of paintings, lithographs and magazine articles. Most Americans, including political leaders, at least knew something about the valley and the great trees. Senator Conness exploited this familiarity with great skill. After referencing Yosemite Valley and the Sierra redwoods as the 'greatest wonders of the world', he reminded his Senate colleagues that the British had once called the Sequoias nothing but a 'Yankee invention'. Spurred on by this key slight to national pride, a Yosemite park bill cleared Congress with ease. On 1 July, 1864, President Abraham Lincoln signed the measure into law.

Under the Yosemite Park Act, California received the valley for 'public use, resort, and recreation', provided the land remained 'inalienable for all time'. The Mariposa Grove of Sierra redwoods, located approximately 30 miles south of the gorge, was transferred to the state under similar conditions. Fortunately, both Yosemite and the Sequoias were on the public domain, a circumstance that allowed tight-fisted federal officials to indulge in the luxury of scenic preservation. Indeed, decades would pass before Congress actually *paid* for scenery. Yet the Yosemite campaign marked the beginning of an idea of nationwide importance: that governments were obligated to protect unique natural areas for the general public. Through the years more and more Americans supported this democratic inspiration, as Farquhar (1965) shows, and by the 20th century the West contained the framework of an extensive park system, both state and federal.

To be sure, Yosemite Valley and the redwoods were not the only great natural marvels in the region considered worthy of preservation. Just a few years after the creation of Yosemite State Park, attention focused on the discovery of another unique wonderland, the Yellowstone wilderness in the north-west corner of Wyoming Territory. For many years this remote portion of the Rocky Mountains was *terra incognita*, known only to Indians and a few hardy trappers who chanced to run across its strange thermal features. Not until the late 1860s, after more weird tales about this region had filtered back to civilization, did white adventurers begin serious efforts at exploration. Between 1869 and 1871 three separate parties of men penetrated the area, the most famous of which were the Washburn Expedition of 1870 and the Hayden Survey of 1871. Through articles, photographs, paintings and other official reports, individual members of these separate ventures introduced the world to the wonders of Yellowstone. Subsequently, concern about the future of these phenomena, as Cramton (1932) has indicated, led some of the explorers into a campaign to preserve Yellowstone, under legislation very similar to the Yosemite Act of 1864.

Like the great waterfalls and cliffs of Yosemite Valley, the natural features of Yellowstone fascinated the American public. Indeed, the reaction to the discovery of the region's geysers, boiling springs, coloured hot pools, bubbling mudpots and other grand attractions took on an almost carnival atmosphere. Descriptions of the territory by its explorers only heightened this 'believe it or not' tone. They portrayed the geysers and other volcanic phenomena as 'freaks' and 'curiosities' of nature, as 'weird' and 'terrifying' marvels.

'We pass with rapid transition from one remarkable vision to another', wrote Dr. Ferdinand V. Hayden of the U.S. Geological Survey, 'each unique of its kind and surpassing all others in the known world.' Nathaniel P. Langford (1871) the most prolific writer of the 1870 Washburn Expedition, was even more enthusiastic.

> 'You can see Niagara,' he wrote, 'comprehend its beauties, and carry from it a memory ever ready to summon up before you all its grandeur. You can stand in the valley of the Yosemite, and look upon its mile of vertical granite, and distinctly recall its minutest feature, but amid the canyon and falls, the boiling springs and sulfur mountain, and, above all, the mud volcano and the geysers of the Yellowstone, your memory becomes filled and clogged with objects new in experience, wonderful in extent, and possessing unlimited grandeur and beauty.' Yellowstone, he maintained, 'is a new phase in the natural world . . ., and, while you see and wonder, you seem to need an additional sense, fully to comprehend and believe.'

During the winter of 1871–72, similar statements provided the impetus for the Yellowstone park campaign, of which both Langford and Hayden were key spokesmen. Once again, preservation advocates were able to profit from the national pride and confidence spectacular landscapes inspired, and from the fact that Yellowstone, like Yosemite, was already government property. Protection merely entailed a land reclassification, and on 1 March 1872, Yellowstone became the first National Park in the United States.

Although the new reserve owed its ideological existence to the Yosemite idea, two major differences set Yellowstone apart from the earlier park. The first of these was with regard to management. Yellowstone, in contrast to Yosemite, remained under federal control. In addition, the national park comprised an area of more than 3300 square miles, while the California preserve, including the Mariposa redwood section, covered only 40 square miles. These differences have led some historians to trace the origins of the park idea to Yellowstone instead of Yosemite, a position that ignores the actual 1864 roots of the scenic preservation ethic. Yet one fact cannot be disputed. Both wonderlands became parks because of their unrivalled magnificence and exceptional curio value. They measured up to what most Americans idealized as scenery—grand, bold landscapes, believed to have no parallel anywhere in the known world. In a nation blessed with an abundance of unspoiled landscapes, only the most unusual, as Ise (1961) shows, were considered worthy of preservation, especially during the 19th century.

Yet aesthetic conservation had one other very practical limitation: economic reality. Unquestionably, both Yosemite and Yellowstone, however magnificent and representative of national pride, could not have become parks if their respective lands had contained valuable natural wealth. Indeed, congressmen were almost unanimously concerned about the economic potential of lands in the proposed reserves. Western senators and representatives were especially anxious to hear assurances like those expressed by Congressman Henry Dawes of Massachusetts, who described Yellowstone as 'rocky, mountainous, full of gorges', and 'absolutely unfit for agricultural purposes'. Nor was his statement that 'even Indians can no more live there than they can upon the precipitous sides of Yosemite valley' mere rhetoric. Unless there had been some solid support for this claim, as was indeed the case, the Yellowstone park proposal would have been defeated. Whenever Americans were given a choice between the preservation of scenery or the use of natural resources, wealth simply commanded more value than wonderlands (Runte, 1972).

The point is that the United States was able to 'afford' Yosemite and Yellowstone. Their creation had no adverse economic impact on western business interests. Of what use were geysers and steep cliffs, except to look at? The value of natural resources, however, needed no justification or explanation. Not until the 1890s was the federal government able to regulate the exploitation of publicly owned timberlands and mineral reserves in the West. The protection of rugged, worthless marvels of nature was one thing, political interference with lumbering, grazing, or mining interests quite another.

The economic concerns of the West during the late 19th century are easily understood when one recalls the observations of writers like Samuel Bowles, Frances Fuller Victor, Joaquin Miller and others. They represented the majority opinion in the United States that natural resources in the West were limitlessly abundant. It is therefore not surprising that few Americans as a whole, and even fewer westerners, accepted the warnings of a few prophets of doom who put a damper on the theory of inexhaustible resources. The economic well-being of the entire nation depended on greater exploitation, and westerners especially wanted to believe in the unlimited truth of their vision, not in abstract statistical predictions.

Limits to Development

Nevertheless, by the 1880s some Americans began to understand the true limits of their nation. The coming of the railroad, a rapidly expanding population and the arrival of big corporations in the West convinced more doubters that the end of the great American frontier was close at hand. In 1883, for example, although deep in the heart of the rugged Bitterroot Mountains on the Montana–Idaho border, a writer for *Lippincott's Magazine* announced the end of the Wild West. 'The Western limit of the Great American Wilderness of our boyhood has been reached', Bailie (1883) claimed. In just a few short years railroads and other technological developments had penetrated every portion of Montana, forcing 'old timers' and 'wild game' either to move aside or to retreat to the most remote corners of the territory. Since the permanent results of such rapid change simply could not be denied, the author was forced to conclude on a sobering note. 'The great Northwest,' he lamented, 'the last remaining stronghold of wild nature, has been invaded, has been vanquished'.

John Muir (1875), the renowned California explorer, naturalist and aesthetic conservationist, backed up this assessment with regard to the Pacific Coast states. During the 1870s he too often had been misled by the popular contention that scenic resources in the West were broadly abundant, yet a few years of widespread destruction, especially in his beloved Sierra, for ever convinced otherwise. In 1875 he was able to describe the High Sierra as a 'vast wilderness of mountains' that remained 'almost wholly unexplored,' penetrated by only 'a few nervous raids . . . from random points adjacent to trails'. In 1890, however, precisely the opposite viewpoint characterized his opinions. By then, he sadly recognized, most of the Sierra had been transformed from a 'garden' to 'rough taluses' devoid of flora and fauna. 'All the flowers are wall-flowers now,' he lamented, '. . . to a great extent throughout the length and breadth of the Sierra.'

The passing away of the great frontier of the West came as an unpleasant shock to many Americans. Ever since 1607, when the first permanent English settlement had been established at Jamestown, Virginia, the United States never had been without a great uninhabited wilderness to the west of the civilized portions of the country. Even as late as the 1870s, it had been possible to believe the nation would never run out of timber, minerals or open lands. The fact that wonderlands such as Yellowstone were not officially discovered until early in the decade only reinforced this conviction. Yet within the space of a few years this belief no longer seemed tenable. Suddenly Americans were forced to concede that their nation had limits and that its remaining wilderness areas would not be an everlasting cornucopia. Blocked by the shores of the Pacific, the American myth of inexhaustible resources at last had run its course.

Although this realization came reluctantly to many Americans, the close of the frontier lent a new sense of urgency to protection programmes. In 1890, following almost two decades of preservation neglect since the creation of Yellowstone National Park, the federal government established General Grant, Sequoia and Yosemite National Parks, all in the High Sierra of California. Nine years later Mount Rainier in the state of Washington received federal park status, and in 1902 Crater Lake, the gem of the Oregon Cascades, joined the growing national reserve system. By 1916 the United States government maintained 11 national parks and over 30 scenic and archaeological monuments, all west of the Mississippi River.

The revitalization of the park idea in the West coincided with the formation of a national forest system throughout the Rockies, the Sierra and the Cascades. In 1891 Congress approved the Forest Reserve Act, designed to protect the great timberlands on the public lands of the western states. Under its provisions, as Hays (1959) has recorded, more than 150 million acres of publicly owned forests were placed under permanent government management between 1891 and 1910.

The West as Refuge

The creation of national parks and national forests on the public lands of the West reflected not only a growing awareness of the need for conservation, but a commitment to protect some of the wonder and majesty of the region. True, much of its mystery and romance was gone, for like the rest of the country the West had become a known quantity. Yet the overpowering uniqueness of the region still fascinated Americans. Even if its hidden wonders had all been discovered, its vastness remained an enduring symbol of national greatness and of growing importance, a reassurance that somewhere the United States was still freshly different. The need for such reinforcement was heightened by the urbanization of the nation. As more Americans began to live in large towns and burgeoning cities, their separation from open country and unspoiled nature rekindled their longing for familiar landscapes of the past. Locked into the drudgery and crowdedness of urban life, many city dwellers looked to the West as a place where some of the old, untrammelled America still held sway. Perhaps John Muir (1898) best summed up this feeling when he wrote:

> 'Thousands of tired, nerve-shaken, over-civilized people are beginning to find out that going to the mountains is going home. Wilderness,' he added, 'is a necessity, and mountain parks and reservations are useful not only as fountains of timber and irrigating rivers, but as fountains of life.'

Initially, of course, the first major parks were established in the West not to preserve its wilderness qualities, but to protect the unique natural features of the region. Striking wonders like Yosemite Valley required little publicity to convince the nation of their rarity and magnificence. It was easy to admire landscapes of great contrast and bold appearance, or waterfalls that were the highest in the known world. Sheer size, whether of trees, mountains, cataracts or geysers, was comprehensible to everyone. Wilderness appreciation was another matter. Wild country depended on the absence of human interference, not merely on the presence of spectacular scenic phenomena. Any unspoiled landscape might be a wilderness, regardless of whether or not it contained the highest waterfall, the widest tree or the deepest canyon in the world. Yet more time was needed before people understood the value of this philosophy. The first wilderness preserve in the nation was not created until 1924, when the United States Forest Service set aside a portion of the Gila River watershed in New Mexico. Forty more years elapsed before Congress passed the Wilderness Preservation Act of 1964. Only then did wilderness protection become law, not merely bureaucratic policy. Nevertheless, the commitment was still not total. Wilderness legislation suffered from the same economic limitations that for years had jeopardized the national park system. Until 1983 backcountry areas are to be open to mineral exploitation, an intrusion glaringly inconsistent with the whole idea of wild country.

Nor was conservation history always comforting for those who wanted to weaken this dangerous legislative loophole. Too many controversies, both past and present, indicated that wilderness preservationists could expect hard fighting ahead before their ideal achieved a truly lasting basis of support. Challenges to the national park system provided especially dramatic examples. Foremost among them was the damming of the Hetch Hetchy Valley, which although a part of Yosemite National Park, was turned over to the city of

San Francisco in 1913 for use as a municipal water supply reservoir. The loss of this spectacular gorge, considered by many a rival of Yosemite Valley, was followed in the 1920s by schemes to divert the lakes and rivers of Yellowstone National Park for irrigation purposes. Although these raids failed, after 1950 the dam builders returned, this time with proposals to construct reservoirs in Dinosaur National Monument and Grand Canyon National Park. In each instance aesthetic conservationists organized to defeat these plans, determined to prevent the re-occurrence of another 'Hetch Hetchy Steal'. To be sure, the hard object lessons of that earlier reversal were never forgotten, and with the successful conclusion of the Grand Canyon dams controversy in the 1960s, scenic preservationists had become a political force of significant and growing influence (Jones, 1965).

Yet modern conservation victories pointed to more than the political emergence of aesthetic preservationists. Of greater importance were indications of a fundamental change in the American environmental perspective. The entire landscape, not merely its great wonders, had become the focus of public concern. Even the role of national parks as museum pieces of nature was being supplemented by a deepening sense of urgency to manage and protect them for their spiritual and restorative qualities as well. Nothing was more indicative of this trend than the renewed popularity of past leaders in the American preservation movement, especially John Muir, and the noted wildlife biologist, Aldo Leopold, often called the father of ecological conservation. Just as Muir had termed a pilgrimage to the mountains 'going home', so Leopold wrote: 'I am glad I shall never be young without wild country to be young in. Of what avail are forty freedoms without a blank spot on the map?' Significantly, this is precisely the question millions of modern Americans are asking themselves as the environmental crisis worsens. Suddenly the Far West has even greater meaning and importance, for here are the last broad and uninhabited stretches of the nation, the last possible havens for 'tired, nerve-shaken, over-civilized people'. The question is whether or not the desire for wealth could permanently be balanced with the ideals of aesthetic and wilderness conservation. One thing is certain: if wild country is to remain a part of the national experience, the West offers the last hope.

References

Avery, B. P., 1874. Summering in the Sierra: up the western slope. *Overland Monthly*, **12,** 81–83.

Bailie, W. A., 1883. Mountain trails of Montana. *Lippincott's Magazine*, **32,** 492–497.

Billington, R. A., 1956. *The Far Western Frontier, 1830–1860*, New York: Harper and Brothers.

Bowles, S., 1869. *Our New West*, Hartford: Hartford Publishing.

Brewer, W. S., 1930. *Up and Down California in 1860–64* (Ed. Farquhar, F. P.), New Haven: Yale University Press.

Catlin, G., 1851. *Illustrations of the Manners, Customs and Conditions of the North American Indians*, London: Henry G. Bohn.

Colvin, V., 1872. The dome of the continent. *Harpers New Monthly Magazine*, **46,** 22, 28.

Cramton, L. C., 1932. *Early History of Yellowstone National Park and its Relation to National Park Policies*, Washington, D.C.: U.S. Government Printing Office.

Evans, R., 1871. Western woodlands. *Overland Monthly*, **6,** 226–227.

Farquhar, F. P., 1965. *History of the Sierra Nevada*, Berkeley and Los Angeles: University of California Press.

Greeley, H., 1860. *An Overland Journey from New York to San Francisco in the Summer of 1859*, New York: Saxton, Barker & Co.

Hayden, F. V., 1872. The wonders of the west: more about the Yellowstone. *Scribner's Monthly*, **3,** 396.

Hays, S. P., 1959. *Conservation and the Gospel of Efficiency*, Cambridge. Mass.: Harvard University Press.

Hutchings, J. M., 1865. *Scenes of Wonder and Curiosity in California*, London: Chapman and Hall.

Ise, J., 1961. *Our National Park Policy: A Critical History*, Baltimore: Johns Hopkins University Press.

Jones, H. R., 1965. *John Muir and the Sierra Club: The Battle for Yosemite*, San Francisco: Sierra Club Press.

Kautz, A. V., 1874. Ascent of Mount Rainier. *Overland Monthly*, **14,** 394.

Langford, N. P., 1871. The wonders of the Yellowstone. *Scribner's Monthly*, **2,** 1–17; **3,** 113–128.

Muir, J., 1875. Studies in the Sierra: mountain building. *Overland Monthly*, **14,** 65.

Muir, J., 1898. The wild parks and forest reservations of the west. *Atlantic Monthly*, **81,** 15.

Nash, R., 1967. *Wilderness and the American Mind*, New Haven; Yale University Press.

Pomeroy, E., 1957. *In Search of the Golden West: The Tourist in Western America*, New York: Knopf.

Richardson, A. D., 1867. *Beyond the Mississippi*, New York: Bliss & Co.

Runte, A., 1972. Yellowstone: it's useless, so why not a park? *National Parks and Conservation Magazine: The Environmental Journal*, **46,** p. 7.

Victor, F. F., 1872. *Overland Monthly*, **10,** 75, 146, 229.

5
The Image of Nature in America

J. WREFORD WATSON

Most Americans in their move into or across America had to pit themselves against nature, go back to nature, or master nature: hence nature played a very important part in their lives. Indeed, one aspect of the frontier theory was that in coming to terms with nature the American had come to terms with himself. He had to strip himself of what was not essential in his European past to become the new man fit for his American future. For generations much of the continent seemed untouched and in its pristine state; a great deal of it was empty and offered room for settlement; a large part was fertile and fit for development; and people could be as free to use it as they knew how.

×

The Second Eden

From the beginning people were fascinated and moved by nature. 'The place they had thoughts on,' said Bradford, 'was some of those vast and unpeopled countries of America, which are frutfull and fitt for habitation. . . .' Edward Johnson went further. 'Know this is the place,' he wrote of New England, 'where the Lord will create a new Heaven and a New Earth in new Churches and a new Common-wealth together'. Here was a chance to 'sow this yet untilled Wildernesse withall', with faith and good works. It is true, reality sometimes belied vision, and Johnson describes the tremendous toils involved in planting the wilderness, but the gains were worth the struggle. Higginson waxed eloquent on the climate of the region, claiming that 'a sup of New-England's Aire is better than a whole draft of old England's ale', while Shepherd vouched for the 'Healthfulness and great increase of posterity' the settlers would earn. Virginia's climate, being much milder in winter, drew more praise. 'The country is not only plentiful,' wrote Hammond, 'but pleasant . . . in regard to the brightness of the weather'. Maryland was described by

Hammond as a country of 'extraordinary goodness'. Pennsylvania seemed to its founder, Penn, 'a fast fat earth', where the 'air is sweet and clear' and the rivers not only abundant but clean and wholesome having 'mostly gravel and stony bottoms'.

Captain Smith perhaps best conveyed the sense of golden opportunity to be found in the new land. Here was a new chance for every one, from master to servant, from gentleman to labourer, most of all in that a man could make good and, more than that, prove himself as a man, whatever he'd been before. As Lewis has pointed out, he could walk the world, a new Adam. Only this was an Adam able from his knowledge and wisdom to improve on the new Eden. 'If he have but the taste of virtue,' wrote Smith, 'what to such a mind can be more pleasant than planting and building a foundation for his posterity, got from the rude earth, by God's blessing and his own industry?' Here a man's pleasures were his gains. 'Here Nature and liberty affords us that freely which in England we want [i.e. lack]. For what pleasure can be more than in planting vines, fruits, or herbs, in contriving their own grounds to the pleasure of their own minds . . . and to recreate themselves before their own doors. . . . ?'

This creative kind of pleasure came not only to the British who founded colonial America, but to Americans or would-be Americans pouring across the Appalachians to settle and develop the Ohio-Mississippi basin. In Frenau's poem on emigration to the American west, he struck the same note. America was not only land, but a condition of land, land wild enough and free enough to be a second chance, the chance for 19th-century man to find the new heaven and the new earth that 17th-century men had sought. Describing Ohio's 'savage stream', he wrote:

> 'Great Sire of floods! whose varied wave
> Through climes and countries takes its way,
> To whom creating Nature gave
> Ten thousand streams to swell thy sway!
> No longer shall they useless prove
> Nor idly through the forests rove;
>
> Nor longer shall your princely flood
> From distant lakes be swelled in vain,
> Nor longer through a darksome wood
> Advance, unnoticed, to the main,
> Far other ends the heavens decree—
> And commerce plans new freights for thee.'

Thus, generation after generation, dwellers in America had this sense of finding in nature the gift of a vast and fruitful and promising land for them to explore, settle and develop.

Eden Despoiled

Unfortunately, the very plentifulness of nature and the new-found freedom in men led many to a swift and thoughtless attack upon the land, that came to waste its resources and beggar the very source of abundance and opportunity. To begin with, men did, of course, have to chop or burn down the forest: this destruction was necessary if they were to improve on the way of life of the Indians and have farming replace hunting. The Indians had opened up a small part for their villages and corn plots, yet they'd not done much to dint the forest front.

In telling of the difficulties the first settlers had in New England, Bradford mentions the continuity and density of the forest that presented an all-but-impenetrable barrier. 'What could they see,' he wrote, 'but a hideous and desolate wilderness . . . the whole countrie, full of woods and thickets, represented a wild and savage hiew'. As people landed and hacked their way forward 'they wandered in the wilderness out of the way and found no citie to dwell in'. The first task then was to de-forest the land, and get it under crops. This was done as rapidly as possible, and trees of all ages and size were cut down, without any thought for their stage of maturity. 'Every one that can lift a hoe', said Johnson, including women and children, was hailed to the task of clearance, 'to teare up the Rootes and Bushes'.

The axe-man became one of the key figures on the frontier, and the sound of the axe rang continuously through the forest. Fenimore Cooper has a marvellous description of an axe-man in Billy Kirby, a character in *The Pioneers*,

> 'whose occupation, when he did labor, was that of clearing lands or chopping jobs. His first object was to learn his limits . . . and then he would proceed to the center of his premises. . . . Commonly selecting one of the most noble trees for the first trial of his power and, wielding his ax, with a certain flourish, not unlike the salutes of a fencing master, he would strike a blow into the bark. . . . From that moment the sounds of the ax were ceaseless, while the falling of the trees was like a distant cannonading.'

Wood was cleared not only to make room for corn and livestock, but to get rid of dangerous animals. Living in the woods was always a threat, as Richter brings out in his novel, *The Trees*. Here a young woman called Genny, left alone one night, became aware of 'something trying to come down her chimney. She could hear the beast giving long snuffs like a hound'. She recalled a story about a charcoal burner, living in the forest, who 'came home one night and found one of those big black wolves inside his cabin. It must have run up a windfall, like the leaning ash ran up her cabin wall, and jumped down the chimney'. She also remembered an old lady, who, when she was a babe, had been picked out of her cradle by a panther, that had come out of the woods into her father's cabin.

One way of getting round these tragedies was to cut the bush well back, even if the land wasn't being used. This wasted land, but saved lives. If not life itself, then health or comfort had to be considered. Fevers broke out, often ascribed to the marshes. The whole of the eastern seaboard is a drowned coast in which the sea has invaded the rivers, often creating marshy lagoons and estuaries. Here mosquitoes and black-flies were a constant nuisance.

They spread up into the woods and made life almost intolerable after the spring thaws and throughout the summer—unless the marshes were drained and the trees cut down. In Bradford's history he mentions that there was a lot of criticism in England of the Plymouth settlement. One of these was 'That the people are much annoyed with muskeetoes.' To this he replied:

> 'They are too delicate and unfitte to begine new-plantations and collonies, that cannot endure the biting of a muskeeto; we would wish such to keepe at home [i.e. in England]. Yet this place is as free as any, and experience teacheth that the more the woods are cut downe, the fewer there will be, and in the end scarse any at all. . . .'

Wood was cut for its own sake, of course. There was no coal on the coast, so wood was everywhere used for fuel. It was burned as logs for cooking and for household fires, or consumed as charcoal in the smelting of metal or work at the smithies. Since winters were long and cold in New England, quite strong in Pennsylvania, and cool enough for fires even in Virginia, an enormous amount of wood was burned. Indeed, something of a boast was made about this by, for example, Higginson.

> 'Though it be here [i.e. New England] something cold in the winter, yet we haue plentie of Fire to warm vs, and that a great deal cheaper than they sell Billets and Faggots in London; nay, all Europe is not able to make so great Fires as New-England. A poor Seruant here that is to possesse but 50 Acres of Land, may afford to giue more wood for Timber and Fire as good as the world yeelds, then many Noble Men in England can afford to doe. Here is good liuing for those that loue good Fires.'

The comparative scarcity of iron also meant a great demand for wood. Wooden pegs were used instead of iron nails in holding house timbers or the struts of furniture together. Many household utensils, like pails, were wood rather than metal. Wood was relied upon on for bridges: indeed wooden trestle-bridges were employed till well on in the 19th century, even when the *chemin de fer* was being carried over them. Wood continued to be used for housing, shops and offices well after it might have been replaced, judging by the European situation, by brick and concrete. But then, Europe was starved of wood; America seemed to have plenty and to spare. Yet plenty can soon turn to scarcity, and in last generation America has woken up to the fact that it has destroyed the greater part of its forest and now has to import from Canada and Latin America to make up its loss.

The soil which was exposed when the forest was cut down seemed very rich and was used year in year out without much manuring. Again, America boasted of its virgin soils, and with some justification, since most of them were very fertile. But the American settler, by and large, did not manage them well, and they became impoverished and eroded, often within a man's own lifetime. We have already noted Jefferson's outburst against the mining of the soil by the tobacco farmer. Understandably few listened to him, because there was yet more land to be mined. People had what we can now see as a pathetically rosy image of their environment. One of the chief writers on the agricultural way of life, John Taylor of Caroline, seemed much more concerned about taxes than erosion as a threat to farming. 'In a climate and soil,' he wrote in 1817, 'where good culture never fails to beget plenty and where *bad cannot produce famine*, ... agriculture can only lose its happiness by the folly or fraud of statesmen, or by its own ignorance.' Presumably here ignorance referred to accountancy rather than agricultural practices; yet it did extend to field practices and, where it did so, led to soil impoverishment. In Burnaby's travels he noted, at the end of the 18th century, 'Virginia is far from being arrived at that degree of perfection which it is capable of. Not a tenth of the land is yet cultivated; and that which is cultivated, *is far from being so in the most advantageous manner*'. Later, he adds, 'There seem to be very few improvements carrying on in the land'. At a time when the Great Improvers were taking in the commons and intensifying the use of arable land in England, America was allowing a prodigal waste in land to occur. And this extended right on to the early 20th century. Part of the problem was that America became so short of labour it could not afford to put in the work necessary to improvement.

> 'The abundance of land and the scarcity of labor made the American farmer the most wasteful of agriculturists. Land was apparently inexhaustible and easily obtained; the normal procedure was to use it up as rapidly as possible and pass on to fresh land. Labor, it was believed, must be conserved, but land might be wasted, a point of view that greatly retarded the introduction of scientific farming.' (Faulkner, 1938, p. 89)

Many farmers were, and still are, also hunters and fishermen, and have kept a large amount of bush on their land to harbour small game. Hunting and fishing have been major interests in America from the beginning. The British were a sporting people and Captain Smith appealed to this in trying to tempt them to the colonies. 'For gentlemen,' he said, 'what exercise should more delight them than ranging daily, using fowling and fishing for hunting and hawking'. At the same time any kind of

> 'man, woman, and child, with a small hook and line, by angling may take divers sorts of excellent fish at their pleasures. Carpenter, mason, gardner, tailor, smith, or what other, may they not make this a pretty recreation, though they fish but an hour in the day, to take more than they can eat in a week.'

Smith's image of the typical American as hunter and fisher was more than justified. People of every walk of life took to this as a common pastime. The result was, of course, an immense and a growing pressure on wild life. Turkeys were shot out of most of the woods by the time Cooper wrote *The Pioneers*. In that book he describes in a most vivid way the slaughter of the carrier pigeon which swiftly led to its extinction. The buffalo then became the main target of mass destruction. Part of this was to fill the very real need for meat of the garrisons of soldiers stationed in the west and of the construction gangs that built the railways. Part, however, was for sport, and sport of a very wasteful kind, in which the head or simply the tongue might be brought back as trophy or delicacy, and the rest of the smitten animal be left to the wolves and the vultures. The immensity of the buffalo herds became one of the wonders of the world. 'As late as 1871,' writes Yorke Edwards, 'one concentration, estimated at 4,000,000 animals, was 25 miles wide by about 50 miles long and required five days to pass a stationary observer'. Perhaps the feeling grew that their numbers could never be exhausted. They had obviously outlasted the attacks of the Indians, though these had been going on for hundreds of years. In any case, they were hunted with gusto. Buffalo Bill Cody reported, with apparent relish, the account of a buffalo killing match he had with Comstock, a noted Western Scout, apparently over a wager at Fort Wallace, Kansas.

> 'The time came to begin the match,' he wrote. 'Comstock and I dashed into the herd, followed by the referees. Comstock took the left bunch and I the right. My great forte in killing buffaloes from horseback was to get them circling by riding my horse at the head of the herd, shooting the leaders, thus crowding their followers till they would finally circle round and round. On this morning the buffaloes were very accommodating and I soon had them running in a beautiful circle, when I dropped them thick and fast, until I had killed thirty-eight; which finished the first run. Comstock succeeded in killing twenty three.' All this was watched by 'an excursion party from St. Louis, consisting of about a hundred gentlemen and ladies who came out on a special train to view the sport, among whom was my wife with little baby Arta. After the results of the first run [i.e. round of the match] had been announced, our St. Louis excursion friends set out a lot of champagne which they had brought with them. . . .'

This story, as extraordinary surely for the manner of its telling as for the tale that was told, showed how little people cared about the wild life already disappearing in such vast numbers from America. It was but one example of how America's wealth of resources was being squandered in what was without question the worst plunder in history, a plunder that extended to fishing and mining, to water and fuel and to the population itself, scarred by riot and killing.

Eden Regained

This very despoilation, however, created a strong reaction and led to a growing wish to respect nature and conserve it. Unfortunately this came late in American history. There is little sign of it during the colonial era. Views of nature did not particularly lead to its respect. Indeed, many of the leaders in early New England were ministers with a profound distrust of nature. Living according to nature could all but be equated with living by the devil. Men were constantly reminded that their own natures and nature itself were under the curse of the Fall; redemption could only be by crucifying the natural instincts. If men had to live in the wilderness they must not give themselves up to the wild. In fact many did, and in fact were not the worse for it. Nevertheless there was always this tendency to try to band men together against the wild which the wilderness might bring out in them. The destruction of the wilderness, then, was part of men's defence against the wild.

By the time of the Revolution such destruction had grown to be disturbing, and some States started to pass laws at least against the excessive killing of game. This was one of the themes in Fenimore Cooper's *The Pioneers*. Leatherstocking, who has made his life by hunting and lived according to the law of the woods, is caught killing game out of season, and punished for it under what must have been amongst the first of the conservation laws. He had been warned by the local judge himself, but had argued his right to kill, if he were in need. This right was not acknowledged: it had been superseded by the need of the community to conserve the game.

> '"The legislature have been passing laws",' Cooper makes the Judge say, '"that the country much required. Among others, there is an act prohibiting the drawing of seines, at any other than proper seasons, in certain of our streams and small lakes; and another, to prohibit the killing of deer in the teeming months. These are laws that were loudly called for by judicious men; nor do I despair of getting an act to make the unlawful felling of timber a criminal offence."
>
> 'The hunter listened to this detail with breathless attention, and when the Judge had ended, he laughed in open derision.
>
> '"You may make your laws, Judge," he cried, "but who will you find to watch the mountains through the long summer days, or the lakes at night. Game is *game*, and he who finds may kill; that has been the law in these mountains for years . . . and I think an old law is worth two new ones."'

But he pleaded a losing cause. Although hunting continued to be allowed, it was as a pastime, rather than an occupation, and gradually became controlled, as population pressed upon the land. But it was a difficult problem, and as the frontier moved west the freedom to shoot went with it. Yet men's concern for protection also moved west. Interestingly enough, in Willa Cather's novel, *O Pioneers*, she too is caught up with this concern. But she shows it as an odd thing in the community. It is one of the crazy obsessions of a Russian religious eccentric, a very humane yet a strangely un-human character called Crazy Ivar, who found contentment in the solitudes of the vast, unbroken prairie.

> 'He disliked the litter of human dwellings. He preferred the cleanness of the wild sod. He best expressed his preference for his wild homestead by saying that his Bible seemed truer to him there. If one stood in the doorway of his cave, and looked off at the rough land, the smiling sky, the curly grass white in the hot sunlight; if one listened to the rapturous song of the lark, the drumming of the quail, the burr of the locust against that vast silence, one understood what Ivar meant.'

Ivar had laboriously dammed a small creek and turned it into a lake visited by thousands of ducks and geese. One day he heard a group of young people, the Bergsons, approach. Thinking that they had come fowling he rushed out of his little sod house shouting, '"No guns! No guns!", waving his hands distractedly'. The eldest Bergson said they had come simply to watch. Ivar watched *them*, nervously.

'"I have many strange birds stop with me here. They come from very far away and are great company. I hope you boys never shoot wild birds." Lou and Oscar grinned, and Ivar shook his bushy head. "Yes, I know boys are thoughtless. But these wild things are God's. . . ."'

The American view of nature was changing; or rather, a view that had actually long been held was getting its due. For had not Thomas Paine, the great revolutionary, written

'The word of God is the creation we behold, and it is in *this* word, which no human invention can counterfeit or alter, that God speaketh universally to man. Do we want to know what God is? Search not the book called Scripture, which any human hand might make, but the scriptures called the Creation.'

How different this was from a Puritan like Edwards who, in spite of his known reasonableness as a Puritan, declared,

'The very thought of any joy arising in me [i.e. independently of God] or any consideration of my own . . . experiences, or any goodness of heart *or life*, is nauseous and detestable to me. And yet . . . I see the serpent [i.e. of natural inclinations] rising and putting forth its head continually, every where, all around me.'

Yet the view that nature was good, indeed that Nature was God, steadily gained ground. In the early 19th century it had more and more champions. Philip Frenau wrote on 'The Universality and other Attributes of the God of Nature'

'All that we see, about, abroad
What is it all, but nature's God.
In meaner works discovered here
No less than in the stormy sphere.

In seas, on earth, this God is seen;
All that exist, upon Him lean;
He lives in all, and never strayed
A moment from the works He made.

This power who doth all powers transcend
To all intelligence a friend,
Exists, the greatest and the best
Throughout all worlds, to make them blest.'

It was from Emerson and Thoreau, however, that the new image of nature sprang, an image that was to have a tremendous impact upon America and the use of its environment. Everywhere men were wrestling with how to make religion accord with science. In Britain, Thomas Carlyle, like Paine, saw nature as the new Scripture. 'That the Supernatural differs not from the Natural is a great Truth', Carlyle wrote. It was his ambition to 'raise the natural to the supernatural'. Emerson strove for this, too. The two men became great admirers of each other and strengthened the new view. 'There are new lands, new men, new thoughts', wrote Emerson (1849/1957 edn., p. 49) introducing his ideas about *Nature*: 'Let us demand our own works and laws and worship.' The wilderness which had put men off, attracted him. 'In the wilderness,' he said, 'I find something more dear and connate than in streets or villages.' Here he discovered an 'occult relationship between man and the vegetable', here he was led to believe that 'Every natural fact is a symbol of some spiritual fact'. These were challenging thoughts. If they were true it meant that nature had a message for man. What was that message to an America that had been so contemptuous of nature as to have grown one of its great despoilers? Development in America had become almost synonymous with ravagement. Love of country had become the rape of the land. The more Americans wanted their country to grow the more they doomed the land to perish. But if they could listen to the voice of nature they might come to find greater meaning for themselves. In a sense, America had lost 'the simplicity of character' of the New Adam coming upon a New Eden, and had given way to 'the prevalence of secondary desires, the desires of riches, of pleasure, of power, of praise'. Thus the power to interpret nature had become lost. This was particularly true where Americans had made such a very 'artificial and curtailed' environment for themselves in the town, that, 'in the roar of cities or the broil of politics', they could scarcely hear 'the call of a noble sentiment'. They had to get back to nature, to where 'the woods wave, the pines murmur, the river rolls and shines, and the cattle low upon the mountains'. Emerson urged men to get beyond the religion, to which so many paid service, that 'put an affront upon nature. The uniform language that may be heard in the churches of the most ignorant sects is—"Contemn the insubstantial shows of the world; they are vanities, dreams, shadows. Seek the realities of religion." The devotee flouts nature.' Yet for Emerson 'The noblest ministration of nature is to stand as the apparition of God.' If men went back to nature they'd get to the heart of truth. To become strangers in nature was to grow alien to God.

Such warnings and appeals had their effect. Men tried, as Emerson urged, 'to understand the notes of the birds', and to recreate an America where 'the fox and the deer would not run away from us'.

In this, Thoreau (1854/1966 edn., p. 5ff) had an exceptional influence. 'The mass of men', he claimed, 'lead lives of quiet desperation.' They did not know what they wanted, and could not get it, even if they were to know. The trouble was they had lost the fundamentals, the 'gross necessaries of life' in their concern about 'the improvements of the age'. Have improvements, of course, but keep the basics. Among these was communion with nature. If we'd lost this, we couldn't be happy until we got it back. 'I long ago lost a hound, a bay horse, and a turtledove', said Thoreau, speaking symbolically, 'and am still on their trail.' Until men got back their scent for nature, their strength in nature, and their peace from nature, they'd never be happy. Thoreau was as good as his word and, building himself a small cabin beside a pond out in the bush, tried 'to make my life of equal simplicity, and I may say innocence, with Nature herself'.

Many of his friends were sceptical. It was certainly something that, with their businesses, possessions and families, they could not do. But what they could do was to live lives as true to their natures as possible, as basic to nature as they could. 'I would not have anyone adopt *my* mode of life', wrote Thoreau. 'I desire that there may be as many different persons in the world as possible; but I would have each one be very careful to find out and pursue his *own* way.' What Thoreau was after was a way, in any kind of occupation or situation, which, like the one he sought in the woods, should 'front only the essential facts of life'. For him this lay in identifying himself with the wilderness, in letting the wild in men have its say. Indeed Thoreau went so far as to say, 'In wildness is the preservation of the world': the motto of the Wilderness Society of America.

That such a Society could have grown up, and that its long uphill battle could have won through to the Wilderness Act of 1964, shows the growing strength of the new image of nature in America, where wildness was seen, at last, as essential to human life—and to civilized life at that. This should not, and indeed does not, in the words of Leo Marx, 'encourage us to believe that we can only solve our city problems by moving into the country . . . or that the recovery of a rural style of life is a genuine alternative to life in our intricately organized, urban, industrial society', but it does mean that sufficient wilderness should be preserved in America so that more and more people may, as Captain Smith put it in 1624, 're-create themselves outside their doors' by making contact with the basic world of nature whenever they feel that need. The conservation of the wild is now one of America's greatest preoccupations, as once the destruction of the wilderness had been. There must be the chance, in Bob Marshall's words, one of America's leading conservationists, to get 'back in a primordial world . . . where only the laws of nature hold sway'.

Eden Divided

America is at present at a crisis of decision: there are those who urge development and growth without too much concern for the land, while there are others calling for conservation and limitation in a desperate attempt to save the land. The extraordinary extent to which America bled itself white of many of its resources during the two world wars in order to become the 'arsenal for democracy' made the country realize, all at once, that it was not an endless cornucopia of good things. The Paley report to the President on Materials Policy, showed to a startled people that America could no longer provide itself with the essential 'sinews of industry' but had already become, and was likely to be increasingly, a net importer of raw materials. This gave great power to the conservation movement; nevertheless, the developers continued to use up the remainder of American reserves at an astonishing rate. Americans themselves demanded a growth in their well-being at home and their power abroad which not only threatened their own basis of prosperity, but began to eat into the basic reserves of Canada, Mexico and other countries—all over the world. It was not alone in this. A world upsurge in growth, in which Germany, the Soviet Union and Japan were all outstripping America in the rate of increase of the gross national product, was putting an equal pressure on resources in every continent. As a result some Americans, Europeans and Japanese have begun to wonder whether the world shouldn't plan for definite limits to growth before resource-depletion and pollution-increase bring everyone to the edge of disaster. As Meadows (1972, p. 184) has said in the rousing survey of *The Limits of Growth* it is time we modified our views of unlimited growth and instead, through limiting excessive or excessively rapid development, 'establish a condition of ecological and economic stability'.

The mineral crisis is particularly acute. Tin and mercury may be used up before the end of the century; lead, zinc and copper by the next mid-century—if present rates of exploitation continue. Oil, too, is being consumed at a fantastic rate and is likely to drop off dramatically in about 50 years. The trouble is that these, and a number of other resources, are being used up much more rapidly than they can be replaced through recycling or through substitutes. Their rate of growth is distinctly higher than that of world population, though this is high enough, because of increasing per capita demand.

To keep up its high standard of living America consumes a far larger proportion of world minerals than any other country or even group of countries. It eats up more oil, natural gas, coal, iron, copper, lead and zinc, nickel, tin and mercury than all of Latin America, Africa and Asia put together, yet it has only 6% of the world's population, whereas these other regions have over 70%. China alone has 20% of the world's people yet uses far less than the United States. Should America slow down: particularly to allow others to catch up? America is already in the post-industrial era, that is, stressing services and welfare rather than mere production, while many countries have not yet hit the industrial age. Can the United States go on at this rate?

Many Americans would say 'Yes' and point to Alaska's untouched reserves, and the nearby resources of Canada and Mexico, only as yet partially exploited. In terms of the continent as a whole, they see no immediate or even early shortage. As has been pointed out, America has already shifted from an *American* to a continent-wide system of development.

Alaska is a case in point. Here great new sources of oil were discovered first near the Kenai Peninsula on the South coast and then at Prudhoe Bay in the North. It was hoped that North slope oil could be piped to the South shore, there to join that being shipped by tanker to continental U.S.A. But the pipe line or lines and the road or roads needed to construct them would have passed through some of the last great stretches of wilderness left to America. What should happen: should oil be developed, or the wild preserved?

Alaska oil is obviously one answer to America's critical oil shortage. Moreover, it is an answer provided from American land, mainly by American companies, over American routes. One of America's problems is that, in using up so many of its wells in conterminous U.S.A., it was getting to be dependent on Canada, Venezuela, Trinidad, Nigeria, N. Africa and the Middle East. The Arab countries have been demanding both an ever greater royalty and an ever greater say in the development and sale of their oil. Would it not be salutary, therefore, to switch to Alaska? The oil-gap between demand and supply has been getting greater. At the time of the Prudhoe Bay discovery, 1968, the gap was of the order of 3·9 million barrels per day. With an expected 2–3 million barrels per day from Alaska the gap might be largely closed. Any additional needs could be met by friendly countries like Canada and Venezuela. The oil-men's case as put to *Fortune* was that 'North Slope oil is U.S. oil—it may make the U.S. no longer dependent on foreign oil. A world power which depends on potentially reluctant or hostile countries for fuel that must travel over highly vulnerable sea routes is by definition no world power.' Besides, Alaska needed the great amount of revenue oil would supply. 'Long little more than a colony dependent on Washington for $1 for every $2 spent within its borders, Alaska is a poor State and suffers the nation's highest unemployment rate.'

The case for development seemed strong. Yet that for conservation was also impressive. The Sierra Club which, along with the Wilderness Society, represents one of the most powerful of U.S. conservation lobbies, at once put pressure on the State and the Federal governments to halt any development that would spoil the only last great area of wilderness left to the country. As Richard Pollak says in *Oil on Ice*, Alaska represents for Americans today the same kind of New Eden that America as a whole did for its first European settlers.

> 'One contemplates Alaska today with a numbing sense of historical perspective. *We have seen this pristine land before*—the United States at its birth two centuries ago! Now, once again, we are playing out the scenario that has reduced so much of the nation to an environmental theater of the absurd. The cake this time is oil—an estimated 100 billion barrels, maybe more—buried beneath the landscape of our last great wilderness. En route the viscous crude would travel the breadth of Alaska's most fragile ecosystems: from the ice-worn coast of Prudhoe Bay, across the lichen-sprinkled tundra of the North Slope, into the glaciated grandeur of the Brooks Range, down to the valleys and forests of the interior highlands, then through the Alaska Range to the waiting ships at Valdez. The 48-inch steel pipe would snake through the once-untrammelled habitats of hundreds and thousands of caribou, through [the last refuge of the] Barren Ground grizzlies, [the unique home of that rare species] the Dall sheep, past peregrine falcon (elsewhere, nearly extinct) and millions of migratory birds and waterfowl.'

The conservationists had all but got the Government to turn much of this area into wilderness preserves, such as the Brooks Range Wildlife Reserve and the Gates of the Arctic National Park, when the oil interests moved in with their road-gangs and rig-crews, and demanded a trans-Alaska routeway for their pipe-line. Thus the last American Eden became divided: an issue that is still rife today. 'The oil companies' to quote the Sierra Club, 'are depending, as the industry always has, on the politics of expedience, and on the traditional American delusion that, as one Alaskan official put it, "This country's so goddam big that even if industry ran wild we could never wreck it. We can have our cake, and eat it, too."' Against this must be put the claim that 'here is the last great wilderness', which is doomed unless a voice is raised. 'There is just one hope of repulsing the tyrannical ambition of civilization to conquer every niche on the whole earth. That hope is the organization of spirited people who will fight for the freedom of the wilderness.'

These concluding quotations show how different are the attitudes, feelings and images which Americans have about their environment. To some it is still a land to be exploited and used up, where nature must be warped to serve man, and in which there is still plenty of room for growth and expansion; to others it is increasingly a land that must be protected and cared for so that people can find in it a last chance to go back to nature and get there the re-creation they believe they so much need. Both these images have profoundly affected the American environment; both are major factors in American history and geography. Both have been at variance for a long time, despoiling the land and building it up; both need constant assessment in that balance between development and conservation which is of such crucial importance to America today.

References

Emerson, R. W., 1849/1957. *Nature: Addresses and Lectures* (1849); *Selections from Ralph Waldo Emerson* (1957) (Ed. Wincher, S. E.), New York: Riverside Press.

Faulkner, H. E., 1938. *Economic History of the United States*, New York: Macmillan.

Meadows, D. D. and Meadows, D. L., 1972. *The Limits to Growth*, New York: Potomac Associates.

Thoreau, H. D., 1854/1966. *Walden, or Life in the Woods* (1854): *Walden and Civil Disobedience* (Ed. Thomas, O.), New York: Norton.

Daniels, P., 1973. *The Stewardship of the Land: a selected bibliography of current readings*, New York: N.Y.S. Office of Planning Services.

Hirth, H., 1972. *Nature and the American: three centuries of changing attitudes*, Lincoln, Nebraska: U. of Nebraska Press.

Kline, M. B., 1972. *Beyond the land itself: Views of nature in Canada and the United States*, Cambridge, Mass: Harvard U.P.

Opie, J. (Ed.), 1971. *Americans and Environment: the controversy over ecology*, New York: Health.

Section II. Perceptions and Policies in the American City

6
The City and the American Way of Life

J. Wreford Watson

As settlers moved across America they rapidly developed a built environment which replaced the natural one. Today, 74% of all Americans are city dwellers, environed not by forest and grassland, but by concrete buildings and tarmac parking lots. They have created a man-impregnated and man-made landscape. In particular, they have made the city central to their life.

Cities are the cultures of their countries made graphic in the landscape. A city's use of site, the layout of its streets, the kind of institutions it has and their place and their importance, the function and facade of its buildings—all tell us whether service or profit, a hierarchical or egalitarian society, continuity or change have been the guiding forces in the lives of its citizens.

American cities strike the European as being quite distinctive; they are rarely dominated by castle or cathedral, their street patterns have not crept out over the centuries testing and reflecting every shade of local relief, they do not often have the anomalies due to a long and complex history, they are less full of the echoes of singular personalities, they are much more changeful, more utilitarian, they attempt to be more businesslike and are more concerned with the masses and the needs and drives of their time. They are, in fact, American: reflecting an American attitude to life and the American way of using the land.

In this they have always reflected the American myth itself: they were part of the frontier spirit, where men could do new things; they spearheaded the growth and progress men were after; they brought men of all races and creeds together and provided a forum for their ideas and ambitions. They expressed American optimism and affluence, brashness and efficiency, love of comfort and show, materialism and yet rectitude, and above all America's sense of freedom and destiny. They were peculiarly American in giving scope for (i) the dynamism of America (ii) the free play of ideals in America and (iii) the humanism of America.

The Dynamism of the American City

The dynamism of the American city is seen in its emphasis on newness, its productivity, its love of change, its mobility, its wastefulness and in the problems of cultural lag where the machine has all but outrun man.

In the growth of America, cities exploded into the landscape with an almost volcanic force. As the bridgeheads of new and swift advance they had to be thrown up quickly. New immigrants pouring into or through them swelled their expansion almost overnight. Virtually all the facilities to which the settlers had been accustomed in Europe—quays, warehouses, shops, offices, homes, churches, schools and hospitals—had to be provided almost at once. Instant cities were the result, often hastily planned and usually erected in haste. Both Jamestown and Plymouth are good examples. In the latter's case, after deciding on Plymouth Harbour as their site, 'rather suddenly the Pilgrims decided to set their plantation upon the ground . . . where the land rose fairly steeply to a hill which commanded the harbour and the country around.' Almost immediately they planned the settlement and set about building it, putting men ashore '"some to fell timber, some to rive, some to saw, and some to carry, so no man rested all day"' Dillon (1973, p. 164). Individual lots and common fields were laid out, a Common House erected, family homes built and a stockade put up, in next to no time.

One reason for this haste was the onset of winter, and certainly throughout the northern states the threat of a long cold winter led to a fever of activity in the summer. This was repeated time and again, as men moved west. With the spring break-up goods could be loaded and sent to a new townsite; the warehouses, shops and homes had to be rushed up in the summer, and then internal improvements added in the winter. Thus towns had to be quick about their business. Billington (1960, p. 304) describes the phenomenally rapid growth of Detroit:

> 'By the 1830's the time was ripe for a rush of settlers [beyond the Lower Great Lakes]. Detroit felt their impact in 1831, as lake steamers deposited the first immigrants at its docks. Buildings were hastily thrown together, and the sleepy little frontier village rapidly emerged as a new metropolis. By 1836 it boasted 10,000 inhabitants, a theater, a museum, a public garden, schools, churches, a library, a lyceum, a historical society, a ladies' seminary, a water and sewage system, and street lights'—all in the space of five years.

Woodward (1965) recalls the equally meteoric rise of Chicago:

> 'For about four decades, 1840–80, Chicago grew more rapidly than any other community in the world. It was like a hearty lad who outgrows his clothes before he has had time to get used to them. This button-bursting expansion led to 93,000 people

twenty years after its incorporation. It was called Slab Town in popular speech because every house in the community was a hastily thrown together box-like structure of split boards.'

Parkman (1849/1961 edn., p. 14) gives a thumbnail sketch of the mushrooming of Independence:

'The town was crowded. A multitude of shops had sprung up to furnish the emigrants and Santa Fé traders with necessaries for their journey. There was an incessant hammering and banging. The streets were thronged with men, horses and mules. While I was in the town, a train of emigrant wagons from Illinois passed through, and stopped in the principal street.'

Horace Greeley (1860/1963 edn., p. 137) described the burst into existence of Denver, on his way west.

'The rival cities of Denver and Auraria front on each other from either bank of Cherry Creek, just before it is lost in the South Platte. Of these rival cities, Auraria is by far the more venerable—some of its structures being, I think, fully a year old, if not more. Denver, on the other hand, can boast of no antiquity beyond September or October.'

Yet it was thronging with men. Oscar Lewis (1963, p. 146) wrote of San Francisco:

'The stream of prospectors from both hemispheres who poured into California during the next two years transformed what had been a sparsely settled frontier into something the likes of which the world had never seen. Changes that in the ordinary course of events would have taken decades were accomplished virtually overnight. The sleepy village of San Francisco changed into a confused bustling metropolis, with substantial business houses surrounding its old Spanish Plaza and acres of tents and shelters covering the nearby sand hills. It was a city unlike all others: few of the rules prevailed that governed the conduct of business elsewhere. Not only bars and gaming rooms, but restaurants, shops, shipping offices and even banking houses remained open twenty four hours a day.'

Even established towns could change overnight. Newness became a habit. The American city is constantly being pulled down to allow newness to take over. Built into the American scene is the 'tradition of the new'. As Rosenburg (1970, p. 78) says, the aim of this is to 'close the doors of history'. America is of the future, not the past. That was certainly the impression Henry James (1907/1946 edn., p. 77) received when he returned to the States after his long European sojourn. Like everyone else he was fascinated and yet appalled by the changefulness of the American scene. This struck him most forcefully in the skyscrapers of New York which

'crowned not only with no history, but with no credible possibility of time for history . . . are simply the most piercing notes in that concert of the expensively *provisional* into which your supreme sense of New York resolves itself. They never begin to speak to you in the manner of the builded majesties of the world as we have heretofore known such—towers or temples or fortresses or palaces—with the authority of things of permanence or even of things of long duration. One story is good only till another is told, and sky-scrapers are the last word of economic ingenuity only till another word be written.'

A major reason for this changefulness is given by James when he describes skyscrapers as 'giants of the mere market' since they are 'consecrated by no uses save the commerical at any cost'. The American city early outstripped its European predecessors in putting profits first. Economic factors came to dominate its development. While the heart of the European city is its CIC, or cultural institutional centre, where the 'authority of things permanent' ruled in castle or cathedral, the heart of the American city lies in the CBD, or central business district. This represents a difference so profound that it is one of kind rather than degree: the American city developed as a new species, springing from the primacy of economic values. Economic rent came to dominate its very structure. Since everyone tried to bid for a central location, here the rents shot up. As finance, shopping, great institutions and industry grew less interested in a peripheral location, away from the point of highest accessibility, rents dropped off. This grading away of rent from a high central point to the lowest point out at the margins, came to grade all major land uses. As Yates and Garner (1971, pp. 238–239) show, within 1 mile of the CBD of a large American metropolis 52% of the land is given up to streets, parking lots and transportation terminals; 23% for offices, shops and hotels; 18% for recreation, culture and administration; 4% for industry and 3% for housing; whereas 14 to 15 miles out from the CBD only 36% is devoted to streets and transportation; 3% for business; 5% for institutions; 15% for industry and 40% for housing.

That so much emphasis is laid on the economic factor is due to America's long-held belief in the competitive system. When Crèvecouer asked, during the American revolution, what was an American, one of the answers he gave was: a person animated with the spirit of an industry which is unfettered and unrestrained, because 'each person works for himself'. President Jefferson himself set his seal upon this by dedicating America, in his Inaugural Address, to 'our equal right to the use of our own faculties, to the acquisition of our own industry . . . resulting from our own actions'. Competition became not only part of the myth but also part of the philosophy of America. Cooley, for example, idealized competition as a major force in making progress. Americans competed with each other for success, which was measured

'chiefly in the development of new means of production. The work at hand has been material work. Men have taken it up, and, with the emulation inseparable from human energy, have striven with one another to excel in it and to gain the power and honor that goes with success. In such work . . . the accumulation of wealth is the proof and symbol of success, and this, accordingly, has more and more become the accepted standard, even for a sort of mind that in another state of things would have risen above it'.

There are many results of this in the city. One is the change due to putting on an ever better front. Face-lifting is a characteristic of the American city. From the very earliest days Americans have been very face conscious. Their pioneer shops, saloons and offices nearly always had false fronts—to show a two-storeyed face when the building was actually only one storey high. The heightening of the facade not merely gave more room for advertisement in larger and bolder signs, but added importance and status to the place. And America was swiftly to become, in Vance Packard's words (1959), the society of the *status seekers*. Status was evidenced in a very concrete way in the American city by making places look 'bigger and better'. In the 19th century upper class homes with their baronial towers, their gothic windows (frequently fitted with stained glass), their large verandahs edged by ornate fretwork, were very typical: so were the big departmental stores with their spacious entrances flanked by corinthian pillars, their windows picked out by white-tile trim, and their walls relieved by carved medallions or cornucopias pouring out fruits and flowers.

Today status lies in moving out to the suburbs. As Vance Packard remarks (1959, p. 83) '. . . the desirability of a [home] address can be determined by its distance from the downtown business district.' People who want to be 'in the swim' have abandoned the tall three-storeyed houses, heightened by an ornamental turret, that used to line the mid-town streets, and have moved into one-storeyed or split-level, 'ranch style' suburban houses, with garden and swimming pool.

Movement to express status dominates the whole appearance of the American city. Few Europeans can conceive of this scale of movement. The drift from country places to the city, going on at a rate nearly four times that of the total immigration into America, is more than matched by the drift from one neighbourhood in a city to another. The mobility of the American population is phenomenal. Moving is part of the American way of life. Washington Irving in his *Legend of Sleepy Hollow* early wrote of the 'shifting throng that forms the population of America.' The American family moves once in every eleven years, or four to five times in the life of every married couple. The contemporary situation is well shown in the study of *Middletown* by the Lynds. The American 'on the make' shrewdly judges the venue of a new push from town centre to suburb, he buys a large lot on the outskirts while land is yet cheap and builds a new home on it; this attracts others and he sells out the rest of his land to these newcomers; finally, he sells his house at a big profit, and moves to a more remote suburb, where he builds a still bigger and better home. Meantime, working class families speculate on the properties left behind, and flit from the inner urban zones which are passing under shops and offices, to middle zones vacated for suburbs. Three out of every five families move out of their original neighbourhood at least once. The majority of them move a second or third time. Many move completely away from the city of their upbringing to find their livelihood in another and yet another city. In fact it is claimed that '33 million Americans now move into a new neighbourhood every year' (Packard, 1959, p. 6); a fact which is staggering to a European who was disturbed at the movement of 9 million Germans out of Slavic-occupied Europe after World War I, or to an Asian hit by the catastrophe of 10 millions moving their homes in the Partition of India. These mass migrations are as nothing compared with the annual flight of Americans from one place to another. The American city, like the old frontier, consists of people on the march—hence the march of their suburbs further and further into the country. 60% of today's urban growth in America is in the suburbs, including 20% in the city–country fringe known as exurbia.

A great deal of change is done without much thought for the land itself, and the American city is marked by a surprisingly great waste in land. Many American cities present a most untidy aspect because of their wide weed-ridden road verges, especially along the expressways, their half-used yards, their all-too-frequently empty plots waiting to be developed but meantime filling up with bottles and tyres and urban bric-a-brac and, above all, the wide circle of waste lands at their margins, held by speculators for development but in the meanwhile lying inert and derelict. On the writer's first visit to Boston he was appalled by the wilderness of weedy fields and farms-going-back-into-bush and other unused overgrown lots with which it was surrounded. A European, he had expected a Von Thunen-like zonation from a circle of beef-fattening farms, through another circle of dairies, to an inner ring of most intensively used land under market gardens, the whole area showing a maximization of the use of the land from the periphery in towards the centre: instead, he felt as though he had wandered into half-cleared Indian land left vacant as the aborigines had gone off for the hunt!

It is now regarded (Berry *et al.*, 1963) as axiomatic that 'land consumed by each household increases with distance from the city centre', but at the city edges land consumed for the non-existent though potential householder is often more than that for actual development. This is a meantime waste, but a waste that, since it will ultimately earn profit, does not seem to worry the American. But then, according to Harry Elmer Barnes, waste is part of the American way of life. America's is a wasteful economy. The whole country, not just the American city, is afflicted with this problem. To produce one drum of oil, America wastes four; to produce one ton of coal it wastes one and a quarter tons; to produce one cubic foot of natural gas it wastes an equal amount: there is hardly any natural product that does not cost America at least half as much in waste to produce. 'Inexcusable waste', says Barnes (Barnes and Reudi, 1942, p. 74) 'has marked our production, distribution, and consumption of goods and services'. In the American city waste includes un-used land, ill-used water and mis-used air. It also embraces poorly used special resources, such as education and medicine. All these may be explained as a part of what America has had to pay for the tremendous dynamism of its life—but they shouldn't be explained away. They are a real cause for concern in that they are polluting and destroying the very environment on which America is nursed.

The Free Play of Ideals

The free play of ideals in American cities rises from their strong democratic spirit and makes itself evident in competition and co-operation, in individuality and conformity, in personal liberty and yet a social conscience.

Problems of wastage lie very much in the freedom America takes with itself in trying to emphasize a free way of life. This freedom does of course have its drawbacks; but it also has its gains. It allows ideas to be tested out against each other. Most of the early immigrants were people in protest, anxious to break free from repressive religious or commercial regulations. The protesters then began to protest against each other. The American city became splintered by sectarianism as one new sect after another sprang up and tried to flourish. The Puritans themselves, who had sought to free themselves 'from their long servitude under usurping Prelacy', had their own usurpation of intolerant power challenged by the New Light movement, soon after they had settled in New England. The leaders of the new movement were banished out of Boston, but set up in Rhode Island. Soon, so many new movements were arising that cities could not cast them out, but had to put up with them. American cities became peppered with competing churches, particularly in the North. This also took place in Britain, of course, where dissent led to such conflict that it broke out in a civil war: but there the distinction between the parish church and the sectarian chapels, although it eroded yet it did not shake the position of the Established Church. In America, ideas of an established hierarchy soon went, and the free churches made free with the city by setting up their own spheres, often in bitter competition with each other. Even the visitor from Protestant Europe must be astonished at the tremendous range in the institutional geography of the American city, not only with different church buildings, but also with different ethnic, social and recreational centres, not a few of which are next door to each other, dividing neighbourhoods into fragmentary sub-sub-communities.

Part of this astonishing sub-division is due to race. There are African Episcopal, African Methodist, African Baptist and other churches, as well as their white counterparts. Indeed, the greater part of the white suite of institutions is paralleled by an African suite, and in many cities also by a Puerto Rican, a Chinese, or a Japanese suite. Of sub-division in the American city there seems no end.

This generally shows itself geographically in physical segregation. This is present in every major American city: it is by no means confined to the South. In the South, it is accentuated because there is a formal and, up until recently, even a legal segregation of black from white. But in the North and West informal segregation, through economic or social pressures, is nearly as effective. Housing segregation by restrictive convenants was long a case in point by which white property owners in a neighbourhood would convenant not to sell out to coloured buyers. These covenants also tended to keep out Jews and Catholics. As a result the American city is a patchwork quilt of Negro, Puerto Rican, Italian, Irish and Jewish 'ghettos'.

To a certain extent this happens everywhere: this writer was born in a city in China where there was a walled Manchu city within the larger walled Chinese city; within this larger enclosure were distinct Mongolian and Chinese quarters. He has visited Indian cities with separate Hindu, Sikh and Mohammedan quarters. Many British cities have long had Irish quarters, and are now rapidly developing coloured ghettoes. But in spite of all this it might be argued that the clash of race, religion and language has gone further in dividing the American city than elsewhere. Certainly the geography of segregation—of coloured areas, or Catholic or Jewish areas, or French-speaking or Italian-speaking areas, each with their own distinctive churches, schools, clubs, eating places and centres of entertainment—is one of the most immediately marked things in the American townscape. The uneven story of the peopling of America lent itself to this kind of distinction. In Boston, for example, following the conflict between Puritans and the New Light movement, came that between Puritans and Baptists, and between Puritans and Quakers. A good bit later, considerable friction grew up between the Protestants combined, especially those of an Anglo-Saxon origin, and the Irish Catholics, just as there next grew up a struggle between the Irish, once they had established themselves, and the Italian immigrants. Subsequently, French-Canadian intruders also found themselves segregated out. Boston (and for that matter any great metropolitan city, New York, Chicago, Los Angeles) is really a congeries of little cities so far as ethnic sensibilities are concerned. It is difficult to get people to think in terms of the city as a whole: each group is out for itself.

These divisions are added to by still another, that of social class. Notwithstanding the American myth that their's is a classless society, Americans do classify themselves into upper, middle and lower groupings. This is due to the free play of two great American ideals, first that a man should be free to gain what recognition he can for himself, yet second, that the recognition of men's potential equality should be admitted. Thus, though an American may be born into a certain class, he need not stay in it but can pull himself up into a higher one: or, of course, fall to a lower. Few scholars have done more work on this than Warner (1962, p. 83).

> 'There is in American communities', he claims, 'a clear understanding of the social differences, values and behaviour which compose a class system. The levels of social class are ranked into superior and inferior according to the values of the community. Classes do not permanently fix the status of either the individual or his family in America. Vertical social mobility, the rise and fall of individuals and families, is characteristic of our class system'.

Nevertheless, even if a man makes his *own* class, he is classified, and this means he is separated.

> 'Lower class people live on the river banks, in the foggy bottoms, in the districts back of the tanneries or near the stockyards, and generally in those places that are not desired by any one else.'

The mental image of class differentiation works itself out in the reality of geographical segregation. Different levels in society equate with distinct zones in geography.

This is very true of Warner and Lunt's Yankee City. Here status areas are marked features of the city scene. How different for example is upper town (Hill Street) from lower town (Riverbrook). On Hill Street population is less than half as dense as in Riverbrook, houses are spaced out with air, light and greenery between them; also the population is only a tenth as mobile, most Hill Streeters are well content with their abode; in it all are native-born as compared with only 57% down by the river, where the foreign-born are congregated. Over 83% of the Hill Street people are in proprietary or professional positions, they are the landlords and owners of factories, the financiers, business managers, lawyers and doctors, but less than 1% of Riverbrook folk are in these occupations. Almost all homes are owned up on the hill, most of those by the river are rented. Less than 1% of Hill Street residents are in houses in poor conditions as compared with 71% of the riverside dwellers. All Hill Street children finish high school, whereas only 15% do so in Riverbrook: less than 0·5% of the former are ever arrested, but 65% of the Riverbrook children have been arrested at one time or another. Here are great disparities, and it is little wonder that Hill Street residents use their influence and, in particular, their disproportionately great participation in the high offices of the city, to keep Hill Street separate, and prevent it from being invaded by lower class communities.

Such hill-side dwellers in another New England town, Boston, have long banded themselves together to segregate themselves out from the rank and file. In 1795, the Mt. Vernon Proprietors developed the south side of Beacon Hill to keep it a select area. 'All the lots were large and suitable for mansion houses for people of some means, and the streets were laid out to minimize the flow of traffic through the area.' In 1922, the Beacon Hill Association was formed to preserve the distinctness of upper class Boston from commercial or other influences that might change its character by lowering its tone.

Every American city has its Beacon Hill: in New York it was Murray Hill at first, then Morningside Heights, now the Heights of the Hudson, in Washington it has become Chevy Chase; in Chicago, Oak Park; in Los Angeles, Beverly Hill. Each of these has its equivalent of Riverbrook. The American city is drawn and quartered by class.

One reason for this is the marked strength of conformity in America. Each man tends to conform to his class. Although many do climb higher or fall lower, most remain in their class and conform to its conventions. Moreover, the person who climbs higher is more anxious than most to behave in a high-class manner, so as to be accepted at the new level; similarly, whoever falls its quick to give up or disguise his former traits. Thus within each class there is an appalling weight of sentiment and tradition to behave in the way of the group. As Warner (1962, p. 83) says, 'the moral principles embodied in parental models inevitably are class-structured so that . . . the child becomes a person in a social-class environment.'

Social conformity is helped by environmental uniformity. Whole areas of a city are the same. It is cheaper and quicker to put up house after house or apartment after apartment from the same plans, using the same materials, with the same skills and equipment. This was never more true than at present, when mass-produced homes are the ideal. There is an astonishing sameness about each part of an American city, and this part is repeated, in almost the identical location, across the American continent, from Florida to Alaska. There is the same flash and glare of motels and filling stations at the entry to every town, singeing the air with their off–on neon signs; the same oppressive repetition of family home/garden/car in suburban crescent upon crescent; the same pressing together of block upon block of tenements, with string-line shopping developments built into their lower storeys, in the mid-town scene; and the same cluster of lofty glass and concrete hotels and office blocks at the city centre.

This is because Americans are sworn to the ideal of a systems-built environment, devoted to models of speed and efficiency. The American is the mass man, mesmerized by *mass* production, *mass* education, *mass* recreation, *mass* standards and *mass* propaganda, 'Built-in systems across the landscape' are the result. Levittown, L. I., is a good example.

> 'Starting from scratch the Levitts converted eight square miles of open farm country into a densely populated community of 70,000 with paved streets, sewer lines, school sites, baseball diamonds, shopping center, parking lots, factory sidings, trunk arteries, garden clubs, swimming pools, doctors, dentists, churches, and a town hall—all conceived in advance and raised by mass production methods right on the building site, to give identical rows of houses along identical street blocks. Levitt owners . . . acquired a certain esprit de corps. Levittown lawns must be mowed once a week and the wash never flaps on Sunday. It's all in the deed. Levittown offers a very narrow range of house to a narrow income range. It is a one-class community on a grand scale.'

There are thousands of Levittowns in America.

> 'In the struggle against monotony houses of the same floor plan have been enclosed by four different types of exteriors, painted in seven varieties of color—so that *your* shape of Levittown house occurs in the same color only once every twenty-eight times. Streets are gently curved for further esthetic effect.' (O'Neill, 1969, p. 41)

To get this accepted (and sold) high-powered advertising plays an enormous role. The advertisers exploit the American ideal of being a good American. This means not stepping out of line, not being exceptional, not going to extremes. Enormous pressures are on a person to do things in the American way. Stereotype images of neighbourhoods attract the search for home location by millions of people and, as a result, influence the form of new towns or the character of urban renewal in older places. For example, the image of the city full of freeways, inviting every family to have a car of its own, to do its banking and its shopping, to go to church or the golf course all on wheels, undoubtedly assists Los Angeles, Houston or Miami to expand, since these largely grew up in the motor-age; but it also imposes on New York, Boston, Philadelphia or Pittsburgh the difficult task of following suit. No large city can now resist this image, each must conform to it, and so there is an astonishing sameness of road layout and traffic flow across the country. Each individual town tries to be the typical American city, and thus attract or retain population and business.

Humanism in American Cities

The humanism in American cities lies in its concern for the common man, its promotion of comfort and well-being, and its desire for immediate betterment. In many of these images of America, man is himself central. And central to that is the man democratic. This is significant. It is man, not land that is important: man believing in an equal chance but having an eye on the main chance for himself. The land, resources, agriculture, industry, living places are thought of as serving man. This is something American. In Europe the individual is still so often expected to serve the institution, the customer is seldom right; in America, institutions scramble over each other to serve the individual; the customer is always right.

The city has reacted to this. More and more it is shaping its growth to reduce labour, to cut down on physical effort, to cater to ease and comfort; in a word, to put personal convenience first. Whether at the centre or at the sub-centre, city design is aimed at reducing the cost and maximizing the profit from the use of land. If it could be shown that a hotel twenty storeys high at the centre would serve more people more rapidly and conveniently than five new four-storey hotels scattered at some remove from the focus of road and rail transport, then there would be little thought for the sky-line, for the physical appearance of the town, or the need to keep the streets sunny and pleasant for pedestrians: development would go ahead. On the other hand, should a twenty-five storey hotel show that the inward-and-upward movement of people was more time-consuming and less convenient than the outward movement, then growth at the centre would be halted, and development be switched to the sub-centre. In each case, the land would be treated for the ease and profit of the individuals likely to be concerned.

One of the best examples of this has been the reaction of the American city to the car. The car has been the greatest single saver of time and labour, it is the principal convenience and comfort of the people, and consequently it has had to be catered for. Even the energy crisis has not shaken its position: Americans will cut out much before the car. Whereas the European city appears to be doing all it can to outlaw the car, the American city has learned to accept it, although, as O'Riordan shows in Chapter 12, this may well be qualified. Take present-day Los Angeles. The town centre is literally lapped around in multi-level, multi-lane flyover expressways. Los Angeles has boldly tackled the problem of access to, and exit from, its central business and social area, by attempting to make the car itself central rather than damping down central urban activities. Since the car is the most vital and revolutionary thing in modern technology and since it allows for the rapid spread of population into country-based suburbs of pleasant, high-value housing, the city has adjusted to it. Whereas many places try to cope with the traffic problem by slowing down private traffic, Los Angeles flushes traffic through, giving millions of people the chance to 'work in' and 'live out', which they want. Smog may have resulted, but it is hoped that a change in cars will overcome this.

In other words, the city is trying to adjust to what its citizens want, it is not forcing the citizens to adjust to the supposed wants of the city. For 9 million people there are 5 million cars—an average of over two cars per family. To enable families to enjoy their cars to the maximum, a vast system of freeways and of parking lots and parking buildings has been planned. By 1980, there will be 1500 miles of freeway in greater Los Angeles. As this system develops the average commuting speed to the city centre will increase from the present 24 m.p.h. to 30 m.p.h., and thus save time, which to the American means money. Already the car capacity at the city centre is over half a million cars per day, and this is to go up appreciably more by 1980, by which time the freeways will handle half of the city's traffic. On the freeways there are no stops, no cross-roads, and traffic is bled off on to normal routeways by an intricate and yet efficient interchange system. This increases the flow of traffic twofold, and increases safety threefold. Within the first two and a half years of operation it was estimated that the commuters had saved 365 million dollars in time, and at least 250 lives.

Increasingly, then, the city is changing to respect what people want, and above all, to supply them with ease and comfort. This is, basically, what many Americans mean by amenity. Northern cities have to keep out the cold, reduce frost, clear the snow, lighten the dark, and in other ways make winter non-existent; while southern ones have to supply shade, offer coolness and suggest that the tropics are never uncomfortable. Comfort is the great factor in house, school, office, hotel and shop, in train and bus, on the road or in the air.

Cities that are clean and open, can be made equable in climate, allow fast movement, and provide easy working and living conditions are attracting more and more business and residents. As Ullman has shown, amenity is the single greatest force in the geography of expansion or decline in the United

States today. 'For the first time in the world's history', he claims, 'pleasant living conditions, instead of narrowly defined economic advantages, are becoming the sparks that generate significant population increase'. In spite of the handicaps of remote location and economic isolation, the fastest-growing states are California, Arizona and Florida. The new 'frontier' of America is thus a frontier of comfort.

What is perhaps really new is that comfort is thought of in terms of the common man. After all, Europe shows many examples of towns or parts of towns built for the ease and convenience of the rich. Edinburgh's 'New Town' when it was planned in the late 18th century was to have been essentially for 'people of fortune and of certain rank'. What America has done is to make the amenity of the rich the perquisite of the common man. Hence the tremendous outward movement into the countryside not only of the upper and middle classes but even of the more ambitious members of the working class. Frank Lloyd Wright makes a strong plea for this. 'The dividing lines between town and country are even now gradually disappearing. The country absorbs the life of the city as the city (proper) shrinks to the utilitarian purpose that now alone justifies its existence.' Why then would America continue to let land 'be parcelled out by realtors in strips 25′, 50′, or even 100′ wide? This imposition is a survival of feudal thinking, of the social economics practised by and upon the serf. An acre to the family should be the democratic minimum if this machine of ours is a success!'

One suspects that the famous architect was exaggerating the point in order to make it: even America would not have that sort of space. Yet if one thinks of the amount of street space, sidewalk, front and back garden, parking area, playing fields and parks the average suburb has to lay aside per family it must consume a lot of ground indeed. To get this ground the city has literally exploded out into the countryside. The era of the exploded city has arrived with fragments of urbanization right out in the country, connected to each other by a shining web of freeways. That this is an enormous drain on space is true, but it is also true that it enormously relieves the strain on man; and *that*, to the American, is the overriding issue. Man must get away from noise and dark, and fumes and overcrowding into light and air and space. In America, where the common man has a short working week and a long weekend, he likes to be out, either enjoying his garden, or engaging in one of a hundred forms of recreation in ball parks, on tennis courts, or at golf courses, and so forth. Why not then move part of the city with him, taking out church, school, or shopping into the country: it will only be an hour's ride by commuter train or by car to his mid-town factory or his down-town office.

However, the common man can ask for so many personal comforts he may embarrass the general ease and clog up the city with his own demands to the point where organization on a community-wide basis is threatened. Private corruption can undermine the public good. The common way may thus frustrate rather than further the overall need, as Professor Higbee points out in Chapter 11.

In reaction to this danger there is a new belief in planning. The pioneer spirit of co-operation has again been revived. Men are acting together to make their living together so much more meaningful. Cities are being expanded or renewed to provide not only greater convenience and comfort, but grace and beauty as well, as the common man reaches out for his full heritage. There is great optimism about this. Amenity is not only thought of in terms of material affluence, of houses with a double garage, spacious lawns, a swimming pool and access to great through-ways carrying people easily and swiftly to work at well-heated and air-conditioned offices: it is also thought of in terms of more play space, pleasant parks, highways centred in a median strip of flowering garden, down-town buildings with more light and space around them, and neighbourhoods planned as social entities where a rich and well-rounded social and cultural life can be enjoyed. Action research and advocacy planning are trying to find out and promote what the common man really wants. His ideal would affect the common lot everywhere, not only in America but in the world. What America is concerned with is not the Welfare State, but a state of well-being, and towards this state of common well-being its cities are moving fast.

References

Barnes, H. E. and Reudi, O. M., 1942. *The American Way of Life*, Englewood Cliffs: Prentice Hall.

Berry, B. J. L., Tennant, R. J., Garner, B. J. and Simmons, J. W., 1963. *Commercial Structure and Commercial Blight: Retail Patterns and Processes in the City of Chicago*, Chicago: University of Chicago, Department of Geography, Research Paper No 85.

Billington, R. A., 1960, *Westward Expansion*, New York: Macmillan.

Dillon, F., 1973, *A Place for Habitation; The Pilgrim Fathers and their Quest*, London: Hutchinson.

Greeley, H., 1860/1963. *An Overland Journey*, 1963 edn. (Ed. Duncan, C. T.), New York: Knopf.

James, H., 1907/1946. *The American Scene*, 1946 edn. (Ed. Auden, W. H.), New York: Scribner's.

Lewis, O., 1963. *Treasure of the American West*. In *The Book of the American West* (Ed. Monaghan, J.), New York: Bonanza Books.

O'Neill, W. L., 1969. *American Society since 1945*, New York: Quadrangle Books.

Packard, V., 1959. *The Status Seekers*, New York: McKay.

Parkman, E., 1849/1961. *The Oregon Trial*. 1961 edn. (Ed. Guthrie, J. B.) Washington Square Press.

Rosenburg, H., 1970. *The Tradition of the New*, London and New York: Paladin.

Warner, W. L., 1962. *American Life, Dream and Reality*, New York: Phoenix.

Woodward, W. E., 1965. *The Way Our People Lived*, Washington Square Press.

Yates, M. and Garner, B. J., 1971. *The North American City*, New York: Harper Row.

7
The Poet and the Metropolis

JOHN PATERSON

In moving into America men had to clear the wilderness, plough the land and set up their cities. Each of these acts created images which have been of great importance. Since America had to wrestle with nature at almost all stages of its development, this always coloured much of its thinking, making it seek a refuge from the wild in the city or, subsequently, long for relief from the city in the wild. As it cleared the forest and domesticated the landscape, it set its values upon the family farm out in the country, or the family business in the market town. However, the small town became a bore as the city grew in attraction, and from the mid-17th century until today the main thrust in America was in the city. Here the outstanding minds met to interpret America to itself and to the world.

The rifle gave way to the plough with the expansion of the frontier, and the plough to the mill with the sophistication of the base. As Max Lerner (1957, p. 155) says, 'while making goods and making money, the American has become in the process a city maker as well.' Yet the city dweller constantly talks of 'grass-roots America', referring to its farms and small towns, and gets 'back to nature' where he can. Thus there is a fascinating interplay which has sometimes confused but more often enriched the American mind.

In the 19th century, with the advance to the West, the Homestead Act and the infilling of the great interior plains, 'middle America' dominated: the people of the farms between the wilderness and the city. The agrarian spirit became all pervasive, making the city out to be a wen of troubles. A Henry George (1930, p. 75) could write, 'Not the desert . . . but city slums . . . are nursing the barbarians who may be to the new (society) what Hun and Vandal were to the old'. Yet by the 20th century the city had come out on top.

Even so, it drew together rather than displaced country and wilderness. The city man might have got his ideas about life from a small-town father;

and his frustration with city life might drive him out to the wilderness: the three images could still be present. Americans have fled, and still flee, the city into the wild; they have vilified the city from the countryside; and they have made the city their beast: in each case the city has been crucial to their mind. Consequently the mind of the city has become of basic importance to America. It has engaged some of the nation's greatest poets, artists, scientists and men of affairs, and so has become in a very real sense the touchstone of American culture. Frequently, the city has been simply the response to and expression of the nation, receiving its image from the realities around. Often, it has created the images of what America wants and hopes for, that have then affected the whole American scene.

During the past century, the proportion of the American population classified as urban has more than doubled: from 34% in 1870 to over 74% in 1970. It is against this background of increasingly urban living that America's artists paint and her poets write. In fact, it is just over a century since this growing awareness of the city as a phenomenon began to manifest itself in the work of the poets.

This study is focused on the poet and the big city. There are, of course, poets whose interest has been focused on the small town; who have been celebrating its virtues or exposing its pettiness. The most obvious case that comes to mind is Edgar Lee Masters and his *Spoon River Anthology*. We are here concerned, however, with the poets of the metropolitan city and, having said that, we can limit the field still further: the poets of the metropolis belong overwhelmingly to only three cities: New York, Chicago and San Francisco.

There is, moreover, a curious distinction to be made even among these three. While virtually all the metropolitan poetry with which we shall be concerned here was written *in* one of these three cities, it was written *about* only two of them, New York and Chicago. San Francisco has, at this particular moment of its rich cultural life, a constellation of poets second to none, but their collective published work contains hardly a line about the city itself, however much it may have inspired them. The local colour they leave to the longshoreman-poet, Eric Hoffer. Rather, the contemporary fashion is, as we shall later see, to withdraw to San Francisco and from there to write about New York and Chicago. Ferlinghetti and Ginsberg recall a New York which they have left behind, but which they cannot forget, and Welch looks back on a Chicago which he found unbearable, and which drove him away to California.

The City in American Life and Thought

In order to appreciate a century or more of the poetry of the metropolis, we must go back and recall the circumstances out of which the city in America and the city poet arose. At the beginning of the 19th century the long series of continental wars was at last over and so, too, was the initial conquest of the wilderness. On the Atlantic seaboard, European-style occupance was by now secure: the threat of Indian attack which, not long before, had been felt even in the coastal cities, had now receded in the consciousness of the settled part of the nation, or had been relegated to that area beyond the Appalachians which only the hardy or the desperate need penetrate. In other words, man had sufficiently established himself in America by this time for there to be room for the development of the romantic view of nature. Romanticism had not marked the writings of earlier generations. To them, nature was something to be conquered. To the New Englander, wildness in nature symbolized Adam's Fall and the curse; the conquest of nature was not only a practical necessity but a moral obligation. But now, with the frontier in Kentucky and Indiana, and moving swiftly westwards, the utilitarian had departed, taking his military metaphors of 'conquest' with him, and there was room and leisure in the east for a fresh appraisal of the environment.

Under these circumstances, we can readily understand the popularity, back east, of Wordsworth and the English romantic poets, the celebrants of the beauty of nature apart from the works of man. They enjoyed a considerable vogue in the period before 1850. It is true that Wordsworth had written, briefly, of the city but what had impressed him, in the view from Westminster Bridge, were qualities not normally associated with urban living: 'The beauty of the morning, silent, bare'.

What was less predictable but, in the event, profoundly influential was the convergence upon New York City of a group of diverse talents who chanced to assemble there about the year 1825 (Bryant, 1970). The two founder-members of this group, soon to become celebrated as the Hudson River School, were Thomas Cole and William Cullen Bryant, painter and poet. To them were shortly added Fenimore Cooper and Washington Irving, as well as the artist and engraver Asher Durand. Out of this chance convergence grew what has been described as 'America's first conscious attempt at an indigenous art style' (Hess, 1970), uniquely integrated as to media (poet and painter would sit together before a landscape, each composing and executing his contribution) and strongly romantic in tone, seeing 'a fundamental opposition of nature to civilization, with the assumption that all virtue, repose, dignity are on the side of "Nature" . . . against the ugliness, squalor and confusion of civilization,' (Miller, 1967, p. 197). When Cole died in 1847, half a million visitors came to see the posthumous exhibition of his work.

In such an atmosphere, the city was seen as something alien: as the encroachment of the urban and man-made upon the beauty of the natural. Miller paraphrases the prevailing attitude of the artist in the following words, with their curiously contemporary ring in the 1970s: 'the artist must work fast, for in America Nature is going down in swift and inexorable defeat' (Miller, 1967).

Nor was this all. The city represented, to the American of the second quarter of the 19th century, an alien environment not only because it was opposed to nature, but also because it represented the antithesis of individualism, and the America of this period was loud with the praise, indeed, the cult, of the individual. It was an attitude which achieved its most felicitous, and perhaps its most influential, expression in Ralph Waldo Emerson's (1850) *Representative Men*. Out of these studies in greatness emerged Emerson's conclusion: 'One man is a counterpoise to a city'.

This is the period of Henry David Thoreau and his retreat to Walden Pond. It was in 1851 that he summed up in a lecture the fundamental opposition, as he conceived it, of nature to the city, and in this same lecture that he uttered perhaps his best-known aphorism: 'In Wildness is the preservation of the World'. This was the period, too, when 'fundamental opposition' between nature and city was finding its most permanent expression in the form of the great city parks laid out, in particular, by Frederick Law Olmstead: Central Park, New York (1858), for which William Cullen Bryant had been crusading for so long; Prospect Park, Brooklyn; the Boston park system and the earliest of them all, Fairmount Park, Philadelphia.

This line of romanticism continues in the United States: painters and poets are its later exponents, denouncing the evils of city life. But, from this point on, it counts among its supporters few major talents. Indeed, much of its work, on the poetic side, scarcely rises above the level of the then popular temperance songs:

> 'Throw down the bottle,
> And never drink again.'

Some commentators see the next major figure in this tradition as T.S. Eliot, his *Waste Land* the city. But this view of Eliot, to which we shall later return, is surely simplistic and unsatisfactory. There was more to him than that.

The City as Stimulation

Rather, our attention is drawn, at the very time that Central Park was being laid out in New York, to the first figure in a new line: to Walt Whitman, the first edition of whose *Leaves of Grass* was published in the same city in 1855. Here, indeed, was a break in the line—a poet who found the city to be a source of endless and varied stimulation; who saw it as a stage, upon which a hundred or a thousand human dramas were being simultaneously played. After him have come many others.

This appreciation by the poet of the city seems to have developed in stages, although these stages are all readily identifiable, even in the first edition of *Leaves of Grass*. In the first place, there was what we may call the aesthetic appreciation; that is, the poet becoming aware, for the first time in America, of the fresh images and sounds of a thoroughly urbanized environment which, nevertheless, afforded stimuli to the poetic imagination. This is the first stage in Whitman:

> 'The blab of the pave, tires of carts, sluff of boot-soles, talk of the promenaders,
> The heavy omnibus, the driver with his interrogating thumb, the clank of the shod horses on the granite floor.

The snow-sleighs, clinking, shouted jokes, pelts of snowballs,
The hurrahs for popular favorites, the fury of roused mobs,
The flap of the curtain'd litter, a sick man inside borne to the hospital,
The meeting of enemies, the sudden oath, the blows and fall,
The excited crowd . . .'

It is, to begin with, the detail, the suddenly captured glimpse, that has fascinated a line of poets from Whitman on. Here, for example, is McLeish, writing of New York in the 1930s:

'Be proud New York of your prize domes
And your docks and the size of your doors and your dancing
Elegant clean big girls and your
Niggers with narrow heels and the blue on their
Bad mouths and your bars and your autobobiles
In the struck steel light and your
Bright Jews and your sorrow-sweet singing
Tunes and your signs wincing out in the wet
Cool shine and the twinges of
Green against evening.'

(Reprinted with permission from A. McLeish, *Collected Poems, 1917–1952*, Houghton Mifflin Company, Boston, Mass.)

This is, recognizably, the New York of Edward Hopper's paintings, of which his 'Nighthawks' is here reproduced. Or there is Denise Levertov, capturing precisely the sensation which is known to everyone who lives in a city laid out on a gridiron street plan: that the sky exists only to the north, south, east and west, framed in the canyon-ends of streets and avenues that look to the cardinal points of the compass:

'As the stores close, a winter light
  Opens air to iris blue,
glint of frost through the smoke,
  grains of mica, salt of the sidewalk . . .
To the multiple disordered tones
  of gears changing, a dance
to the compass points, out, four-way river.
Prospect of sky, wedged into avenues,
left at the ends of streets,
  west sky, east sky, more life tonight.
A range of open time at winter's outskirts.'

(Denise Levertov, *With Eyes at the Back of Our Heads*. Copyright © 1959 by Denise Levertov Goodman. Reprinted by permission of New Directions Publishing Corporation, New York.)

The City as Structure

But now the poet moves on, beyond the range of the purely sensual. The next stage is the recognition of the city as structure. This was the period of what Kenneth Clark (1969) has identified as 'Heroic Materialism', a period in which gigantic feats of construction and engineering were attempted, with

Figure 7.1. 'Nighthawks', by Edward Hopper. Courtesy of the Art Institute of Chicago

technical means often wholly inadequate, and in which the constructor triumphed, much more often than not, over his technical limitations. The city in mid-century America could with justice be regarded as the modern counterpart of the heroic structures of the classical world:

> 'Mightier than Egypt's tombs,
> Fairer than Grecia's, Roma's temples,
> Prouder than Milan's statued, spired cathedral,
> More picturesque than Rhenish castle-keeps,
> We plan even now to raise, beyond them all,
> Thy great cathedral, sacred industry, no tomb,
> A keep for life for practical invention.' (Whitman)

Not only had the young republic overtaken the Old World in its technical skills; its purposes, too, were superior: practical, benevolent, forward-looking.

In the pre-skycraper era of Whitman, the structures which attracted most attention or admiration were the bridges. Perhaps because in shape they held a particular aesthetic appeal, or perhaps because in their function they were the most altruistic of man's structures, uniting rather than separating men from one another, they have exercised a particular fascination upon the poet. Hart Crane's *The Bridge* is probably the outstanding example:

> 'Through the bound cable strands, the arching path
> Upward, veering with light, the flight of strings—
> Taut miles of shuttling moonlight syncopate
> Up the index of night, granite and steel—
> Transparent meshes—fleckless the gleaming staves—
> Sibylline voices flicker, waveringly stream
> As though a god were issue of the strings.'

(From 'The Bridge', Hart Crane, *Collected Poems*, 1930, Liverright Publishing Co., New York.)

And of all the bridges in 19th-century America, one stood out above the others in the poet's vision: Brooklyn Bridge. Many years in the building, it was opened in 1867 and was designed by a German engineer, Roebling. Of it, Kenneth Clark (1969) says: 'all modern New York, heroic New York, started with Brooklyn Bridge'. Ever since, it has exercised upon poets a particular fascination—not only upon Hart Crane, but also on Marianne Moore:

> 'Enfranchising cable, silvered by the sea, of woven wire,
> greyed by the mist, and Liberty dominate the Bay—
> her feet as one on shattered chains,
> once whole links wrought by Tyranny.
> Caged Circe of steel and stone, her parent German ingenuity.
> "O catenary curve" from tower to pier,
> implacable enemy of the mind's deformity.'

(From 'Granite and Steel', Marianne Moore, Faber and Faber.)

Even Ferlinghetti has felt obliged to take notice of this strange attraction. In far-off San Francisco, he declares:

'I once started out
to walk around the world
but ended up in Brooklyn.
That bridge was too much for me.'

(Lawrence Ferlinghetti, A Coney Island of the Mind. Copyright © 1958 by Lawrence Ferlinghetti. Reprinted by permission of New Directions Publishing Corporation, New York.)

The City as Catalyst

The next stage in the poet's exploration of the city comes with the recognition of man as the builder of cities. The structures grow larger, but man does not. Alongside his structures, he appears insignificant; yet it is he who has brought them into being. There is an interplay here in which the city acts as a catalyst to bring out man the maker. Struck by this contrast in scale between the creator and the creation, Carl Sandburg can only smile:

'In the evening there is a sunset sonata comes to the cities.
There is a march of little armies to the dwindling of drums.
The skyscrapers throw their tall lengths of walls into black
  bastions on the red west.
The skyscrapers fasten their perpendicular alphabets far across
  the changing silver triangles of stars and streets.
And who made 'em? Who made the skyscrapers?
Man made 'em. the little two-legged joker, Man.
Out of his head, out of his dreaming, scheming skypiece,
Out of proud little diagrams that danced softly in his head—
  Man made the skyscrapers.'

(From 'Good Morning, America', Carl Sandburg *Complete Poems*, 1928, Harcourt Brace Jovanovich, Inc., New York.)

Beyond this, again, lies another level of recognition: that although the city is man's creation, it may eventually take on a character of its own, so that now it becomes legitimate for the poet to speak of the city's 'soul'. In its extreme form, we find this consciousness expressed by Whitman in extravagant terms in his 1856 poem *Crossing Brooklyn Ferry:*

'Thrive, cities—bring your freight, bring your shows, ample
  and sufficient rivers,
Expand, being than which none else is perhaps more spiritual,
Keep your places, objects than which none else is more lasting . . .
We fathom you not—we love you—there is perfection in you also,
You furnish your parts toward eternity,
Great or small, you furnish your parts toward the soul.'

The City as Hero

Obviously, by this time the idea will have suggested itself to the poet of making the city, rather than its people, the focus of his imaginative effort. This is, in fact, the city as hero: the drama lies in the reaction of its inhabitants to it and, since it can outlive them all, it is the drama of succeeding generations reacting to the city, which is the permanent foundation, the link between people who are physically linked in no other way.

We have no difficulty in identifying the classic example of this *genre:* it is William Carlos Williams' *Paterson*. The city is neither New York nor Chicago, but it resembled New York in its origins and it lies not far away from the metropolis, on the other side of the Hudson. Leaving aside for the present such fascinating layers of symbolism as the sexual inwardness of the opening lines which some critics have discovered, we have here an epic poem of book length, interspersed with extracts from guidebooks and tit-bits of local history, that holds the reader all through to the city as the continuing reality:

> 'Paterson lies in the valley under the Passaic Falls
> its spent waters forming the outline of his back. He
> lies on his right side, head near the thunder
> of the waters filling his dreams! Eternally asleep,
> his dreams walk about the city where he persists
> incognito. Butterflies settle on his stone ear.
> Immortal he neither moves nor rouses and is seldom
> seen, though he breathes and the subtleties of his machinations
> drawing their substance from the noise of the pouring river
> animate a thousand automatons.'

But this, of course, was exactly what Emerson and his anti-urban contemporaries had feared: that the city would impose its own soul on its inhabitants, and thus rob them of their individuality. It is interesting, therefore, to notice how the young Carl Sandburg takes this argument and stands it on its head. Of all the big-city poems, his *Chicago* is probably the most widely known. But we need to identify what, precisely, is the poem's drift. It is that if a few outstanding individuals found a city (and since Chicago was founded only in 1833, there seems no reason why Sandburg should not have known some survivors of the first generation in his early days), then the city they create takes on their own character and, in due course, imbues others with it. Far from stifling individualism, Sandburg argues, the city multiplies it; it transmits the spirit of the founders to the generations which succeed them.

> 'Hogg Butcher for the World,
> Tool Maker, Stacker of Wheat,
> Player with Railroads and the Nation's Freight Handler;
> Stormy, husky, brawling,
> City of the Big Shoulders:
> They tell me you are wicked and I believe them, for I have seen
> your painted women under the gas lamps luring the farm boys.
> And they tell me you are crooked and I answer: Yes, it is true
> I have seen the gunman kill and go free to kill again.
> And they tell me you are brutal and my reply is: On the faces
> of women and children I have seen the marks of wanton hunger.
> And having answered so I turn once more to those who sneer at this
> my city, and I give them back the sneer and say to them:
> Come and show me another city with lifted head singing so proud
> to be alive and coarse and strong and cunning.

Flinging magnetic curses amid the toil of piling job on job,
here is a tall bold slugger set vivid against the little soft cities;
Fierce as a dog with tongue lapping for action, cunning as a
savage pitted against the wilderness, . . .'

(From 'Chicago' in *Chicago Poems* by Carl Sandburg, copyright 1916, by Holt, Rinehart and Winston, Inc., copyright 1944 by Carl Sandburg. Reprinted by permission of Harcourt Brace Jovanovich, Inc.)

The City as Destination

As Sandburg very well knew, of course, not everyone could stand up to Chicago. The character-imparting process did not make 'bold sluggers' of all its inhabitants; this was an educational process in which only the fit survived. And in particular the problem of survival confronted the immigrant. The century between 1815 and 1914 saw the arrival in the United States of 30 million Europeans (including Sandburg's own parents, who migrated from Sweden). By processes which have been sufficiently studied elsewhere to be passed over in the present context, most of these millions in due course found themselves inhabiting, at least for a period before they moved on, the immigrant quarters of New York, or Boston or Chicago.

It is significant that the first generation of immigrants produced little or no poetry which has survived. That they should fail to produce poetry in English is not surprising: many of them could not speak it, and to establish beyond dispute the proposition that the immigrant did not write poetry it would be necessary to carry out research in all the languages of Europe. What is, however, clear is that the children of those immigrants included among them some notable poets: both Sandburg and Ginsberg, for example, fall into this category. It was left to them to record the experiences of the immigrants as they saw and heard them through their parents. That both saw the task as a duty laid upon them is suggested by the following extracts. First, Sandburg in his *Onion Days*:

'Mrs. Gabrielle Giovannitti comes along Peoria Street every
morning at nine o'clock.
With kindling wood piled on top of her head, her eyes looking
straight ahead to find the way for her old feet.
Her daughter-in-law, Mrs. Pietro Giovannitti, whose husband was
killed in a tunnel explosion through the negligence of a fellow-servant,
Works ten hours a day, sometimes twelve, picking onions for
Jasper on the Bowmanville road.
She takes a street car at half-past five in the morning,
Mrs. Pietro Giovannitti does,
And gets back from Jasper's with cash for her day's work,
between nine and ten o'clock at night.
Last week she got eight cents a box, Mrs. Pietro Giovannitti,
picking onions for Jasper,
But this week Jasper dropped the pay to six cents a box because so
many women and girls were answering the ads in the *Daily News*.'

(From 'Onion Days', Carl Sandburg, *Complete Poems*, 1928, Harcourt Brace Jovanovich, New York.)

Ginsberg matches this with his *Kaddish*, the poem in which he seeks to recreate the impact of the new continent on his own mother, arrived from Russia as an immigrant, and confronted by a struggle for survival which ended in her mental collapse. He thinks of her

'. . . as I walk toward the Lower East Side—where you walked fifty years ago,
  little girl—from Russia, eating the first poisonous tomatoes of America—
  frightened on the dock—
then struggling in the crowds of Orchard Street toward what?—
toward candy store, first home-made sodas of the century, hand-
  churned ice cream in back room on musty brownfloor boards—
Toward education marriage nervous breakdown, operation, teaching
  school, and learning to be mad, in a dream—what is this life?'

The City as Challenge

But mention of Ginsberg brings us to changed times. The lift and the excitement have gone out of city poetry. The spirits that once soared with the skyscrapers and the bridges are today crushed by them: the era of heroic materialism has ended; indeed, it ended several decades ago. Reading over the poetry of Whitman–Sandburg, one is impressed by the Camelot-like quality of the vision. One can only ask: what went wrong?

We may attempt an answer to this question in various forms. First, and most obviously, it has always been the case that, even when the poet himself was being inspired and enthused by the city, the inspiration was not universal; it did not touch Ginsberg's mother and her kind. To *most* of their inhabitants, bent only on survival, the American metropoli were no Camelot but places where the weak and the unfortunate were crushed and the remainder might count themselves fortunate to be merely bored by the humdrum nature of their existence. Recalling early days in New York City, Ferlinghetti catches a glimpse of an existence in which to have survived is already an achievement, but in which, also, the question: survived for what? receives no clear answer:

'The El
        with its flying fans
and its signs reading
                SPITTING IS FORBIDDEN
the El
        careering thru its thirdstory world
with its thirdstory people
                        in their thirdstory doors
looking as if they had never heard
                                of the ground
an old dame
            watering her plant
or a joker in a straw
                    putting a stickpin in his peppermint tie
and looking just like he had nowhere to go but coneyisland
or an undershirted guy
                        rocking in his rocker

watching the El pass by
as if he expected it to be different
each time'

(Lawrence Ferlinghetti, *A Coney Island of the Mind*, Copyright 1955 ©1958 by Lawrence Ferlinghetti. Reprinted by permission of New Directions Publishing Corporation, New York.)

And from his San Francisco vantage point, Ferlinghetti is struck by the same emptiness: there may be drama in the city, but it is drama in which men are objects rather than subjects; drama in which the fifth act never seems to have been written:

'But now we come
to the lonely part of the street . . .

the lonely part of the world
And this is not the place
that you change trains
for the Brighton Beach Express
This is not the place
that you do anything
This is the part of the world
where nothing's doing'

(Lawrence Ferlinghetti, *A Coney Island of the Mind*, Copyright 1955 © 1958 by Lawrence Ferlinghetti. Reprinted by permission of New Directions Publishing Corporation, New York.)

The City as Conflict

But all this is general and merely intuitive: there is a much more proximate cause of the loss of joy in celebrating the city, and that is the steady take-over of the central city by the Negro ghetto. The ghetto is not, of course, a phenomenon of the post-war years alone, although our consciousness of it may have increased enormously during the past three decades. There were ghettos in American cities as soon as immigrants began arriving in large numbers. In this period, there were even arguments *in favour* of ghettos: that a newcomer from Russia or Greece, who knew no English and had no preparation for life in America, should be able to spend a transitional period in a community of his own people, where the cultural shock of the New World could be moderated by at least an echo of the familiar. But the word 'transitional' is the key. There have always been ghettos but, for the European immigrant, there has also always been a way out. However distinctive his appearance may have been on arrival at Ellis Island, a couple of years in the ghetto, a superficial wash and brush-up and he was indistinguishable from American Americans, whose life-story he was, in any case, re-enacting.

Not so the Negro: 10 or 11% of Americans had and have no way of disguising themselves; in this sense, there is for them no 'way out'. With the congregation of blacks in the industrial cities of the United States there has been formed a *permanent* ghetto population. It is a population which contains, in itself, a full range of socio-economic differentials. It is a city within a city, for even wealthy blacks live in it: moreover, it is to the advantage of the blacks to live in it, at least in a political sense. The black Americans, scattered evenly throughout the population, could nowhere hope to muster a majority but together, in the ghetto, they have at least a base for political power—the opportunity of electing one of their own people to the city council, or even to Congress. If they are there for good, they may as well make what they can of their segregation. But this is a far cry from the heroic materialism with which we began.

To the growth of the ghetto (Osofsky, 1966; Rose, 1971) there is, of course, a counterpart; the withdrawal of the ethnic majority to the suburbs, leaving the central cities in the hands of minority groups (for there are, besides the blacks, such groups as the Puerto Ricans and the Chicanos to consider). For a growing number of whites, the cities have become *terrae incognitae;* more, they have become *terrae periculosae*. Small wonder, then, that hymns in their praise have been conspicuously absent in recent white poetry.

The City as Separation

This leads us, appropriately, to the subject of Black poetry of the ghetto. And here we confront something of a paradox. For there have been, at least in New York, two main schools of ghetto poets, separated from each other in time; one dates from the twenties and therefore ante-dates the Civil Rights movement, while the other is contemporary and is the voice of a minority now growingly self-aware and assertive. And of these two there can be very little question that the former is, poetically, superior to the latter. The 1920s saw the so-called Harlem Renaissance (Huggins, 1971) and the emergence of such major figures as Langston Hughes and Countee Cullen; they were the poets of an oppressed minority in a period when to aspire to equality with the whites was beyond imagining. The Renaissance ended with the Depression (Osofsky, 1966) and ghetto poetry revived, on a significant scale, only in the post-war drive for conscious ethnic upgrading. Of this revival, however, it has to be said that little of the poetry which it has produced appears to have enduring quality; at least, there is little to match the quality of Baldwin in prose, on the one hand, or the traditional art form of the Negro, song, on the other. This is not so much a criticism of the present generation of poets themselves as a recognition that their sensations are too immediate; they have no room to withdraw into artistic detachment from the experience undergone.

In any case, the ghetto poet has formidable obstacles of the mind to overcome. The ghetto is, in numerous cities, an area of once-palatial housing (this is certainly the case with Harlem), in which five or ten black families occupy the shell of a mansion that fifty or seventy years ago housed a single wealthy, white family. There is a long tradition of poets living in poverty and dreaming of riches, but what can one reasonably expect of a poet who lives in squalor, but in a palace whose splendour has gone? The irony is caught precisely by the gentlest of all the black poets, Countee Cullen:

Yet Do I Marvel

'I doubt not God is good, well-meaning, kind,
And did He stoop to quibble could tell why

The little buried mole continues blind,
Why flesh that mirrors Him must some day die . . .

Yet do I marvel at this curious thing:
To make a poet black, and bid him sing.'

(From 'Yet Do I Marvel', Countee Cullen, *On These I Stand*, 1925 and 1953, Harper and Row Inc., New York.)

There is yet another factor to consider in the changing attitude to the city. On the poetic side, the influence of T. S. Eliot on the generation immediately past was great and Eliot, as noted above, has been seen by some commentators as the anti-urban, or arcadian, poet of the mid-20th century. But what Eliot was protesting, and very specifically in *The Waste Land*, was not merely the loneliness and sense of alienation that the city may induce but, from his own, deeply religious, standpoint the reduction in the stature of man himself.

The City as Desolation

The structures of the era of heroic materialism expressed the outlook of the men who built them; men who were optimists and, despite any Emersonian fears, individualists in the spheres of creation and technology. Sandburg understood this, that the structures in themselves were nothing without the people who occupied them:

'By day the skyscraper looms in the smoke and sun and has a soul.
Prairie and valley, streets of the city, pour people into it
and they mingle among its twenty floors and are poured out again
back to the streets, prairies and valleys.
It is the men and women, boys and girls so poured in and out
all day that give the building a soul of dreams and thoughts and memories.
(Dumped in the sea or fixed in a desert, who would care for
the building or speak its name or ask a policemen the way to it?).'

(From 'Skyscraper' Carl Sandburg, *Complete Poems*, 1928, Harcourt Brace Jovanovich, New York.)

The metropolitan structures of the present era, by contrast, tell us very little of the men who built them: they merely express the high levels of ground rent in the Central Business District. The poet senses that the structures are now more important than the men who build them: in the final analysis, that the roles of creator and creation have somehow been reversed and it is now the structure, rather than the user, which is the animate thing. Byron Vazakas has this very sensation on entering a skyscraper:

Skyscraper

'The purpose of its sex is obvious, the body articulate and masculine.
It concedes nothing, yet the weak flock toward its entrances arterial
With tubes. Light and heat are incidental to its chemistry.
Even the sky is invaded. Looking upward from the street, the
window spacing is absolute
As time and frequent as death. None deviate . . . I enter the

elevator, and mention names, but impersonality's infectious
glaze calls numbers only.
In the outer office, the wind moans by steel window-sash,
echoing the terror of my being here. No other hand constructed
this; no brain conceived
So monstrous a detachment. I want nothing in this place. I want
the forest crackling underfoot . . .'

(Byron Vazakas, *Transfigured Night*, 1942, MacMillan, New York.)

So we can plot the footsteps of the urban poet, away now from the city; away from Chicago, the poet's city *par excellence*. Lew Welch, safely now in San Francisco, looks back to the metropolis of the Midwest with a shudder, his pessimism highlighting the optimism of Sandburg's *Chicago*:

Chicago Poem

'Driving back I saw Chicago rising in its gases and I knew
that never will the
Man be made to stand against this pitiless, unparalleled
monstrosity. It
Snuffles on the beach of its Great Lake like a blind, red
rhinocerus.
It's already running us down.
You can't fix it. You can't make it go away.
I don't know what you're going to do about it,
But I know what I'm going to do about it.
I'm just going to walk away from it.'

(LewWelch, *On Out*, 1965, Oyez Berkeley Publications.)

And will the last poet to leave (as they said about Seattle) please turn off the lights?

So we come, in conclusion, to what is, up till now, the most violent protest of all: Ginsberg's *Howl*. Here the poet is moved to great heat by the way in which the city consumes and obliterates the individual like the ancient god Moloch, to whom human sacrifices by fire were made.

'What sphinx of cement and aluminum bashed open their skulls and
ate up their brains and imagination?
Moloch! Solitude! Filth! Ugliness! Ashcans and unobtainable
dollars! Children screaming under the stairways! Boys
sobbing in armies! Old men weeping in parks!
Moloch! Moloch! Nightmare of Moloch! Moloch the loveless!
Mental Moloch! Moloch the heavy judger of men!

Moloch whose eyes are a thousand blind windows! Moloch whose
skyscrapers stand in the long streets like endless Jehovahs!
Moloch whose factories dream and croak in the fog! Moloch
whose smokestacks and antennae crown the cities! . . .'

Moved as we may be by the intensity of feeling, we look in vain to Ginsberg or his contemporaries for any positive solution to the complex of urban problems which have forced themselves on the American consciousness in the past half-century. It is difficult to overstate the magnitude of these problems, and certainly Ginsberg is not guilty of understatement: he and his contemporaries are passionate but negative, and now we need to know what to do next.

The century since Whitman has seen a remarkable flowering in what might be called the social philosophy of cities—in their planning and in the analysis of their character as communities. Now we need, but do not have, a moral philosophy to accompany this (Harvey, 1973), and the present generation of poets has accurately diagnosed this lack. In particular, we may say, we lack a view of man, lofty in the Emersonian manner or, at the lowest estimation, worthy of the structures he is capable of erecting. But if life is really meaningless, then it is 8-million-or-so times as meaningless in New York City as it is outside.

It has been the achievement of the poet that he has, in the past, sometimes been able to act as both the prophet and the moral philosopher of his society, diagnosing weakness and pointing the way ahead. In the urban sphere, the problems challenging American society are enormous, and so the challenge to the poet to rise to the situation today is enormous also. What will the message of the next generation of urban poets be?

References

Bryant, W. C., II, 1970. Poetry and painting: a love affair of long ago. *American Quarterly*, **22,** 859–882.

Clark, K., 1969. *Civilization*, London: Murray.

Emerson, R. W., 1850. *Representative Men*, Boston.

George, H., 1930. *Social Problems*, New York: Doubleday.

Harvey, D., 1973. *Social Justice and the City*, London: Arnold.

Hess, J. A., 1970. Sources and aesthetics of Poe's landscape fiction. *American Quarterly*, **22,** 177–189.

Huggins, N. L., 1971. *Harlem Renaissance*, New York: Oxford University Press.

Lerner, M., 1957. *America as a Civilization*, New York: Simon and Schuster.

Miller, P., 1967. *Nature's Nation*, Cambridge, Mass: Belknap Press.

Osofsky, G., 1966. *Harlem: The Making of a Ghetto. Negro New York* 1890–1930.

Rose, H. M., 1971. *The Black Ghetto: A Spatial Behavioural Perspective*, New York: McGraw Hill.

8
Black Culture, Violence and the American City

TIMOTHY O'RIORDAN

The following three chapters revolve around the themes of black culture, violence and the urban landscape. The linkage of these three themes in no way implies that black culture and violence should be equated. The Kerner Commission (1968) made it abundantly clear that violence in American cities is not a product of culture, but is associated with deeply rooted social injustice, frustration, bitterness and a general feeling of hopelessness. The chapter by Florence Ladd portrays how the black sees the American city and it is clear that, given a sense of fellowship, and the opportunity to enjoy a variety of constructively stimulating experiences, the city can be an exciting and meaningful place to live in regardless of ethnic origin.

Nevertheless, the hard fact remains that much of urban America is moving toward two societies, one black, one white, separate but unequal (Kerner Commission, 1968). In fact this trend has moved Gold (1970) to describe the American city as a defensive structure composed of separate cells each bent upon protection of their sub-cultures and property values. The consequence of this pattern, implies Gold, is a series of quite separate urban villages, all isolated from each other, and each denying itself the necessary activity space to make the city truly liveable.

The physical manifestation of black inequality is the ghetto, a distinctive urban residential form spreading in a sectorial manner from the fringes of the downtown core. Morrill (1965) has described its location and form in some detail and Rose (1972) has provided firm evidence that the concentrations of blacks in such areas are becoming steadily more pronounced. Adams (1972) has shown how, relative to the expansion of the total metropolis, the ghetto is diminishing in size even though its population is increasing. The inevitable implication is that ghetto inhabitants are trapped in an ever more closely confined activity-space with little hope of release.

The ghetto exists because forces are in operation to ensure that it remains. Morrill (1965) documents these forces. These include external pressures such as the policies of the Federal Housing Administration, up until 1949, which openly encouraged segregation of urban housing; the current practices of realtors and financial institutions to discourage blacks from moving into white residential neighbourhoods except when 'blockbusting' is intended; pressures from white vigilante groups aimed at intimidating pioneering black families in white neighbourhoods; the location of freeways and urban renewal schemes which form deliberately designed barriers to ghettos; and insufficient legal and political aid to help blacks fight for their rights. But there are also positive forces holding the ghetto together: the need for protection against whites, especially white policemen; the opportunity to gain political representation at various levels of government; and the desire to maintain viable social institutions, such as the church.

The ghetto is also a locale for violence, and in America today violence has become almost a necessary way to command attention over grievances and to bring about change. Violence is so much part of American city life that it must be seen as a major force shaping urban residential form. Gold (1970) has discussed this matter and Higbee develops this theme in Chapter 11. Crime is primarily associated with large cities. One-third of all reported crimes in the United States take place in six cities, each with a population exceeding one million yet with a combined population that accounts for only 12% of the nation's inhabitants. Compared with communities of 10–15,000 there are eight times as many arrests in cities with populations exceeding 250,000. This same group of cities reported increases of 90% in robberies, 51% in homicides and 46% in assaults between 1961 and 1967. This was a period that was also marked by an outburst of savage social violence which accounted for a number of lives and damaged much property in a number of cities. The history of racial violence in America is discussed in Chapter 9 by Lewis, who also analyses the geographical distribution of the serious riots of the mid-60s.

Why did these riots take place? The answer is not entirely clear but it has much to do with the discrepancy between image and reality which forms the central theme of this book. It is quite probable that much of the migration of blacks from the south to northern mid-western and western cities during the 20th century was the result of the push of reality, (rural poverty resulting from tenancy and changed agricultural economics), and the pull of illusion, (the bright lights and the opportunities for making money in the big city). Morrill and Donaldson (1972) document the migration patterns of American Negroes from 1770 to 1970. Brown (1965, pp. vii–viii) describes the feelings of the Negro pioneers who began the big migration to northern cities following 1915.

> 'These Negroes were told that unlimited opportunities for prosperity existed in New York and that there was no "color problem" there. They were told that Negroes

> lived in houses with bathrooms, electricity, running water and indoor toilets. To them this was the "promised land" that Mammy had been singing about in the cotton fields for many years . . . [But in fact] there was a tremendous difference in the way people lived up north. There were too many people full of hate and bitterness crowded into a dirty, stinking, uncared for closet-sized section of a great city . . .
>
> The children of these disillusioned colored pioneers inherited the total costs of their parents—the disappointments, the anger. To add to their misery, they had little hope of deliverance. For where does one run to when one is in the promised land?'

This northward migration was encouraged partly by the supply of low-income jobs which followed the First World War. But it was also made possible by the process of 'filtering down' whereby old houses in the central city were vacated by the last of the European immigrants who were becoming upwardly mobile and who were looking for higher-income residential areas towards the fringes of the city. The rise of urban public transportation assisted this trend. For 40 years or so this filtering down process seemed to work fairly effectively, and even though incoming blacks were occupying sub-standard housing at least they had a roof over their heads.

However, in the '50s and '60s the expectations of the urban Negro began to rise. The success of the Civil Rights Movement in the South brought with it hope for the embattled ghetto resident of the north. The charisma of John Kennedy, the Great Society pledges of Lyndon Johnson, numerous historic supreme court rulings denouncing segregation as unconstitutional, all added to the illusion that the famous 'dream' of Martin Luther King, so poignantly expressed by him immediately prior to his assassination, might be fulfilled. Outwardly there were signs that this was indeed the case. During the prosperous years of the early '60s black incomes rose relative to white incomes and the number of non-whites falling officially below the poverty line dropped from 55% to 35% during the period 1961–1967. The 1973 Annual Report of the President's Council of Economic Advisers indicated that there has been no significant widening of the gap between the rich and the poor in the United States since 1947. However, this has only been achieved at the cost of billions of dollars in welfare programmes such as income supplements, food stamps and unemployment insurance. Of more significance perhaps, there has been no narrowing of this gap, and the statistics do not take into account the effects of inflation (calculated at 10·8% in 1974) which disproportionately impinge upon the poor who have less room to manoeuvre when prices rise. Nor do these statistics reflect that, as incomes rise, aspirations rise even faster.

Adams (1972) hypothesizes that much of the racial disturbance documented by Lewis can be related to the fact that blacks were frustrated in achieving what they regarded as legitimate and attainable goals, namely decent housing, a decent job, a reasonable degree of protection against crime, drugs and disease and the opportunity to get a proper education. Instead they were increasingly trapped in the hopelessness of the ghetto, where housing was neither available nor of any reasonable quality, where unemployment rates rose to as much as 40% among high school drop-outs, and where, as already documented, crime and violence were a fact of life. The Kerner Commission (1968) showed that many more crimes were perpetrated by blacks against blacks than against whites, with the exception of robberies. Indeed, the chance of a ghetto black being assaulted is 1 in 77 compared with 1 in 10,000 for a white suburban resident. Droettboom and colleagues (1971) noted that those most affected by crime and violence are those least able to do anything about it. While many of them long to move out to escape urban violence they are unable to relocate. As one woman put it to the U.S. Commission on Civil Rights (1967, p. 6), 'The first thing I would do is move out of the neighborhood. I fear the entire neighborhood is more or less a trap'.

Frustrated aspirations are exacerbated by numerous impediments to the job market caused either by the hatred or the fear of the white working-class groups (including policemen) who stand to lose most by policies aimed at improving the social and economic conditions of non-whites, or by restrictive union practices which deny non-whites the opportunities to enter well-paid jobs for which they are suited (e.g. the construction industry), or by inadequate education. This is an explosive mixture. Adams (1972) found that those cities which recorded the worst racial violence were those which were large with packed and compact ghettos, with increasing black populations but a virtually static housing supply, and with extensive affluent white suburbs. The major riots occurred in the inner areas of the ghetto where mobile blacks with high aspirations found themselves trapped in the company of dispirited people in overcrowded tenements.

Lewis makes the point that most of the violence was directed against property, not individuals, except possibly policemen and national guardsmen. Gans (1972, p. 336) interpreted many of the riots almost as 'carnivals'.

> 'This is not because the participants are callous, but because they are happy at the sudden chance to exact revenge against those who have long exploited or harassed them. The rebellion becomes a community event, a community activity; for once the ghetto is united and people feel they are acting together in a way that they rarely can. But most important, the destruction and looting allows ghetto residents to exert power . . . for once they have some control over their environment, if only for a little while'.

The only solution to the problem seems to involve radical new concepts in urban living. A stimulating education, the opportunity to create and develop imagination, a sense of belonging in a viable community which is able to design its own urban space and a varied urban habitat are generally regarded as necessary ingredients (Rappoport and Hawkes, 1970; Rose, 1972; Adams, 1972). Today the situation is superficially more peaceful but little changed. While school integration is now more widespread, the bussing of children from one community to another arouses hatred among the whites and induces fear and alienation among blacks who are used to operating in tight social spaces. Racially mixed housing is more popular but it really only works among middle-income blacks and whites.

One ray of hope lies in the political sphere where black power is finally beginning to be felt. Seven major cities including Washington, Los Angeles and Detroit now have highly respected black mayors and Congress and state legislatures are increasingly filled with black representatives. The growing politicization of the American Negro may lead to the emergence of political control over the ghetto and with it the prospect of improved economic and social conditions for its inhabitants (Goldsmith, 1974). Political skill plus a growing public willingness to support programmes of reform offer a cautiously promising prospect, but few observers are sanguine that current social injustices will be eradicated in the foreseeable future (Warren, 1969; Harvey, 1972). Meanwhile, the currents that shape such injustices also leave their mark upon the American urban landscape, as documented in part by the chapters that follow.

References

Adams, J. S., 1972. The geography of riots and civil disorders in the 1960's. *Economic Geography*, **48,** 24–42.

Brown, C. 1965. *Manchild in the Promised Land*, New York: The New American Library.

Droettboom, T., McAllistar, R. J., Kaiser, E. J. and Butler, E. W., 1971. Urban violence and residential mobility. *Journal of the Institute of American Planners*, **37,** 319–325.

Gans, H. J., 1972. *People and Plans: Essays on Urban Problems and Solutions*, Harmondsworth, Middlesex: Penguin.

Gold, R., 1970. Urban violence and contemporary defensive cities. *Journal of the American Institute of Planners*, **34,** 146–159.

Goldsmith, W. W., 1974. The ghetto as a resource for black Americans. *Journal of the American Institute of Planners*, **40,** 17–30.

Harvey, D., 1972. Revolutionary and counter revolutionary theory in geography and problems of ghetto formation. *Antipode*, **4**(2), 1–13.

Kerner Commission, 1968. *Report of the National Advisory Commission on Civil Disorders*, New York: Bantam.

Milgram, S., 1970. The experience of living in cities. *Science*, **167,** 1461–1468.

Morrill, R. L., 1965. The Negro ghetto: problems and alternatives. *Geographical Review*, **55,** 339–361.

Morrill, R. L., and Donaldson, O. F., 1972. Geographical perspectives on the history of black America. *Economic Geography*, **48,** 1–24.

Rappoport, A. and Hawkes, R., 1970. The perception of urban complexity. *Journal of the American Institute of Planners*, **36,** 106–111.

Rose, H. M., 1972. The spatial dimension of black residential subsystems. *Economic Geography*, **48,** 43–65.

U.S. Commission on Civil Rights, 1967. *Final Report*. Washington, D.C.: Government Printing Office.

Warren, R. L. (Ed.), 1969. *Politics and the Ghetto*, New York: Atherton Press.

9

Geographical Aspects of Race-related Violence in the United States

G. MALCOLM LEWIS

The following study is concerned with the areal distribution of race-related violence in the United States. It has two primary objectives: first, to examine the general areal distribution of race-related incidents during the violent years of the middle 1960s in relation to the distribution of incidents during earlier periods of racial violence and second, to discover regional differences in the race-related incidents during these years according to type of place and type of incident, and region.

The Areal Distribution of Incidents During Earlier Periods of Racial Violence

Walter Lippmann is reputed to have observed that a cardinal weakness of Americans is their unwillingness to read the minutes of the previous meeting and this would seem to apply just as much to the less completely recorded events which have taken place outside committee rooms as to the carefully minuted discussions within. 'Because Americans tend to suffer from chronic, historical amnesia, it is important to stress that riots are not a new phenomena in our country' (Speigel, 1967, p. 1). The history of American riots in general and of race riots in particular was first presented in detail by Grimshaw (1959) and has subsequently been reviewed by others (Waskow, 1966; Kerner Commission, 1968). There have also been several definitive studies of specific urban riots (Rudwick, 1964; Chicago Commission on Race Relations, 1968; Gilbert, 1968; New Jersey Select Commission on Civil Disorder, 1968). However, there would appear to have been no direct attempt to consider the areal patterns of incidents in each of the fairly recognizable periods of racial tension (Deskins, 1969). This chapter attempts to fill this gap and to provide a perspective against which to assess the distinctiveness of the pattern of race-related violence in the middle 1960s.

The first settlement within the present borders of the conterminous United States which is known to have contained Negroes was a short-lived colony near the mouth of the Pee Dee River in what is now South Carolina. In 1526, within a few months of settlement, several of the Negro slaves rebelled and defected to the Indians (Aptheker, 1963, p. 63). However, Negroes were not brought to the English colonies until 1619, in which year 19 arrived at Jamestown, Virginia. But it was only towards the end of the 17th century that Negroes began to be brought in large numbers to the tidewater areas of Maryland, Virginia and the Carolinas (Greene and Harrington, 1966). After 1718 the Company of the Indies began to introduce Negro slaves into Louisiana, where they were concentrated in New Orleans and on the levee lands of the Mississippi for 100 miles or so upstream from there.

Almost from the beginning slave plots and insurrections would seem to have been fairly common in those areas containing relatively large numbers and/or high proportions of Negroes in the population. In the English colonies south of Chesapeake Bay almost all of them occurred within a very short distance of the coast (Figure 9.1). There were relatively few incidents in the New Orleans area, partly because there were fewer Negro slaves, and partly because it was easier for Negroes to obtain their freedom (Aptheker, 1963). During the late 17th century and the first half of the 18th century some of the

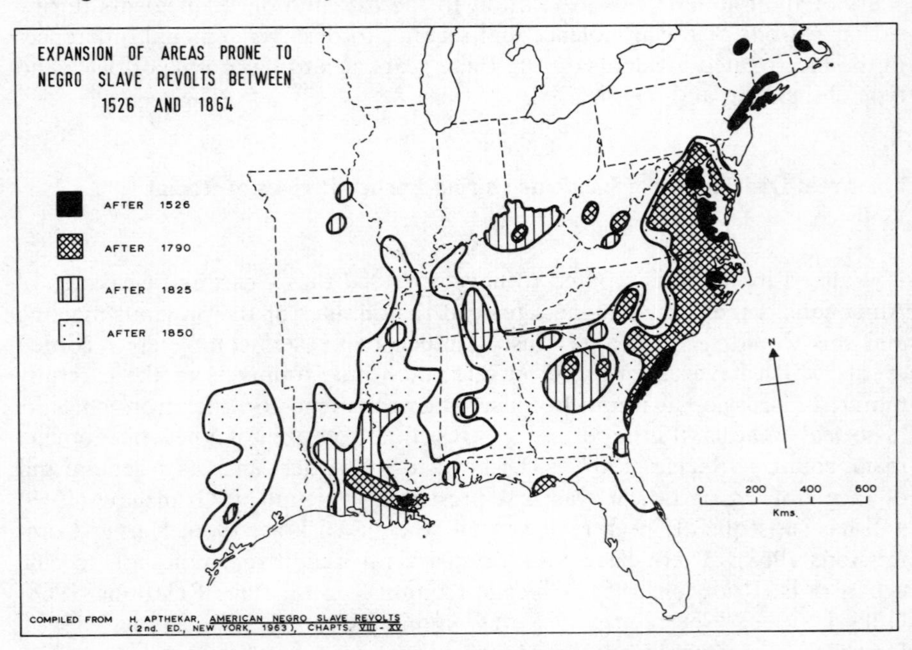

Figure 9.1. Race-related violence in the United States

worst incidents and plots occurred in the towns of New England and the Middle Colonies, where slavery did not begin to be abolished until 1774 (Aptheker, 1963). In 1681 two Negroes in Massachusetts were convicted of arson and executed. One of these incidents occurred in Roxbury, thus anticipating the riot of June 1967 by almost 300 years. Other cases of arson in which Negro slaves were either known or believed to have been implicated occurred in New York City (1712 and 1741), Boston (1723), New Haven (1723), Philadelphia (1737) and Hackensack (1741). Occasionally, Negroes in the Middle Colonies actually attacked their white masters, as in 1708 when a group of slaves in Newton, Long Island, rebelled and killed 7 whites.

Plots and rumours of plots were fairly common in northern towns during the first half of the 18th century and, together with actual incidents, they would appear to have occurred in cycles. In particular, the period 1737 to 1741 was one of widespread discontent among Negro slaves. The peak of this crisis was reached with an outbreak of mass hysteria in New York City in 1741, as a result of which 31 Negroes were executed and 70 more banished. The 101 thus disposed of must have represented a high proportion of the total Negro population since, four years before, the City and County of New York together contained only 1283 Negroes over the age of 10 (Greene and Harrington, 1966, p. 97).

A mid-18th century period of relative tranquillity was followed in the 1770s and 1780s by a renewal of disturbances in the South, for which there was no equivalent in the North. By this time Negroes were becoming numerous in the South and were beginning to comprise approximately one-third of the region's population (U.S. Bureau of the Census, 1918). Conversely, in the towns of the North overt tensions decreased as more and more Negroes attained freedom and as they became increasingly outnumbered by whites.

Following the American Revolution plantations began to be established away from the original coastal plain areas. Between approximately 1790 and 1820 many were created on the Piedmont in Virginia, South Carolina and eastern Georgia, and across the Appalachians in central Tennessee and in the Bluegrass region of Kentucky. These, together with the older coastal plain areas, became the centres of Negro population (Grey and Thompson, 1933). Between 1790 and 1820 the number of Negroes (including free Negroes) in the South increased by almost one million, of which approximately half occurred in the newer areas. During this period there were two cycles of discontent among the slaves, the first from approximately 1790 to 1802 and the second from 1810 to 1816. During the first of these, the distribution of suspected plots and disturbances was similar to that in the latter part of the colonial period, with a high frequency in the coastal plain areas of Virginia and the Carolinas and occasional incidents in the towns of the North and on the Mississippi delta. During the second cycle inland disturbances became more frequent and occurred for the first time on the Piedmont in South Carolina and Georgia and in the Bluegrass region of Kentucky (Figure 9.1).

The second quarter of the 19th century witnessed two further cycles of rebellious activity: 1829 to 1831 and 1835 to 1842. The first of these was associated with an economic depression, a temporary pause in the areal expansion of the plantation system and the consequent 'piling up' of the Negro population in the already densely settled areas. During the 1830s plantation agriculture became established in several new areas, the most notable of which were the Black Belt of western Alabama and north-eastern Mississippi, the Piedmont area of eastern Alabama and the Mississippi bottom lands as far north as south-west Tennessee. Consequently, during the 1835 to 1842 periods of rebellious activity a high proportion of the incidents occurred in areas which hitherto had had no tradition of racial violence (Figure 9.1). This was particularly so in the Mississippi bottomlands, where incidents were numerous as far north as the Yazoo. Indeed, rumours of conspiracies occurred as far up the Mississippi valley as Memphis, Tennessee (1839) and even Fulton, Missouri (1842).

The period from 1843 to 1849 was one of relative quiet but the decade leading up to the Civil War was marked by increasing tension throughout the South. For the first time, disturbances would seem to have been widespread throughout the lowland South, peninsular Florida and, less explicably, the Alabama–Mississippi borderlands being the main exceptions (Figure 9.1). Beyond the Mississippi, disturbances occurred in the Red River valley lands of central Louisiana but the most significant change during this period was the increasing frequency of disturbances in eastern Texas. In the summer of 1860 this culminated in plots, arson and a general reign of terror in at least 14 counties. The region had only been effectively settled by planters and their slaves after the Texas Revolution of 1835 and particularly after the annexation of Texas by the United States in 1845. Thereafter, settlement on the alluvial lands of the Red, Trinity, Brazos and Colorado rivers marked the last major phase in the spread of slavery in North America (Grey and Thompson, 1933, p. 907). As in earlier phases of planter–slave settlement, plots and disturbances followed almost immediately.

The Civil War brought about the end of slavery in the United States and, consequently, the end of slave plots and insurrections. The last known slave conspiracy was planned in December 1864 in Troy, Alabama and in the South this was followed by almost 20 years of relative calm. However, during the Civil War a new type of racial violence occurred in the North when several towns and cities experienced riots involving both whites and free Negroes (Figure 9.2). The first of these occurred in 1862 in Cincinnati and arose from a dispute between Negro and Irish American riverboat hands. Similar riots occurred in Newark, New Jersey and in New York City, Buffalo and Troy, New York. Generally referred to as the 'draft riots', they resulted from fears on the part of recent poor white immigrants (particularly those of Irish origin) that Negroes would take their jobs whilst they themselves were drafted to fight in the war. In terms of loss of life, the New York City race riot of 1863 was by far the worst in American history. White fatalities were estimated at

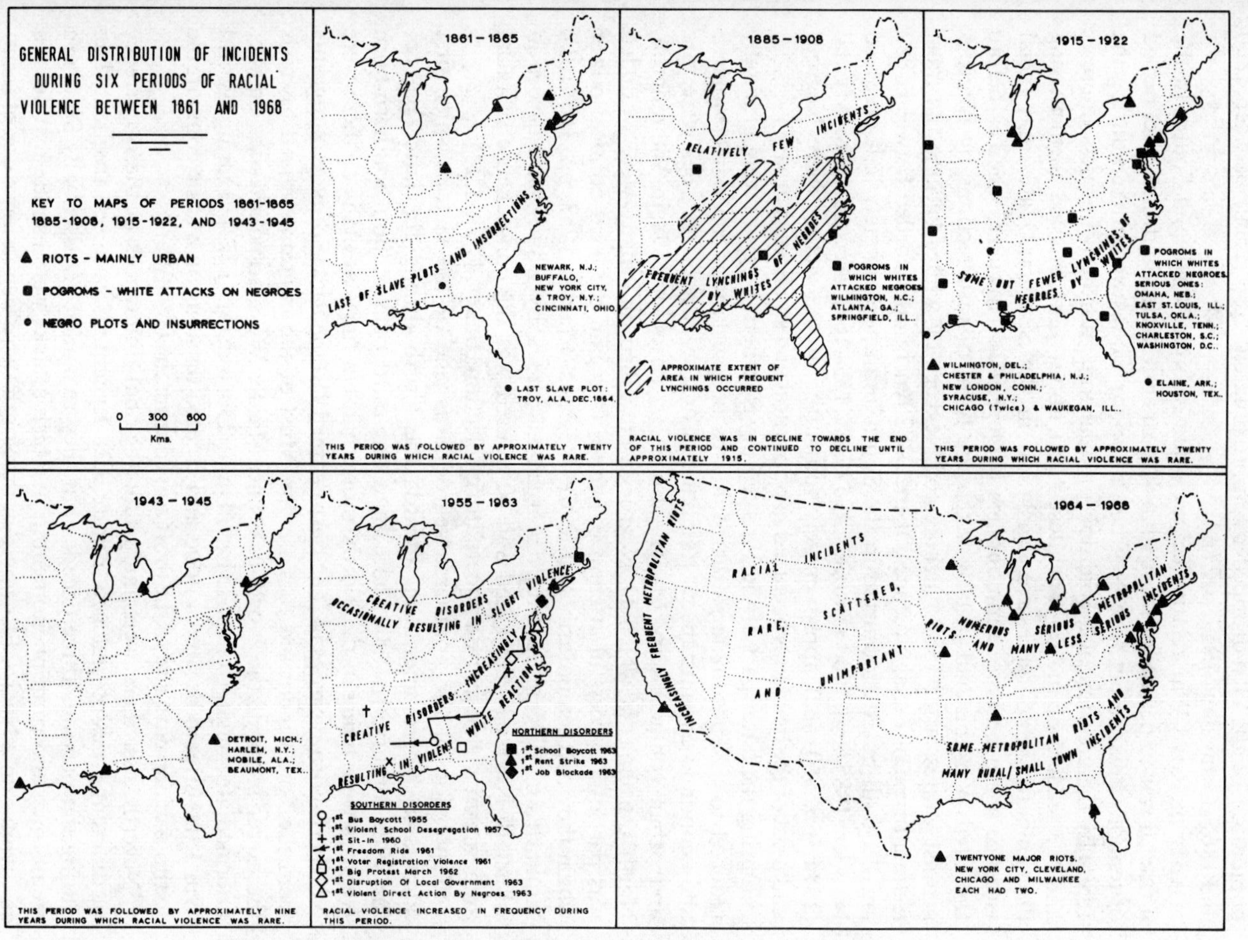

Figure 9.2. Race-related violence in the United States

1500. The number of Negro fatalities and injuries is not known but the Negro population of the city dropped by approximately 20% (2500) between 1860 and 1865 (Grimshaw, 1959, p. 27).

After the Civil War relative calm existed in both North and South. In the former it was to persist for more than 40 years but in the South the brief period of tenuous alliance between Negroes and poor whites was broken during the 1880s by the creation of a coalition of Bourbons, northern business interests and the leaders of the poor whites. After about 1885 this led to increasing discriminatory and repressive legislation in the Southern states and racial violence again became commonplace (Figure 9.2). However, whereas before the Civil War this had usually taken the form of insurrections by Negroes against whites, after 1885 lynchings of Negroes by whites became the most frequent type of racial violence. Of the 3382 lynchings recorded between 1882 and 1936 all but 108 occurred in the South and approximately half occurred between 1885 and 1900 (Frazier, 1957, pp. 159–160). According to Myrdal (1944, pp. 560–562) approximately four-fifths of the victims were Negroes and most of the lynchings occurred in poor rural and small town areas.

During the latter part of the 19th century Southern Negroes were gradually reduced to a subordinate status by means of legislation, repressive violence and economic circumstances. Retaliations on the part of Negroes were rare and urban conflicts were few (Figure 9.2). A riot in 1898 in Wilmington, North Carolina, in which 11 Negroes were killed and many were wounded was part of a successful attempt by whites to regain the local political power which they had temporarily lost to the Negroes (Frazier, 1957, pp. 160–161). Similarly, in 1906 in Atlanta, Georgia, an attack by whites on Negroes in which 10 Negroes were killed and 60 injured was the consequence of racial antagonism stirred up by a gubernatorial election (Baker, 1964, p. 15).

Repression of the Southern Negroes was one of the factors which first stimulated Negro migration towards the North during the latter part of the 19th century. In 1900 more than one-third of a million Southern-born Negroes were living in the North, more than ten times as many as there were Northern-born Negroes in the South. After 1910 the trickle of Negro migrants from the South increased to a stream. Most of the migrants moved towards the larger cities of the North-east and Middle West, sometimes via smaller intermediate staging cities (Morrill and Donaldson, 1972). During World War I this stream became a flood and created tensions on a scale comparable to those which gave rise to the draft riots during the Civil War.

Nevertheless, there were important regional differences in the occurrence of different types of violence (Figure 9.2). Lynchings were less frequent than during the previous three decades but they remained an essentially Southern small-town and rural phenomenon, in which attacks were usually by one or a few whites on a Negro. Elsewhere in the South there were a few cases of insurrections by Negroes, of which a civil insurrection in Elaine, Arkansas (1919) and a military mutiny in Houston, Texas (1917) were the most serious. After the war there were numerous riots in Southern cities, most of which were really pogroms, being one-sided attacks by whites on Negroes, with little or no attempt at retaliation by the latter.

A new regional element in the pattern of racial violence was the reappearance after a lapse of more than half a century of rioting in Northern cities. Some were Southern in style, being precipitated by violent mass action on the part of whites, e.g. Springfield, Illinois (1908) and Omaha, Nebraska (1919) (Chicago Commission on Race Relations, 1968, pp. 68–71; Waskow, 1966, pp. 110–119). However, most of them involved violence on the part of both races and were the result of increasing tension consequent upon the massive in-migration of Negroes. These tensions were most apparent in the spheres of housing, employment and public facilities, especially in those parts of cities where Negroes came into residential contact with recent working-class immigrants from Europe. Towards the end of and immediately after World War I riots occurred in cities as far apart as Chester (1917) and Philadelphia (1917), Pennsylvania; New London, Connecticut (1919); Wilmington, Delaware (1919); Syracuse, New York (1919); and Chicago (1920) and Waukegan, Illinois (1920). However, the two most serious riots of this period occurred in East St. Louis, Illinois, in 1917 and Chicago, Illinois, in 1919. In East St. Louis there were 39 Negro deaths, 9 white deaths and several hundred injuries. More than 300 buildings were destroyed, 44 railroad cars were burned and the fire loss was later estimated to have been almost $400,000 (Chicago Commission on Race Relations, 1968, p. 73; Rudwick, 1964, pp. 52–53). In the Chicago riot of 1919 fatalities were almost as great as they had been in East St. Louis (23 Negroes and 15 whites), injuries were again very high (at least 342 Negroes and 178 whites) and approximately 1000 people were rendered homeless (Chicago Commission on Race Relations, 1968, p. 1). In terms of fatalities and injuries these two riots were as serious as any of the urban riots of the middle 1960s, though less serious than the New York City draft riot of 1863. However, even allowing for the vagueness of the estimates and the changed value of the dollar it would seem that the value of property damaged was in both cases less than that in Watts (1965), Newark and Detroit (1967) or Baltimore, Washington, D.C. and Chicago (1968), (U.S. Senate Committee on Government Operations, 1967; Lemberg Centre for the Study of Violence, 1968).

The interval from 1923 to 1942 was one of relative calm, though the migration of Negroes to Northern cities continued and both Negro and anti-Negro organizations slowly built up their strengths. Between 1915 and 1944 the Ku Klux Klan registered approximately two million members, of which one-half lived in metropolitan areas with more than 50,000 inhabitants. Half the new members lived in the Northern States and only about one-third in the South. Indeed, the largest concentrations of Ku Klux Klan members were all outside the South: Chicago, 50,000; Indianapolis, 38,000; Philadelphia–Camden, 35,000; and Detroit, 35,000. Of 20 cities housing Klan memberships of more than 10,000, only Atlanta, Birmingham and Memphis were in the South (Jackson, 1967, pp. 236–239). Between 1918 and 1942 the National Association for the Advancement of Coloured People increased in membership from less than 10,000 to approximately 85,000 and the number of branches from 85 to 481. Though the numerical increase was small compared to that of the Ku Klux Klan, it included a higher proportion of capable, middle-class people in the cities of the North, within which, for the most part, the more active branches were located (Myrdal, 1944, pp. 822–826; Kellogg, 1967, pp. 236–239). The only significant riot during this period occurred in 1935 in Harlem, New York. It was an unorganized outbreak by Negroes against continuing discrimination by whites and, as such, presaged the riots of the middle 1960s (Myrdal, 1944, p. 568).

Twenty years of relative calm were terminated early in World War II when, as in World War I, there was an increase in the rate of Negro migration to cities both beyond and within the South. A major riot in Detroit in 1943 resulted in 34 deaths (including 25 Negroes) and property damage in excess of $2,000,000 (Lee and Humphrey, 1943, pp. 82–86; Kerner Commission, 1968, p. 104). A riot in Harlem in the same year resulted in 6 deaths, more than 500 injuries and damage to property estimated at $5,000,000 (Lee and Humphrey, 1943; Kerner Commission, 1968). Negroes were attracted not only along established paths of migration but also to the newest centres of industry, e.g. the shipbuilding centres of the South (Hampton Roads, Charleston and Mobile) and the industrial sections of Southern California, the San Francisco Bay area and the Pacific north-west. Not surprisingly, therefore, some of the riots during this period had occurred in cities which had hitherto experienced little or no violence, e.g. Mobile, Alabama, and Beaumont, Texas (Figure 9.2).

The racial disturbances of the World War II period were followed by a further interval in which violence was rare but during which attitudes changed.

> 'White opinion in some quarters of America had begun to shift to a more sympathetic regard for Negroes during the New Deal, and the War had accelerated that movement. Thoughtful whites had been painfully aware of the contradiction in opposing Nazi racial philosophy with racially segregated military units. In the postwar years, American racial attitudes became more liberal as new non-white nations emerged in Asia and Africa and took increasing responsibilities in international councils. Against this background, the growing size of the Northern Negro vote made civil rights a major issue in national elections and, ultimately, in 1957, led to the establishment of the Federal Civil Rights Commission . . .' (Kerner Commission, 1968, p. 105).

The establishment of the Kerner Commission was part of the first Civil Rights legislation to get through Congress since 1875. Two further Civil Rights in 1960 and 1964 still further widened the gap between rights and reality. During the early 1950s the National Association for the Advancement of Coloured People had achieved a series of legal victories but these successes had not been followed by comparable economic and social progress. Between 1953 and 1962 the unemployment rate among non-whites (92% of whom were Negro) increased both absolutely and relatively compared with that among whites. Between 1952 and 1959 the median family income of non-whites fell from 56·8% to 51·7% that of white families (U.S. Bureau of Labour Statistics, 1966, pp. 20, 138). (However, non-white incomes were to rise relative to white median income in the 1960s).

For the first time direct action techniques designed to accelerate social and economic progress began to be used by Negroes and their supporters: at first in the South; later, after about 1962, in the larger cities of the North-east and Middle West; and from 1964 onwards in the large metropolitan areas on the West coast. The one-year-long bus boycott in 1955 and 1956 in Montgomery, Alabama, was the first direct action to capture the attention of the nation as a whole and to give confidence to tens of thousands of Negroes. It led in the South to the invention and widespread use of a variety of direct action techniques: attempts by Negroes to register at hitherto all-white schools and colleges (1957 onwards); sit-ins and their various derivatives, swim-ins, stall-ins, etc. (the first sit-in was at Greensboro, North Carolina, in February, 1960); freedom rides (the first major freedom ride was from Washington, D.C., to Jackson, Mississippi in the spring of 1961); voter registration drives (from 1961 onwards); large-scale protest marches (the first was in Albany, Georgia in 1962); and the disruptions of the working of local government (first attempted at Danville, Virginia, in 1963) (Figure 9.2; Lewis, 1966). Whilst non-violent in concept, direct action by Negroes almost from the beginning resulted in violent retaliation by whites. Thus, the Library of Congress Legislative Reference Service listed six incidents in 1961 in which white mobs attacked Negroes, five of which were in the South. There were no incidents, however, in which Negro mobs attacked whites (*Congressional Record*, 1967, S. 10,260).

In the early 1960s moral and political attitudes in the United States underwent a sudden change. The political correspondent Andrew Kopkind (a self-styled product of the 'silent' generation of Americans, which came of age in the late 1950s) returning to the United States in 1961 after only two years abroad found a 'new America I did not know had changed' (Kopkind, 1969, preface). Federal legislation, an upsurge of white liberalism and the growing popularity of direct action campaigns combined, among other things, to give the Negroes a new sense of self-respect. Many were no longer prepared to accept the humiliation of second-class citizenship. This led to growing impatience and some direct action on the part of Negroes became less non-violent than before. Several well-established civil rights organizations became more militant and new and still more militant Negro movements began to emerge. In 1963 the Library of Congress Legislative Reference Service listed eight riots in the South, in at least five of which Negroes would seem to have been the first to use violence (*Congressional Record*, 1967). This was also a significant year in that it marked the beginning of direct action in the North (Figure 9.2). Violent incidents also occurred in several cities in the North. Escalation of tension eventually led to the Civil Rights Act of July 1964 and to yet another cycle of racial violence. This surpassed all other cycles, not only in the number of serious, damaging riots but in the variety, frequency and areal dispersion of lesser disturbances.

Race-related Violence in the Middle Sixties

It would seem that the cycle of race-related violence which began as a consequence of direct action by Southern Negroes and their supporters in the spring of 1964 probably came to an end with the series of often violent student protests in the schools and colleges of the North and West during the fall term of 1968. By the end of that year it was becoming increasingly difficult to distinguish between violence which stemmed from race-related civil disorders and that which stemmed from non-racial incidents. According to the American Insurance Association more than half of the relatively minor amount ($14,000,000) of damage to insured property resulting from riots and civil disorders during the first eight months of 1969 occurred as a result of demonstrations on college campuses and in high schools (*Daily Telegraph*, 24 September 1969, p. 22). Indeed, after the widespread violence which followed the assassination of Dr. Martin Luther King, Jr., in April 1968 the summer was surprisingly uneventful. By the fall there was a growing feeling in many major Negro communities that 'large scale summer violence that pitted Negroes with rocks and bottles against policemen could be a thing of the past' and this began to be seen as a consequence of 'the growing black consciousness of the Negro and his desire to take over and run his own community' (Johnson, 1968, p. 1). During the first eight months of 1969 a few small towns were troubled by racial violence but there was an estimated 50% reduction in riots compared with 1968. Discussing this trend, one commentator recognized a wide range of factors, including the improved training and equipment of the police and National Guard, a general decrease in the unemployment rates in Negro ghettos, a somewhat cooler summer than usual, the success of some remedial measures, the growing interest and pride in Negro culture, and, perhaps most important of all, the increasing tendency for ghetto Negroes to involve themselves in indigenous community organizations (Heren, 1969, p. 6).

It is virtually impossible to measure with any degree of precision the magnitude of race-related violence or to compare its consequences accurately with those of other types of disaster. However, it is possible to estimate the cost of race-related violence in the United States during the middle 1960s in terms of deaths, injuries and certain types of economic loss. According to the data compiled for the purpose of this research, there were 579 violent incidents during the five-year period from 1964 to 1968 inclusive, involving 270 deaths and almost 8700 injuries. Property damage resulting from race-related violence for the period March 1965 to September 1967 was estimated to be $210 million (U.S. Senate Committee on Government Operations, 1967, pp. 14–15). These figures compare with 50,000 deaths annually from automobile accidents, 18·2 million injuries annually in the home and annual property damage of $2·25 billion due to fire and storm.

Most of the deaths, injuries and economic losses resulting from race-related violence seem to have occurred in a relatively few places and within brief periods. Thus, for the period 1964 to 1968 inclusive, more than 80% of the estimated property damage, more than 60% of the injuries and approximately 45% of the fatalities occurred in five major riots: Los Angeles (1965), Newark (1967), Detroit (1967), Washington, D.C. (1968) and Chicago (1968). However, even these, with 121 fatalities, approximately 5300 injuries and property damage estimated at between $210 million and $220 million, were relatively minor compared with other types of localized disaster. For example, the explosion and subsequent fire in Texas City in April 1947 resulted in 468 deaths, approximately 3500 injuries and damage to insured property alone estimated at $150 million.

Race-related violence during the middle 1960s might therefore appear to have received publicity and attention out of proportion to the losses of life and property. Conversely, however, unlike any other type of disaster, race-related incidents have involved and influenced millions of both black and white Americans in many places and within various parts of the United States. Certainly, there are and will be moral and political consequences which should prove to be of importance in the development of the nation. (These implications are touched upon briefly at the introduction to this chapter.)

The Distribution of Race-related Violence in the Mid-sixties

The percentage distribution of race-related incidents of all types during the period 1964 to 1968 inclusive was remarkably similar to the percentage distribution of Negroes in 1960. Incidents would seem to have been most numerous in those regions which experienced large absolute increases in the number of Negroes between 1930 and 1960. Thus, those regions which together experienced 85·5% of all the incidents, accommodated 7,540,000 additional Negroes between these years.

Almost 72% of the race-related incidents which occurred in the United States during the five years from 1964 to 1968 took place in one of the 211 Standard Metropolitan Statistical Areas. However, since 62% of the nation's population and 64% of its Negroes lived in such areas in 1960 this proportion was perhaps to be expected. More significantly, 42% of all the incidents took place in the nation's two Standard Consolidated Areas (New York–north-eastern New Jersey and Chicago–north-western Indiana) and the twenty Standard Metropolitan Statistical Areas with populations each in excess of one million in 1960. Together these contained 35·2% of the nation's population and 38·4% of its Negroes. So there was a close coincidence between the percentage of incidents which occurred in S.M.S.A.s and the percentage of the Negroes living in them.

However, some regions had much higher incident rates per S.M.S.A. than others. Thus, northern regions experiencing Negro immigration had almost three times as many incidents per S.M.S.A. than southern regions reporting increases in black populations. This probably reflects the fact that on average northern S.M.S.A.s had approximately three times the population of southern S.M.S.A.s but may impart a bias in the newspaper reports surveyed.

Incidents occurring in cities (incorporated places with populations of 25,000 and over) but outside Standard Metropolitan Statistical Areas were most frequent in the South *sensu lato*. However, they comprised barely more than one-twentieth of all incidents and were of less significance than incidents taking place outside both cities and Standard Metropolitan Statistical Areas. Though these small-town and rural area incidents represented less than one-quarter of all race-related incidents they were also dominant in the South, particularly so in northern and western Mississippi, one of the most reactionary parts of the nation, where civil rights workers from outside were actively opposed by whites. In fact the South *sensu lato* experienced almost 90% of all incidents occurring in such places as against just over 29% of all incidents taking place in Standard Metropolitan Statistical Areas.

The Timing and Nature of Incidents

Racial incidents can be categorized by time as well as by severity. Over 85% of all events took place between March and October, with the worst disorders being reported during the sultry months of July and August. Three-quarters of major disorders (where there was widespread damage and loss of life) and 56% of serious disorders (less disruptive involving fewer people and security forces) took place during the last three weeks in July and the first three weeks in August. These were also weeks during which teenagers and young adults were either on vacation or seeking their first jobs after graduation. Unemployment rates among young people were therefore particularly high, especially in the inner suburban areas of the larger cities. The absence of the daily school routine, of adequate communal recreation facilities and of money with which to engage in creative or sustained activities, together with overcrowded dwellings and sidewalks, created tensions which tended to build up as the vacation season progressed.

In terms of diurnal time, large-scale disorders generally commenced during the evening, individual attacks took place at night, while demonstrations were often held during the day. Characteristically, about one-fifth of these events (except demonstrations) took place on a Sunday, a day when the tensions of the working week most frequently came to a head. Monday was also noticeable for racial tension, possibly as a sequel of weekend activities.

Minor disorders (involving relatively few people, little property damage and only local police) were the most frequent of the race-related incidents, closely followed by isolated incidents involving attacks on or fights between individuals or small groups. In contrast, major disorders were relatively rare, comprising a mere 3% of all the race-related incidents that occurred between 1964 and 1968 inclusive. However, the proportion in the various categories varied considerably from region to region. The North experienced

81% of all major disorders, 77% of the serious disorders and 73% of all minor disorders. In contrast, the South *sensu lato* was more prone to lesser incidents, experiencing 53% of attacks involving small groups, 70% of disorders involving damage to property, 54% of incidents involving demonstration and 84% of demonstrations involving arrests but little violence.

In terms of damage to human life and property the 21 major disorders were more destructive than all the other 640 incidents analysed. Together they resulted in more than 200 deaths (approximately two-thirds of all race-related deaths), nearly 6000 injuries (approximately two-thirds of all race-related injuries) and more than 30,000 arrests (more than half of all the arrests in race-related incidents). The aggregate property damage in major disorders was approximately $150 million, several times greater than the aggregate property damage from all other race-related incidents. Each of the major disorders took place in a Standard Metropolitan Statistical Area, of which only three (Tampa–St. Petersburg, Memphis and Wilmington) had less than one million inhabitants in 1960. All but one of the incidents (Memphis, February to March, 1968) occurred between early April and late August and all but four occurred beyond the South *sensu lato*. Major disorders were characteristic of the Negro ghettos of the inner areas of large cities, particularly those beyond the limits of the South *sensu lato*. In such cities the main Negro communities were large and increasing in population, compact but expanding in area and segregated from adjacent non-Negro communities. Socio-economic conditions were poor; family, group and community tensions were high and a large proportion of the population consisted of teenagers and young adults. The 17 counties within which the 21 major disorders occurred together experienced an increase of approximately one and one-third million Negroes between 1950 and 1960, Negroes accounting for almost one-third of the total population increase.

In the middle 1960s the large Negro ghettos were still a relatively new element in the social structure of most large cities outside the South *sensu lato*, usually dating back to the first two decades of the 20th century in the North-east and Middle West and to the 1940s and 1950s in the West. As such, they still constituted a relatively new potential threat by a subordinate group on the established accommodation pattern. Here, more than anywhere else, flourished the more radical and militant black organizations, partly awakening and partly satisfying an awareness among Negroes of their inferior status in relatively new environments, of which only a decade or so before they or their parents had hoped for something better. In these ghettos, the chance of a minor incident precipitating a major disorder was therefore far greater than elsewhere. Even so, 10 of the 21 Standard Metropolitan Statistical Areas in the North with more than one million inhabitants in 1960 failed to experience a major disorder between 1964 and 1968. These included three with very large Negro ghettos: New York, Philadelphia and St. Louis. (Subsequent analysis by Adams (1972) has shown that in St. Louis at least the availability of housing was equal to the rate of growth of the Negro population, indicating that overcrowded housing conditions were in part a contributor to the dissatisfaction expressed by urban blacks.) In the South *sensu lato* only three Standard Metropolitan Statistical Areas had more than one million inhabitants. Of these, Baltimore and Washington, D.C. each experienced a major disorder but Atlanta did not.

Serious disorders, though more than four times as numerous as major disorders, were in aggregate less destructive, only causing approximately one-fifth as many deaths, less than a third as many injuries and arrests and a relatively small amount of damage to property. In several important respects, however, they were similar to major disorders: 93% of them occurred in Standard Metropolitan Statistical Areas; 77% occurred in the North; 87% occurred between early April and late August; and they occurred in counties in which 29% of the aggregate population increase between 1950 and 1960 consisted of Negroes. One important difference, however, was the tendency for serious disorders to occur in smaller Standard Metropolitan Statistical Areas and smaller cities. Thus, whereas only 14% of the major disorders occurred in S.M.S.A.s having less than one million inhabitants, 54% of the serious disorders occurred in S.M.S.A.s or other cities with less than this number.

Minor disorders were almost twice as numerous as serious disorders and almost nine times as numerous as major disorders. However, in aggregate they resulted in only 17 deaths, less than 600 injuries and probably not more than 3500 arrests. Although 82% of all minor disorders took place in Standard Metropolitan Statistical Areas, many of the latter were in the small-to-medium size range. Indeed, 15% occurred in rural areas or in places with populations of less than 25,000. Consequently, they were somewhat more evenly distributed than either the major or serious disorders. Even so, 61% of all minor disorders occurred beyond the South *sensu lato*.

Therefore major, serious and minor disorders were essentially non-Southern phenomena. Thus, whereas 58% of the incidents in all other categories occurred in the South *sensu lato*, only 25% of the more serious disorders occurred therein. Disorders were also essentially metropolitan phenomena, 86% of the incidents in the three categories occurring in S.M.S.A.s as against only 59% of the minor racial incidents. Disorders were also a markedly spring–summer phenomenon, 83% taking place between early April and late August as against only 55% of the minor racial incidents. Furthermore, disorders were particularly associated with counties which had experienced large population increases between 1950 and 1960, with a concomitant expansion of black populations. The corollary to this is that minor incidents tended to be found in smaller centres and rural areas generally in the South (75% of all such incidents). Incidents which involved violent attacks in small groups followed an interesting pattern.

Most racial demonstrations (which did not result in violence) occurred during 1964 and the first eight months of 1965. Some were spontaneous and indigenous but, particularly in 1965, they were usually organized or assisted from outside by civil rights' organizations and deliberately centred in the very communities where Negroes had remained most oppressed and hitherto had protested least. Their concentration within areas in which there were few or no larger scale incidents reflects the important fact that they were non-violent in concept and occurred in communities within which the native Negroes were passive by tradition. They also met with strong measures from control forces, together leading to some 11,600 arrests. The relative absence of minor and small-scale incidents beyond the South *sensu lato* was partly due to the fact that demonstrations in North-eastern, Middle Western and Western cities frequently escalated into more serious types of disorders (the major disorders in Cincinnati in June 1967, in Milwaukee from August to November 1967 and in Memphis during February and March 1968 each started as demonstrations).

Even minor incidents were by no means characteristic of the whole of the South *sensu lato*. In particular, those regions with a long history of Negro occupancy, such as Virginia and the Carolinas, were singularly free from these and, incidentally, other categories of incidents. So were peninsular Florida, the outer delta of the Mississippi and the Gulf Coast of Louisiana and Texas. Since most parts of these latter areas experienced healthy economic growth during the 1940s and 1950s it is probable that the Negroes had less to demonstrate about than in most other parts of the South. Furthermore, with the possible exception of the outer delta of the Mississippi, these areas lacked the repressive tradition of the grosser forms of Negro oppression found in the former plantation areas.

The Historical Perspective

Racial violence occurred more frequently during the middle 1960s than ever before. The period was the most recent of a series of relatively short cycles of violence which have occurred at intervals of a few decades during the past 300 years. However, it differed from even the most recent of the others in several important respects: for the first time, serious urban riots were numerous and widespread; a higher proportion of the violent incidents commenced as demonstrations; the total damage to life and property was greater than ever before; and the location of the incidents established a new geographical pattern. In the South there were more urban riots than in any previous cycle and a higher proportion of the incidents were precipitated by Negroes (albeit often with non-violent intentions) than during any cycle since the Civil War. Pogroms were still a characteristic element of the violence in this region but they were proportionally less important than in previous cycles. In the North-east and Middle West the differences from previous cycles were even greater. Serious rioting in the ghettos of large cities was more frequent, widespread and damaging than ever before. Unlike many of the urban riots of the Civil War and World Wars I and II they rarely, if ever, stemmed from open conflicts over labour issues but were spontaneous expressions of deeper frustrations and resentments; the culmination, in most cases, of more than half a century of growing pressures and little progress in the Negro communities. For the first time, race-related violence began to occur in the larger cities of the West. Here, in the newer Negro ghettos, the physical conditions of life were perhaps somewhat better than in the older cities of the east of the Mississippi. However, traditions were even shorter and, in consequence, their restraining influence even less.

In one important respect the areal pattern of race-related violence did, however, conform to that of all the previous cycles. Outside the South, small towns and rural areas experienced singularly little race-related violence and in the absence of Negro communities within these places and areas this is likely to remain the case. Most of America, by areas, never has had, does not now have and is unlikely in the foreseeable future to experience race-related violence. Conversely, the tendency for an increasing proportion of both Negroes and whites to concentrate in or near large cities (coupled with the increasing efficiency of the mass communications media) is resulting in an ever increasing proportion of the population being acutely aware of, at risk from, or involved in, this type of violence. Hitherto, these two fundamental and essentially geographical conditions do not seem to have been adequately appreciated.

Appendix

Classification of Incidents

Each incident was classified according to an eightfold classification based on intensity, duration, crowd size and number of control forces. The categories were as follows.

Category A. Major Disorders. Involving all or most of the following elements.

(1) Many fires, serious looting and some sniping.
(2) Violence over a period of more than two days.
(3) Several sizeable crowds.
(4) Deployment of federal forces and/or the National Guard as well as local control forces.

Category B. Serious Disorders. Involving all or most of the following elements.

(1) Some fires, isolated cases of looting, some throwing.
(2) Violence over a period of one or two days.
(3) One sizeable crowd or many small groups.
(4) Use of state police though generally not federal forces or the National Guard.

Category C. Minor Disorders. Involving all or most of the following elements.

(1) A few fires and broken windows.

(2) Violence usually lasting for less than one day.
(3) Small numbers of people involved.
(4) Use only of the local police or police from a neighbouring community.

Category D1. Isolated Incidents Involving Attacks on/or Fights between Individuals or Small Groups of People. Involving all or most of the following elements.

(1) Violence generally occurring at a moment in time or lasting for only a few minutes.
(2) Violence committed by one person or a very small group of people.
(3) Control forces not usually present.

Category D2. Isolated Incidents Involving Attacks on Specific Properties. Involving all or most of the following elements.

(1) Attacks occurring at a moment in time.
(2) Attacks committed by one person or a very small group of people.
(3) Control forces not present.

(In the case of bombings it was sometimes difficult to decide whether the attack was directed primarily against a person or persons (Category D1) or at a property (Category D2) and in consequence some rather arbitrary distinctions had to be made.)

Category E. Violent Incidents Associated with Protest Demonstrations. Including all or most of the following elements.

(1) Violence between individuals participating in the demonstration and the control forces or ideologically opposed onlookers.
(2) Violence generally lasting for only a few minutes.
(3) Very small numbers of people involved in the violence, although the number of demonstrators may be large.

Category F. Demonstrations Without Violence but Resulting in Arrests.
Category G. Sundry and Generally Minor Disorders for which Little Information is available.

References

Adams, I. A., 1972. The geography of riots and civil disorders in the 1960's. *Economic Geography*, **48,** 24–42.

Aptheker, H., 1963. *American Negro Slave Revolts*, New York: International Publishers.

Baker, R. S., 1964. *Following the Color Line*, New York: Harper and Row.

Chicago Commission on Race Relations, 1968. *The Negro in Chicago: A Study of Race Relations and a Race Riot in 1919*. London, Cass. (Photo litho prints of the Chicago edition of 1922).

Deskins, D. R. Jr., 1969. Geographical literature on the American Negro: A Bibliography. *The Professional Geographer*, **21,** 145–149.

Frazier, E. F., 1957. *The Negro in the United States*, New York: Macmillan.

Gilbert, B. W., 1968. *Ten Blocks from the White House: Anatomy of the Washington Riots of 1968*, London: Pall Mall Press.

Greene, E. B. and Harrington, V. D., 1966. *The American Population Before the Federal Census of 1790*, Gloucester, Mass.: P. Smith.

Grey, L. C. and Thompson, E. K., 1933. *History of Agriculture in the Southern United States to 1860*, Washington, D.C.: Carnegie Institution Publication No. 430.

Grimshaw, A. D., 1959. *A Study of Social Violence: Urban Race Riots in the United States*, Philadelphia: unpublished Ph.D. Dissertation, Department of Sociology, University of Pennsylvania.

Heren, L., 1969. Quiet in the ghettos. *London Times*, August 26, p. 6.

Jackson, K. T., 1967. *The Ku Klux Klan in the City, 1915–1930*. New York: Oxford University Press.

Johnson, T. A., 1968. Negroes see riots giving way to black activism in the ghetto. *New York Times*, October 21, p. 1.

Kellogg, C. F., 1967. *A History of the National Association for the Advancement of Colored People*, Baltimore: Johns Hopkins University Press.

Kerner Commission, 1968. *Report of the National Advisory Commission on Civil Disorders*, New York: Bantam Books.

Kopkind, A., 1969. *America: The Mixed Curse*, Harmondsworth Middlesex: Penguin.

Lee, A. M. and Humphrey, N. D., 1943. *Race Riot*, New York: Dryden Press.

Lemberg Centre for the Study of Violence, 1968. *Riot Data Review*, Waltham, Mass.: Brandeis University.

Lewis, A., 1966. *The Second American Revolution*, London: Faber.

McConnell, R. C., 1968. *Negro Troops of Antebellum Louisiana*, Baton Rouge: Louisiana State University Press.

Morrill, R. L. and Donaldson, O. F., 1972. Geographical perspectives on the history of black America. *Economic Geography*, **48,** 1–23.

Myrdal, G., 1944. *An American Dilemma: The Negro Problem and American Democracy*, New York: Harper.

New Jeresey Select Commission on Civil Disorder, 1968. *Report for Action*, Trenton: Office of the Governor.

Rudwick, E. M., 1964. *Race Riot in East St. Louis, July 2, 1917*, Carbondale: Southern Illinois University Press.

Speigel, J. P., 1967. Race Relations and Violence. Paper presented to the Annual Meeting of the Association for Research in Nervous and Mental Diseases.

U.S. Bureau of the Census, 1918. *Negro Population 1790–1915*, Washington, D.C.: Government Printing Office.

U.S. Bureau of Labor Statistics, 1966. *The Negro in the United States: Their Economic and Social Situation*, Washington, D.C.: U.S. Department of Labor, Bulletin 1511.

U.S. Senate Committee on Government Operations, 1967. *Riots, Civil and Criminal Disorders*, Washington, D.C.: Government Printing Office.

Waskow, A. I., 1966. *From Race Riot to Sit In, 1919 and the 1960's*. Garden City, N. Y.: Doubleday.

10
Black Perspectives on American Cities

FLORENCE LADD

Tuan (1974) has used the term 'topophilia' to mean 'all of the human being's ties with the material environment'. Topophilia couples sentiment with place, though sentiment will include negative as well as positive feelings. In seeking to review the topophilic experiences of black Americans to their cities I am equally struck by the paucity of relevant material and the tremendous importance of the task. Today the centre city essentially belongs to the American black who has inherited it by default. The future of the city, its liveliness, its cultural interest, certainly its social importance, will in part depend upon how its inhabitants regard it at present and imagine its future. Let me begin by recounting some personal experiences.

A Childhood Perspective

As I reflect on the development of my own sense of cities and the images their names evoke, I am continually reminded of one of my grandfathers, my father's father, through whose eyes I came to look at my first city. His attachment to Washington, D.C., my birthplace and the city of my childhood, and his generous sharing of his Washington gave me an early consciousness of the fact that cities are perceived differently by people of different backgrounds and generations. Before I introduce some generalizations and speculation about black perspectives on American cities, I want to acknowledge this debt to my grandfather through a brief description of him and his Washington.

He was called 'Pop' by his children, grandchildren, nephews, nieces, neighbours and by a number of other Washingtonians who were neither relatives nor neighbours, but casual acquaintances who had come to recognize and greet a giant-like, elderly acorn-brown man who occasionally passed their way on his walks and wanderings through the streets of Washington. His figure was unmistakable and unforgettable. He was tall (6′3″ which was very tall for his generation), enormous (weighing about 220 lb) and pear-shaped. He was also gentle and genial.

Pop was born in a hamlet in Caroline County, Virginia. He was four years old in 1870 when his parents, former slaves, moved to Washington, presumably in the spirit of others who made that northward trek. It was thought to be 'better up there'—opportunities, a better way of life.

Pop was a printer's assistant at the U.S. Government Printing Office for most of his working life, during which time he married and subsequently reared eight children. He was retired from the Printing Office in 1932, the year of my birth. From then until his death in 1954, he occupied himself with hunting small game in his native Virginia, creating printer's knives, wine-making, shopping and walking the streets of Washington. He never owned a car or even learned to drive. He walked. It is those walks that are significant in this context.

Pop would set out early in the morning on a day-long odyssey. Dressed in a navy wool cardigan, dark and rather baggy trousers, and comfortable, high-topped shoes (their cracking leather appeared moulded around his immense feet), and carrying a black leather shopping bag, Pop set out. When I was six or seven years old, Pop began inviting me to join him on his walks on Saturdays, school holidays and during summer vacations. Our first stop would be a covered market (the O Street Market) not far from his house. After he had inventoried the day's stock and inspected the prices, we would move on. I never knew what determined our itinerary. Sometimes with a grocery list he and my grandmother had prepared, we would visit a large open market he called 'Camp Meg' and other covered markets, but more often, the shopping seemed incidental. While the itinerary was never revealed to me, we always set out in a purposeful fashion as if we had a specific mission. Now I think that perhaps my grandfather did have a specific purpose: showing his first grandchild his Washington. One day we would walk toward Dupont Circle, to sit there while observing the pigeons and the people and then walk home again. Another day we would walk toward the White House, viewing it from Lafayette Park, or to the Capitol grounds or the Washington Monument. We would visit the neighbourhood near the Capitol and the Government Printing Office where he and my grandmother had lived when their children were young, or visit his favourite spots in Rock Creek Park and favourite streets in Georgetown. Sometimes we would go and sit on the steps of the Library of Congress, where we might share a bag of peanuts while we watched tourists and scholars come and go. As we walked, he talked about how Washington had changed in his lifetime, how times had changed.

At an early age I knew (perhaps Pop had told me) that Washington was a planned city, that a Frenchman named L'Enfant, assisted by a black man named Benjamin Banneker, was responsible for its design. Its wide avenues and impressive federal buildings made for a grandeur appropriate for pageants, parades and demonstrations. Pop always knew when there would be a major parade and what perch near a downtown street corner would afford us an excellent view of it. Even now as I walk or drive along Washington's Pennsylvania Avenue or Constitution Avenue, it occurs to me that it was here that Pop and I were standing when George VI and his Queen passed by with the Roosevelts, or there that we saw the President of Brazil, recalling the pageantry of the occasion. Between segments of a parade, Pop would point out significant buildings, the Department of Justice or the Treasury Department building, and I would look up in awe at their white marble majesty. My Lord, what a city!

As a child I realized that Washington enjoyed international attention. I was keenly aware of its importance and prominence during World War II when Washington's significance, in terms of references to it in radio broadcasts, seemed much greater than that of other U.S. cities I was aware of: Baltimore, Philadelphia, New York, Boston, Pittsburgh, Chicago and Detroit. It was through reports of air attacks abroad and through the experience of air raid drills and blackouts that I began to recognize the visual distinctiveness of Washington, so easily identifiable as a target. I began to pay attention to its landscape, contours, landmarks and boundaries. I began to recognize the pace and the ambience of the city which gave Washington its special character. It was becoming my city in the way that it had become Pop's city.

What about our being black, Pop and me, black Washingtonians in the 1930s and '40s? Because I had led a sheltered early life there seemed to be little conscious connection between my being black and my image of Washington until I was fifteen years old, when I had my first personal encounters with its segregation and discrimination.

These then are the perspectives of a young black girl growing up in Washington. How did other blacks feel about the city? I often wonder about Pop's contemporaries' views of Washington, especially the notable black men and women such as E. Franklin Frazier, the sociologist, Carter G. Woodson, the historian and Mary McCleod Bethune, the reformer. Their roles and their work took them to numerous other cities in the U.S. and abroad which provided them with a wide range of urban skylines and elements. It is within a much wider range than Pop's that their views of the characteristics of Washington fell. They went through city streets with brief cases, not with black leather shopping bags; and it is likely that they drove or were driven, rather than walked. Needless to say, there were thousands of other blacks of their generation, when the social structure of black society was considerably simpler than it is now, who had different perspectives on the same city. In other words, neither then nor now was/is there an image of a city held by blacks *en masse*. Rather, there is a complexity of individual, ethnic group and class-related experiences which determine how people view cities and the phenomena associated with urban life.

The Perspectives of Recent Migrants

Recently there has been a considerable amount of interest in the experience and perspective of poor urban blacks, southern-born migrants to northern cities. When World War II began, there was a substantial increase in the number of blacks who migrated from the rural south to urban areas. For earlier black migrants, those who moved from south to north during World War I and in the 1920s and '30s, the bourgeois prospects of life in Washington and the visions of a 'promised land' for blacks in New York's Harlem had attracted a significant black population, as had Chicago and St. Louis. Just as New York had been a symbolic city of hope for European immigrants, it was the symbolic destination of black immigrants. Langston Hughes (1963) captured aspects of the pattern of black immigration to Harlem and some of the problems in these lines:

I was born here, he said
watched Harlem grow
until colored folks spread
from river to river
across the middle of Manhattan
out of Penn Station
dark tenth of a nation,
planes from Puerto Rico,
and holds of boats, chico,
up from Cuba Haiti Jamaica,
in busses marked New York
from Georgia Florida Louisiana
to Harlem Brooklyn the Bronx
but most of all to Harlem
dusky sash across Manhattan
I've seen them come dark
wondering
  wide-eyed
  dreaming
Out of Penn Station—
but the trains are late.
The gates are open—
but there's bars
at each gate.

The creation of a black urban culture and the impact of urban problems for black migrants was experienced in microcosm by Harlem's blacks, whose situation anticipated the circumstances of other migrants to other black urban settlements. The problems have been created by prejudice and discrimination against blacks in two critical areas of living: employment and housing. The process and its result is well described by Warner (1972):

'Beginning with World War II, the blacks of the Southeast ... poured out of the old Confederacy into Northern and Pacific cities. There they have faced in our own time, as in previous periods, two special obstacles that have never confronted their white counterparts. Job prejudice consistently held down newcomers and older residents alike and excluded even the skilled and qualified from jobs commensurate with their abilities. Moreover, prejudice blocked Negroes from the traditional practice of one man's using his established position to make room for his friends, relatives, townsmen, and fellow ethnics. Employment restrictions closed down the historical process of urbanization whereby newcomers advanced either through job improvement or accumulation of property. Housing prejudice, far in excess of any that existed in respect to Jews or poor families of any sort, closed vast areas of the city to Negroes, and the black ghettos could often only expand by violence or by the purchase of housing at exorbitant prices. There had been ghettos and prejudice before in American cities, but the rapid growth of communities of Negro migrants in the North and the relentless job discrimination heightened the segregation. The outcome has been the emergence of an unprecedented situation in American cities: vast quarters are occupied exclusively by the members of single race or origin.' (Reprinted with permission from San Bass Warner, *The Urban Wilderness: A History of the American City*, Harper and Row Publishing Company.)

What does one know of a city if upon arrival he finds his first home behind the walls, so to speak, of the ghetto of that city? In this regard, Coles (1971) reports on the oppressive living conditions newcomers find and the anti-city sentiments some express. He conveys a sense of the imagery and astonishment of blacks who are newcomers to northern cities. 'Some may have seen or been in a city before: Greenwood and Greenville in Mississippi, or Selma and Montgomery in Alabama, or Lexington in Kentucky or Charleston in West Virginia or Atlanta and New Orleans, those big, big cities. For many, though, the cities up North are the first cities they have looked at and lived in. "Lord, I never knew there were so many buildings." "Lord, I never knew what a street was, not really, not streets like we have up here, not miles and miles of them."'

Some of those new migrants were aware of the artifacts of the city: sewers, locks, letter boxes, mail keys and street lamps. The noise level and the pace of city life were attended to especially by the children 'who are sensitive to noises and curious about where they come from. The streets never seem to quiet down. Even in the middle of the night there are cars moving, people walking, things going on. What is that? Why is the whole rhythm of life so different in New York or Philadelphia or Boston or Chicago?' (Coles, 1971).

While the rhythms, sights and sounds of cities hold a certain mystique for new migrants, that mystique is remote. Of much greater meaning to poor newcomers are the streets where they live and the neighbourhoods which surround them. Coles (1971) notes that 'the streets have received, continue to receive men and women and children who have had to leave their homes elsewhere. . . . There is the address of a brother to find or the address of kin, not close kin, but above all kin.' Hannerz (1969) and Liebow (1967) have reported the use of specific streets and neighbourhoods in Washington, D.C. as settings for significant occasions among lower-class and working-class blacks. The neighbourhood was far more significant culturally then the 'image' city of the whites. Liebow (1967) describes the function of 'Tally's Corner'—a carry-out shop on a corner in north-west Washington, D.C. 'It would be within walking distance of the White House, the Smithsonian Institution, and other major public buildings of the nation's capital, if anyone cared to walk there, but no one ever does.' It is as if the city beyond their neighbourhood has no relevance or purpose in their lives. Liebow's subjects were not newcomers to Washington. Rather, they had been residents of Washington for a long time. Liebow's observation of his subjects' lack of awareness of 'official' Washington is replicated elsewhere. In a map of part of Boston drawn by a black adolescent boy, he delineated his 'neighbourhood' clearly; it consisted of the public housing project in which he lived, a school and a few nearby shops. Beyond that neighbourhood (which is only two or three blocks from Boston's Museum of Fine Arts and not far from the shopping area surrounding Prudential Centre), the city does not exist for him. The rest of the city is *terra incognita*, undefined and presumably unexplored. What is significant is that for the working class and the poor, there are few occasions and little time for them to lift up their eyes to the skyline and enjoy a panoramic view of their cities. Instead, their attention and energy must necessarily be riveted to the environment at hand with its many debilitating problems.

A Literary Perspective

In autobiographical and fictional works of a few black authors, the geography and imagery of aspects of U.S. cities have been depicted vividly. For example, the autobiographical works of Claude Brown and Malcolm X have given us some personal views of New York; the latter also included some impressions of parts of Boston. If one attempts to map Claude Brown's New York, the most highly differentiated section would be his Harlem. There are several references to 8th Avenue in the 1940s. He mentions speakers' corner (125th Street and 7th Avenue), Harlem Hospital, the Harlem River and Washington Irving High School. His picaresque experiences took him to other parts of New York: downtown to Eldridge Street, to school on Forsythe and Stanton Streets, to rob a nightclub on Delancey Street, to Times Square, and to Bellevue Hospital. For a while he lived in a Greenwich Village loft on Cooper Square. There is a marvellous sequence in which his underground image of New York is stirred upon his return from the South. Riding the subway, he recalls his experiences at various stations:

> 'The train ride uptown was the longest train ride I ever took. As the train came to each station, I remembered something about that station. I remembered shaking down the two white shoeshine boys at the 42nd Street station. And I remember the time the cops saw me beating a gum machine at the 59th Street station . . . I must have done something at 72nd Street, but I couldn't remember it, and that bothered me. I remembered the way that lady screamed when I snatched her pocketbook at 81st Street . . . at every stop, I wanted to get off the train and yell that I was back . . . When I came out of the subway at 145th Street and St. Nicholas Avenue, I thought there had never been a luckier person in the world than me. I wanted to grab the sidewalk and hug it tight.' (Reprinted with permission from *Manchild in the Promised hand*, Claude Brown, Macmillan Publishing Co. Inc. and Jonathan Cape Ltd. Copyright © Claude Brown, 1965.)

Certain places in New York remained warmly memorable and vivid for Claude Brown because of what he did or experienced in those places.

Malcolm X came to the North-east and its major cities from Lansing, Michigan. It was in the summer of 1940 that he took a Greyhound bus from Lansing to Boston where he lived for a while with a half-sister 'on Waumbeck Street in the Sugar Hill section of Roxbury, the Harlem of Boston'. He soon began to explore Boston and its environs, aware of its historic buildings, plaques, markets and statues 'for famous events and men'. He wrote: 'One statue in the Boston Commons astonished me: a Negro named Crispus Attucks, who had been the first man to fall in the Boston Massacre'. He walked to Boston University. A subway ride took him to Cambridge and Harvard. He watched people come and go at North Station and South Station and wandered among the piers and docks along the harbour. He noted 'winding, narrow cobblestoned streets, and the houses jammed up against each other'. In downtown Boston he was impressed by 'the biggest stores I'd ever seen, and white people's restaurants and hotels'. He took a job as a railroad cook on a Boston-to-Washington train. He was shocked to find 'just a few blocks from Capitol Hill, thousands of Negroes living worse than any I'd ever seen in the poorest sections of Roxbury . . .' Then Malcolm went to New York where his first stop was a well-known Harlem night stop, Small's Paradise. ('Within the first five minutes in Small's I had left Boston and Roxbury forever.') In his early period in New York, Malcolm X recalled living at the Braddock Hotel near Small's and hustling at the Astor Hotel 45th Street and Broadway. Even at this stage, the settings of Malcolm X's activities were subordinate to the events in which he was involved. It must be noted that where Malcolm X has described Boston, Washington, New York and especially Harlem, he adds a social and historical comment, or a statement which reveals his awareness of the socio-political significance of the place.

Black fiction, poetry, drama, music and films also offer views of American cities which would interest the researcher. I know of no research on what all these media have revealed about blacks' images of cities; however, there is an essay which considers the physical settings of some black fiction. In his essay, 'The Negro's Image of the Universe as Reflected in His Fiction', Blyden Jackson (1968) examined the Chicago of Richard Wright's *Native Son* and the New York of Ralph Ellison's *Invisible Man*. He concludes:

> 'That the two best-known Negro novels should put the emphasis, which they do, upon the two largest Negro ghettos (Harlem and Chicago's Southside) is an informative circumstance . . . All Negro fiction tends to conceive of its physical world as a sharp dichotomy, with the ghetto as its central figure and its symbolic truth, and with all else comprising a non-ghetto which throws into high relief the ghetto itself as the fundamental fact of life for Negroes as a group.'

Jackson laments the lack of variety in the settings used in Negro fiction.

> 'It is, in short, the limited universe of a literature of protest, a universe that, with its quality of *statis*, can well be seen as the same universe which Negroes see but only

from two different angles, when they see first in it either the ghetto or the irony of color caste.'

Jackson was writing in the 1960s when the U.S. was being urged by an active civil rights movement to recognize and ameliorate the living conditions of ghetto dwellers and the socio-economic disadvantages imposed because of one's race or colour. The circumstances and problems engulfing the lives of most poor blacks were and *are* so distressing that black authors who choose black people and themes as their subject matter, no doubt, feel compelled to make their contribution to the improvement of the lives of their people by writing of those in poorest circumstances. The background for those characters is a ghettoscape. As long as poor blacks are forced to live under economic and residential conditions which are physically and psychologically dangerous and depressing, it is essential that writers and researchers remind us of their discomfort and despair. The stark reality of the horrors of day-to-day living in Harlem and the micro-Harlems of the U.S. must be presented repeatedly until the resources of the nation are directed toward improving the living conditions of poor blacks trapped in urban slums.

A Current Perspective

During World War II there was an increase in the number of blacks who migrated to cities to find wartime jobs in government and industry. That war, like World War I and the recent wars against South-east Asia, took black men to cities abroad, London, Paris, Berlin and Rome; Tokyo, Seoul and Saigon, as well as to cities in the United States. They returned with remembrances and images of far-away places, images that are not foreign to those who never left home who, through radio, television and films, have had some of the sights and sounds of distant places brought home to them.

It is difficult to assess the impact of the mass media in the U.S. on viewers' images of urban settings and urban life. To be sure, witnessing through television the dramatic historical events of the 1960s, which included civil rights' demonstrations, political assassinations and urban riots, had a profound effect on blacks' views of their sociopolitical position as a group in the U.S. and heightened awareness of the distribution of the population and power (and powerlessness, alas!) of blacks. For decades New York, Philadelphia, Detroit, Chicago, Cleveland, Newark, Washington and Baltimore were the reference cities for blacks. In the wake of the riots of the '60s, Watts and Oakland gained prominence as West Coast centres with ghetto problems and politics. Now smaller cities across the country have gained substantial populations of blacks who lived ghettoized in their poorer sections. Rochester, Gary, Hartford, Dayton and Dallas appeared in the array of cities of consequence in the lives of black people. A network of urban areas with numerically significant black populations now stretches across the nation.

During the '60s blacks in the U.S. were recognized as an urban people.

Indeed, where the term 'urban' appears in some contexts one should read 'black'; the term 'inner city' is used to denote the area of a city where blacks predominantly reside. The 'ghettos' of the U.S. are black enclaves.

The urban riots of the late 1960s afforded cities an opportunity to rebuild sections of their ghettos in a style which reflects the values, aesthetics and preferences of its disenchanted black citizens. But except for the 'walls of respect', murals depicting aspects of black life painted on ghetto buildings in riot-torn areas, there is little or no evidence (except in Watts, perhaps) of the clear recognition of the presence of black people in a prominent form, on a grand scale (though the decor of many small black-owned businesses has given America's black main streets a distinctive quality). We can only speculate about what forms might emerge if American blacks were encouraged to search their culture, experience and imagination for their particular urban images and forms. We first should learn about the shape and content of blacks' images of U.S. cities and understand how the social and political experiences of urban blacks influence their images and ideals of city life.

Concomitant with the growth of black settlements in cities has been the flight of whites to suburban areas. One political consequence of their flight is the ascendance of blacks into high elected political positions in several cities. In 1974, there were black mayors in six major cities in the U.S.: Los Angeles, Detroit, Atlanta, Newark, Raleigh and Gary. But as Williams (1974) has noted, the cities of the U.S. have become the black man's burden. Now that it appears that U.S. cities may not be economically and socially viable entities, the blacks and black politicians who have inherited them are handed the seemingly insurmountable challenge of rescuing them from impending disintegration.

Why should the challenge and the responsibility of the future of U.S. cities be thrust upon blacks, most of whom reside in cities because of lack of options with respect to other places to live? What do U.S. cities mean to black urban residents in the '70s? Blacks, particularly poor blacks who live in cities and are confronted daily by the problems of city life, can hardly be involved in romantic images and futuristic plans for urban places. Instead, they must necessarily be concerned with the immediate conditions surrounding their homes and neighbourhoods. Given a choice, would they want to renew, restore and revitalize sections of the cities they live in? (Commercial sections which served black residents prior to the riots of the '60s when they were damaged or destroyed have not been redeveloped. Given the shift toward shopping centres, it is likely that many such commercial areas will not be re-established.) Or do they want to flee to the suburbs as whites have or perhaps find more satisfying residences elsewhere?

Preferences for suburbia on the part of some blacks have been indicated. For example, a group of urban black adolescents who served as research subjects expressed a clear preference for future homes in suburban settings (Ladd, 1972). Rose (1971) has described a 'national ghetto network' in a

> 'number of suburban ring communities ... evidenced by the increase in the magnitude of the black population outside central cities in a few select locations ... Subur-

banization is thus underway, but it is occurring essentially in the same way as did residential development in the central city. Suburban ghetto clusters can now be added to the list of ghetto centers which have facilitated escape from central-city locations and which are beginning to serve as direct magnets of attraction'.

Those blacks who have been able to move to suburbs are middle class. New towns are claiming some middle-class blacks as residents. And a few young blacks are seeking new possibilities in semi-rural areas. A few middle-class professional blacks are leaving northern cities for better opportunities in cities and towns in the 'new South'.

Despite this dispersion of segments of the black population in the U.S., the least mobile are those who comprise the underclass in urban areas. As long as they remain confined to the poorer quarters of their cities, for some the city, or at least its ghetto, will be regarded as a 'trap'. Tuan (1974) reminds his readers of the importance of terms used to characterize cities. An extensive study of the labels people use to describe their cities and their places in those cities would reveal how individuals at different socio-economic levels see cities. At this point in history, a greater exposure of the perspective of the urban poor might in some way contribute to a movement to improve their living conditions. Improving the circumstances of the urban underclass, black and white, will improve the quality of urban life and, in turn, change views of cities: not only the viewpoints and views of poor residents, but the views of city dwellers who are better-off and of other users-of-the-city who might accord cities more respectful use and award them more of each tax dollar. This economic analysis seems circular. But a start should be made in assessing the topophilia of blacks for their urban surroundings and subsequently searching for means to make their surroundings most meaningful and inhabitable. To make cities liveable in and lovable is surely the major challenge of the final quarter of this century.

References

Brown, C., 1965. *Manchild in the Promised Land*, New York: The New American Library.

Coles, R., 1971. *The South Goes North*, Boston: Little, Brown.

Du Bois, W. E. B., 1961. *The Souls of Black Folk*, Greenwich, Conn.: Fawcett.

Hannerz, U. S., 1969. *Soulside: Inquiries into Ghetto Culture and Community*, New York: Columbia University Press.

Hughes, L., 1968. *Selected Poems*, New York: Knopf.

Jackson, B. 1968. The Negro's image of the universe as reflected in his fiction. In *Black Voices* (Ed. Chapman, A.), New York: Mentor.

Ladd, F. C., 1972. Black youths view their environment: some views of housing. *Journal of American Institute of Planners*, **38,** 108–116.

Liebow, E., 1967. *Tally's Corner: A Study of Negro Streetcorner Men*, Boston: Little, Brown.

Rose, H. M., 1971. *The Black Ghetto: A Spatial Behavioral Perspective*. New York: McGraw-Hill.

Shelton, F. C., 1967. A note on 'The World Across the Street', *Harvard Graduate School of Education Bulletin*, **XI,** 47–48.

Tuan, Y. F., 1974. *Topophilia*, Englewood Cliffs, N. J.: Prentice-Hall.
Warner, S. B., 1972. *The Urban Wilderness: A History of the American City*. New York: Harper and Row.
Williams, R. M., 1974. America's black mayors: are they saving cities? *Saturday Review/World*, **1**(17).
X., M., 1965. *The Autobiography of Malcolm X*, New York: Grove Press.

11
Centre Cities in Canada and the United States

EDWARD HIGBEE

Although American broadcasts dominate Canadian airwaves, American periodicals swamp Canadian newstands, and Canada's economy rocks on waves created by American corporations, Peter C. Newman, editor of the Canadian monthly *Maclean's*, rejects the notion that his country is a cultural suburb of the United States. In just the past few years, according to Newman, 'Canada virtually reinvented itself'. Among other reasons there seems to be a mounting disenchantment with events in the United States. 'Our quarter-century old admiration of all things American', writes Newman, 'exhausted itself in the blazing villages of Vietnam, the dark labyrinth of Watergate, and the long-overdue realization that the United States was crowding out not just our industries, but our way of life' (quoted in *New York Times*, 14 October 1973).

A growing divergence in the life-styles of Americans and Canadians is nowhere more evident than in their centre cities where a substantial portion of Canada's middle class remains by choice but from which, in America, the white middle class feels forced to retreat. American tourists returning from visits across the border comment enthusiastically about the safety and cleanliness of Canadian cities as compared with their own. They are impressed not only by the new skyscrapers, which they have at home, but more so by the superior maintenance of old buildings, old neighbourhoods, and pleasant parks which are more characteristic of Europe. The efficiency and comfort of public transportation facilities are admired. Traffic jams generated by automobile commuters are as aggravating in Canada as in the United States but mass transit systems, including railroads, offer superior service. Street crimes, which are commonplace in America, are far less frequent. 'In Toronto I feel safe wearing my mink stole in the subway', comments a middle-aged socialite, 'at home I'd be crazy to try it'. Toronto has one-fourth the population of New York City but in 1971 New York had 62 times as many robberies, and 38 times as many murders. On New York's subways alone there were 1315 robberies and 304 felonious assaults in 1972 (*New York Times*, 11 November 1973).

Until recently, the number of Canadians migrating to the United States exceeded the number of Americans settling in Canada. Ten years ago four times as many Canadians migrated to the United States as vice versa. A popular joke once defined a Canadian as 'someone who hasn't yet been offered a job in the United States' (*Canada Year Book*, 1971, pp. 268, 274). Now Americans moving to Canada outnumber Canadians moving to the United States by two to one. While higher wages paid for similar work have long attracted Canadians to the United States (and still do), Americans now migrate to Canada expecting to earn less but convinced that they will live in more wholesome environments. They are willing to take a cut in personal income in order to experience a better public community. A New York banker who moved recently to Montreal declared, 'I had to make the switch for my family's sake. I could see Manhattan getting worse and it wouldn't have been fair to the children to stay'. Of the 60 thousand persons granted Canadian citizenship in 1969 almost 89% took up residence in urban centres (*Canada Year Book*, 1972, p. 280).

The question arises as to why it is that city life is safer and more gratifying for the middle class in Canada than in the United States where the per capita gross national product is approximately 30% higher. Part of the answer seems to lie in the way governments in the two countries spend public wealth, reflecting political priorities regarding the well-being of people and the quality of their cities. Racist attitudes are a major factor in determining the social structure of American cities, but play only a minor role in Canada.

The frontier-agrarian notion that the individual should be self-reliant and responsible for his own fate still dominates the social psyche in the United States, although farmers are less than 5% of the population and the unemployed who are looking for work are more than 5%. In frontier days it was possible for a family on its own land to earn a subsistence living by its own labour. In an industrial society the option of self-employment, if all else fails, does not exist. In Canada the English concept of government's role in providing social security is more seriously considered; consequently, welfare is a respectable term in Canada whereas most Americans are conditioned to despise it as charity rather than a right. The environmental consequences of such attitudinal differences are profound. No urban poor in Canada are as wretched as their counterparts in American cities. Now, with mounting inflation, the American middle class finds itself a victim of its own immature social institutions, particularly in education, medical care, comprehensive social security and maintenance of the common public environment. A change in attitude toward these environmental services will come as the middle-class majority finds that its real income is beginning to diminish.

Public Spending and Public Service

In 1970 the gross national product of the United States was $974 billions (*Statistical Abstract*, 1972, pp. 312). Total spending by all levels of government (federal, state, local) was $27·5 billions or 28% of the total (*Statistical Abstract*, 1972, p. 411). Spending on military affairs was $84·2 billion or 30% of the sum-total spending by all levels of government. This is equivalent to some $400 for each man, woman and child in the nation. (In 1975 this spending could increase to $100 billion (*New York Times*, 20 January 1974, p. 1).) The $84.2 billion disbursed by the Pentagon almost equalled the $89·9 billion (some $430 per capita) spent by all levels of government for the most crucial human needs: education ($55·7 billions), public welfare ($17·5 billions), health and hospitals ($13·5 billions), and housing and urban renewal ($3·2 billions) (*New York Times*, 20 January, 1974). The public environment of cities in the United States is in stiff competition with the military establishment for fiscal support. Mayors and governors have been put on notice by the Federal Government that they cannot expect a change in Washington's priorities. In September 1973, President Nixon delivered a 'State of the Union' message to the Congress in which he indicated that he would veto legislation that proposed either significant cuts in military spending or increases in spending for civilian social services (*New York Times*, 11 October 1973). Paradoxically, the United States withdrawal from Vietnam was followed not by a decrease in military spending but by an increase.

'I have always thought', Richard Nixon once commented, 'this country could run itself domestically without a President—All you need is a competent Cabinet to run the country at home'. To this he added, 'You need a President for foreign policy; no Secretary of State is really important. The President makes foreign policy' (*New York Times*, 17 July 1973, p. 23). In this respect Nixon is no different from his predecessor, Lyndon Johnson, whose rhetoric proclaimed the 'Great Society' but who spent the national wealth to escalate the Vietnam war. As New York's former Mayor, John Lindsay, commented: 'All budgets are political documents'. Lindsay said on many occasions that there is no genuine hope for New York or any other large American city unless state and federal governments, which control the greatest share of tax revenues, also assume the major responsibility for funding public services and for maintaining the quality of the public environment.

Since the population of Canada is roughly one-tenth that of the United States its aggregate public spending is not as impressive. What is significant is the higher priority which the Federal Government of Canada places upon services to its people and their environment. The gross national product in 1967 was $65·7 billions (*Canada Year Book*, 1971, p. 1179). Total spending by all levels of government (federal, provincial and municipal) was $21·8 billions or 33% of the GNP (*Canada Year Book*, 1971, p. 1125). Outlays for military affairs came to $1·78 billions (or $80 per capita), less than 3% of the GNP and less than 8% of the sub-total spending by all levels of government (*Canada Year Book*, 1971).

By comparison, the aggregate spending by all levels of government in Canada for civilian services was high: education \$4·2 billions, social welfare \$3·2 billions, health services \$2·2 billions, housing and urban renewal \$0·1 billion, a total of some \$470 per capita (*Canada Year Book*, 1971). Whereas in the United States military outlays approximate public spending on the four most important domestic services, those same four services in Canada receive over five times as much public funding as the military. This has made possible more comprehensive social security systems in Canada than in the United States and more equitable educational and medical care facilities. When planning Canada's federal budget up to the year 1974, Prime Minister Trudeau placed the military in a category of 'low-growth functions' whereas health and welfare, education and regional economic development emerged as 'high-growth functions'. 'This will permit the people of Canada to see what priorities we have', declared the Prime Minister. Canada is more of a welfare than a warfare state and this favourably affects the quality of civilian life and the character of the nation's cities.

Despite these obvious fiscal advantages many Canadians are critical of their urban environments. They see them not in comparison with those in the United States but in terms of what they might be, considering the more favourable ratio of natural resources to populations which prevails in Canada. Other Canadians feel that whatever happens in the United States today will happen in Canada tomorrow; so they anticipate that their cities, too, are headed for more slums, frequent street violence, domination by the automobile and more pollution. An unemployed widow, Mrs. Beth MacDonald, lives on a pension in a shabby Toronto rooming house. 'The poor people who live in the cities', she declares, 'are the ones who are forgotten in Canada. The politicians worry about everyone else but not us' (*New York Times*, 23 October 1972). (For an analysis of Canadian urban politics and the plight of the poor, see two recent books by Lorimer (1970, 1972).)

While politicians deny it, the fact remains that, as Canadian cities grow, the disagreeable aspects of urban life do become more prominent, even if they are not as visible to American tourists as they are back in the United States. Urban critic, Boyce Richardson, formerly an associate editor of the *Montreal Star*, foresees Canada heading for a degraded environment unless it deviates even more than it has from American models and greatly increases social and environmental services as has been done in Britain, Scandinavia and the Netherlands. In the introduction to his book on Canadian urban problems, Richardson (1972, p. 8) raises such questions as, 'What kind of cities do we want? What sort of lives do we want to live? How should our wealth be distributed? And what stands highest in our list of priorities? . . . What is the proper balance between the public and the private interest?' Unless questions such as these guide citizen thinking about cities, Richardson believes, it will be impossible to make them liveable as they grow more crowded.

It appears that these questions are beginning to be raised in Canada. In both Toronto and Vancouver, civic elections in 1972 replaced 'old guard' governments connected to urban developers and the ethic of more growth, with a younger and more radical group of city politicians pledged to incorporate public opinion in the urban planning process. Growth is no longer uncritically accepted, even though no Canadian municipality has the power to limit growth or control the pace of development (Greer, 1973). Currently, plans are being formulated in both these cities to limit the growth of the urban core, but how successful they will be remains to be seen.

As a rule Europeans envy the low taxes and high private incomes of middle-class America and Canada while some North Americans envy the higher quality of the public environment and more comprehensive social welfare systems operating in Sweden, the Netherlands and Britain, which are funded with higher taxes. Sweden, says Richardson, constructs housing with government aid at a faster rate per capita than Canada. 'Stockholm', he notes (Richardson, 1972, p. 64), 'has cleared its whole population out of slums, and has found ways to make it possible for slum dwellers to live in decent houses, like everyone else. It is a piece of social engineering, if that is the expression, that has seen few equals in the world'. The reaction of many middle-class Swedes to this achievement is that they would rather have lower taxes, less public welfare and more personal spending money. In the modern industrial state there may be little compassion for the poor, the handicapped and the elderly. Their needs are commonly neglected unless, in contemporary social institutions, there persists the tradition of reciprocal rights between government and citizen that in Europe goes back to the feudal manor of medieval times.

Housing in Canadian and American Cities

In the matter of housing there is not much difference between the aggregate statistics for Canada and the United States. In 1970 Canada had approximately one dwelling unit for every 3·6 persons while in the United States there was one dwelling unit for every 3·2 persons. In the decade of the 1960s the numbers of houses grew faster than the human populations in both countries. The major difference between housing in Canada and the U.S. lies in the amount of public subsidy of shelter for the poor, handicapped and elderly. In 1969 some 28,000 housing units, or just over 13% of all new housing built in Canada, were financed under the National Housing Act and designed for occupancy at low rents by those with sub-standard incomes (*Canada Year Book*, 1971, p. 834).

In the United States, with its vastly larger population, only 33,000 units of low-rent public housing were built in 1969 and most of these were for lower-middle-income people rather than for the poor. These units represented only 3% of all new housing units built that year (*Statistical Abstract*, 1972, p. 681). In 1970 two-thirds of the Canadian federal housing budget of $854 millions was allocated to shelter low-income families. It was enough to build 35,000 dwelling units (*Canada Year Book*, 1971, p. 834). While it is more comfortable to be middle class in the United States it is harder to be poor. This fact has its social consequences and is creating larger urban ghettos which threaten to become a permanent feature of the American landscape.

The housing situation affects the quality of life in the city not only because urban housing is proportionately older than suburban housing, but because the poor are increasing in the cities while older housing is being demolished and rents are raised on what remains. When what is called 'urban renewal' destroys old tenements to make way for commercial shopping centres, office towers and middle-class high-rise apartments it aggravates the housing situation for the poor unless there is compensatory construction of low-rent, subsidized public housing.

In his exposé of middle-class ideology, William Ryan (1970) illustrates how the poor themselves are blamed for their misfortune rather than the structural inequities of American society. The Housing Act of 1949 was passed ostensibly to provide safe, sanitary, decent housing for every American family. It provided for federally subsidized urban renewal by which the government would pay up to two-thirds of the cost of bulldozing a slum and turning the recovered land over to a developer for reconstruction. In other words, developers and not the poor were subsidized. By 1967 some 1400 urban renewal projects had been executed in over 700 cities. A total of 383,000 slum dwelling units were demolished in what were regarded as prime sites for 'higher uses'. Ryan, who was Urban Renewal Chairman of the Fair Housing Federation of Greater Boston, reveals that in the whole United States only 107,000 new dwelling units eventually replaced those destroyed and only about 10,000 of these were for low-income people. 'The net effect', he says, 'was a loss of over 350,000 homes for low-income people'. Today, after a quarter of a century of urban renewal, the housing situation of the urban poor is more desperate than ever. Because four out of five families displaced by urban renewal have been Negroes, urban renewal has been called 'Negro removal'. Yet according to prevailing middle-class ideology it is the blacks' culture of poverty which is responsible for their being in slums in the first place.

In nearly every major American city the poor eventually spread out into old middle-class districts as they were ejected from demolished slums. They managed to do this by doubling-up to share rents and the resultant crowding of limited facilities produced more slums. Inasmuch as a high proportion of the urban poor are black this process aggravates the racism that is pervasive in America and is exploited, sometimes inflamed, by self-seeking politicians from city ward to the White House. Because black slavery never gained a strong foothold in Canada, modern urban Canadians are spared the racial antagonisms which rack American society. White discrimination against minority blacks intensifies and sharpens economic segregation.

In 1970 there were 22·6 million blacks in the United States. This was slightly more than the entire population of Canada. Blacks are 11% of all Americans. Of the total, 13·1 million live in central cities, 3·6 million live in suburbia while 5·8 million live in small towns and rural areas (*Statistical Abstract*, 1972, p. 26).

58% of all American blacks live in central cities where they average 20% of the population (Statistical Abstract, 1972, pp. 16, 28). Washington, D.C. is 71% black, Atlanta is 51%, Detroit 44%, Chicago 33%, St. Louis 41%, Los Angeles 18%, Cleveland 38%. The suburbs of all these cities are predominantly white. Taken as a whole, the suburbs of the United States are 95% white and their populations are wealthier than those of central cities. Eleven and one-half million families, or 22% of all Americans, had income exceeding $15,000 in 1970. Only 3·3 million of these families live in centre cities.

Discrimination and its Motives in the American City

Stamford, Connecticut, is an urban hub in upper middle-class suburban Fairfield County, which ranks as the eighth richest county in the United States on a per capita basis. Stamford is within commuting distance of New York and has a relatively small number (5%) of poor people, who are mostly black. The city's mayor, Julius H. Wilensky, and its school board are opposed to any publicly subsidized school lunch programme because they believe that warm lunches served to poor children encourage their parents to remain in Stamford. Thus for the past three years the city schools have not served a noon-time meal, whereas elsewhere in Fairfield County it is standard practice. 'It's mostly incompetence, plus a feeling that the poor people are getting too many handouts', says Mrs. Vivian White, president of the Committee on Basic Human Rights. 'In other cities, children get a free or reduced-rate meal that is supposed to provide one-third of their daily nutritional requirements'. In Stamford, Mrs. White adds, '. . . the Mayor frankly says he tries to get welfare people out of town—there's a lack of compassion for poor people who live in Stamford and this is part of it' (*New York Times*, 21 October 1973, p. 45). When it is realized that the per capita income in Stamford and Fairfield County is above the national average and that people resent giving food to a small minority of school children then it can be understood why social frictions and frequent street violence flare in larger urban centres where the middle class is not so well off, while the poor and black are more numerous.

In the New York borough of Queens, not far from the J. F. Kennedy International Airport, 30 men and teenaged boys using axes and picks attacked and smashed a home in the Rosedale community because it had been rumoured that a black family was about to buy the house in their all-white neighbourhood. After breaking windows, punching holes through interior walls, and smashing appliances, the vandals flooded the basement to a height of six feet and wrote on the outside walls of the house, 'Stamp out Niggers'. One white woman bystander told police investigators to 'Send a copy of a picture of this to Harlem to show them what happens if niggers try to move into this neighbourhood'. As a matter of fact the house was being bought not by a black, but by two Chinese, who after the riot, decided they were no longer interested in it (*New York Times*, 12 June 1971, p. 31).

In Boston, where there is intense white resistance to letting black children be bussed to better white schools, there was a rash of racially motivated murders and assaults in 1973 and 1974. In one instance a gang estimated at 40 to 50 black teenagers stoned to death a 65-year-old white man who came into their area to fish behind a supermarket. Two days before, another gang of blacks poured gasoline upon a white woman then lit a match and burned her to death. When a respected member of the black community was interviewed and asked the inevitable questions, he replied, 'The young people of this community have no resources and no hope' (*New York Times*, 7 October 1973).

For years a rabble-rousing white woman, Louise Day Hicks, has tried with considerable success to inflame the white community of Boston with her segregationist rhetoric, and at one time was nearly elected mayor of the city on her pledge to keep black children in black schools. When her ideological colleague, William O'Connor, became chairman of the School Committee, he declared there was no need to improve black schools in the Roxbury ghetto. 'We do not have inferior schools', O'Connor proclaimed, 'We have been getting an inferior type of student' (quoted in Ryan, 1970, p. 31). In such a way has Boston defended its reputation as a distinguished centre of American culture.

Education and Equality

Primary and secondary education in the United States is financed chiefly by taxes levied on real estate by local governments or school districts. As a result, the quality of education in most communities depends upon the quality of real estate within their jurisdiction. In 1972 a Presidential Commission on School Finance documented what has long been known: that the assessed value of property per student may be many times higher in one district than in another, even within the same state. In Kansas, for instance, the ratio of the assessed valuation per pupil between the wealthiest and poorest districts in the state was 182·8 to 1. In North Dakota the situation was more equitable, with a ratio of 1·7 to 1. Only in Hawaii did the state assume the costs of primary and secondary education so that the ratio was in effect 1 to 1. In New York the ratio was 84·2 to 1 and in California it was 24·6 to 1. The state supreme courts of both California and New Jersey have declared that the property tax is unconstitutional in those states because it discriminates against pupils in poor areas. Nevertheless, as yet, neither state has developed an alternative system of school finance. The United States constitution does not specifically guarantee equality of education.

As a national average, localities in the United States pay 52% of the costs of primary and secondary education, states pay 41% and the Federal Government pays 7% (Schultze, 1972, p. 324). In Canada the provinces pay the major costs of primary and secondary schools. In 1967 provincial governments paid 52% of the costs, localities 42% and the Federal Government 4% (*Canada Year Book*, 1971, p. 431). By carrying the chief costs of public schools Canadian provinces reduce somewhat the grosser disparities that exist in the resources of school districts. Middle-class parents are not under quite as much self-inflicted pressure as they are in the United States to move from centre city to suburbia in order to secure a better education for their children (*Canada Year Book*, 1971, p. 1124). Economic segregation naturally leads to educational segregation, and this fact is basic to the bussing controversies in the United States. Parents of the poor and the black want their children bussed to better schools while the parents of children in better schools want to keep such children out. What the underprivileged call 'enrichment' the privileged call 'dilution'. The issue is bitterly debated, and despite numerous court cases in favour of bussing, remains largely unresolved. Education is too closely related to economic opportunity to be acceptable on a region-wide basis in America at present. Yet equality of educational opportunity is a critical element in the drive toward greater social equality.

Welfare Programmes

Inasmuch as the poor tend to concentrate in cities, welfare is commonly regarded as an urban problem even though the causes of poverty are not generally related to place but rather to old age, childhood dependency, unemployment, injury or racial discrimination, which are universal and systemic. The public revenues of cities are insufficient to cope with what is the proper burden of the total society. In Canada the Federal Government assumes somewhat more responsibility for health and social welfare programmes than in the United States. Also the Canadian Federal Government sets more equitable national minimum standards with which provinces and municipalities must comply. In 1969 Canadian municipalities paid only 2% of the total costs of health and social welfare programmes, provinces paid 37% and the Federal Government paid 60%. In the United States the Federal Government covered only 50% of the costs of health and public welfare programmes and there was great inequity in the minimum standards of states (*Statistical Abstract*, 1972, p. 411).

Mothers with dependent children receive an average of $54·00 a month in welfare aid in Mississippi whereas in Minnesota the average is $268·00 (Schultze, 1972, p. 187). Some localities make no contribution to welfare programmes whatsoever while 23% of the budget of New York City is so allocated. Other aspects of the public environment are bound to suffer when cities are sufficiently concerned with human agony to assume a welfare role that is more properly the obligation of the Federal Government.

Canada is more advanced than the United States in its federal programmes for universal medical care, hospital insurance, family allowances and guaranteed income supplements for the aged. These programmes, which are administered by the provinces and in part financed by them, differ significantly from what might be regarded as their counterparts in the United States. All qualified persons, regardless of where they live, are entitled to comparable benefits under the Canadian system. There are increasing pressures in the United States to correct the discrepancies that exist between states and localities but as long as they persist they will serve to foster the economic and racial segregation which penalizes the American city and perpetuates the social discord for which it has become notorious.

In 1968 the National Advisory Commission on Civil Disorders (Kerner Commission, 1968) recommended that massive federal aid be given to central cities to finance programmes that would rehabilitate the disadvantaged, both black and white, so as to enable them to have truly gainful employment and to enter the mainstream of American life. The Commission concluded that otherwise 'Within two decades, this division could be so deep that it would be almost impossible to unite'. In the eventuality of a permanent racial split, the Commission warned, America will face '. . . the danger of sustained violence in our cities. The timing, scale, nature, and repercussions of such violence cannot be foreseen. But if it occurred, it would further destroy our ability to achieve the basic American promises of liberty, justice, and equality' (Kerner Commission, 1968, pp. 407–408). To this the Commission added, 'We cannot escape responsibility for choosing the future of our metropolitan areas and the human relations which develop within them. It is a responsibility so critical that even an unconscious choice to continue present policies has the gravest implications' (Kerner Commission, 1968).

It is five years since the Commission filed its report. Recently the Nixon Administration refused to expend funds for human welfare programmes which the Congress had appropriated. Simultaneously, federal funding of police surveillance has been increased. At the moment the Administration apparently believes that repression is preferable to amelioration because it is cheaper and because its prejudices both shape and reflect the national mood. To escape the inevitable blow-up when it comes the white middle class has continued for the past five years to retreat from centre city. With the recent devaluation of the dollar, inflation and escalating interest rates it has become more difficult for persons of moderate means to buy a home anywhere. Racial and class antagonisms can be expected to fester in centre city as the suburban escape-hatch closes for all but the more affluent.

Life and Lifelessness in the Centre City

In both Canada and the United States the sprawl of suburbs during the past two decades has created similar changes in metropolitan landscapes. The social consequences, however, are quite distinct. In general, the populations of centre cities in both countries have grown very little; in some cases there has been a net decline. The composition of centre city populations, however, is quite different in Canada and the United States. In America the more affluent young families have almost unanimously opted for suburban living. In Canada the same group is divided in its preferences; some migrating to the suburbs while others of equal means remain because centre city alone accommodates their cosmopolitan life style. Jerome Friedland is a young advertising executive who three years ago migrated from Detroit to Toronto with his wife and three children. 'We love city life', Friedland remarked recently, 'and I can't think of too many cities in the United States where it's possible any more' (*New York Times*, 11 November 1973).

In Montreal, Toronto and Vancouver the middle class pays premium rents to remain in centre city apartments. They pay substantially more for a single family residence within the city limits than for equivalent house space in suburbia. Gantry cranes perched above new high-rise apartment buildings under construction are the most conspicuous feature of Vancouver's centre city skyline, and have been for the past 15 years. The city proper, not its suburbs, is where people want to live if they can afford it. So popular is centre city living in all of Canada's 16 largest metropolitan areas that apartment construction downtown outpaced single family residence construction in suburbia in 1969 (*Canada Year Book*, 1971, p. 836). Between the census years 1956 and 1966 there were net population increases in the centre cities of Montreal, Ottawa, Winnipeg and Vancouver. In central Toronto there was a decrease of less than 3000 in a total population of two-thirds of a million. Major growth, however, occurred in the fringe areas of all of these top five metropolitan areas (*Canada Year Book* 1971, p. 233).

Since World War II many more aliens have immigrated to the United States than to Canada but their impact upon urban life styles has been far greater in Canada, where the ratio of recent immigrants to the native population is higher. During the decade 1961–70 some 3·3 million immigrants entered the United States but they constituted only 1·6% of the total American population of 203 million in 1970. During the decade 1960–69 Canada received slightly more than 1·0 million immigrants but in 1969 they constituted roughly 5% of the total Canadian population of 21 million.

More important than numbers is the fact that alien immigrants to both countries are now chiefly professionals and skilled white-collar workers. Engineers, scientists, physicians, teachers, artists and other professionals made up 35% of all heads of family who migrated to Canada in 1967–69. Common labourers made up 3% (*Canada Year Book*, 1971, p. 270). A majority of these immigrants were from Europe where, traditionally, professional people prefer to reside in central cities. They are conditioned to and thrive in a cosmopolitan atmosphere. The civilized amenities of centre cities which are patronized after dark are as important to their life-style as are working accommodations during daytime. In Europe, as in Canada, violent street crime arising out of economic deprivation and racial segregation is minimal. People generally go about their affairs day or night without fear. Today all major Canadian centre cities reflect this European preference for urban rather than suburban living.

Vancouver's Robson Street, affectionately dubbed *Robsonstrasse*, testifies to the compatibility of contemporary Canadian and European urban cultures. It is 'human scale' yet cosmopolitan: a melange of restaurants, coffee shops, apartment buildings, hotels, boutiques and other specialty stores that cater to a clientele seeking everything from staples to high fashion, gourmet foods from all continents and imported toys. Scandinavian shops vie with those operated by recent immigrants from Germany, England, China, Australia, Turkey and elsewhere around the globe. At its north-west end Robson Street is only a short walk from Stanley Park and the public beaches on English Bay. A fine zoo, tennis courts, children's playgrounds and picnic areas are not only available but they are clean and otherwise well-serviced. At its south-east end Robson Street enters the city's commercial core of office towers and department stores. Because the city itself offers a setting in which to satisfy a great variety of tastes, the suburbs and their conventional shopping malls have not been able to top the commercial, recreational or cultural appeal of downtown. Big department stores have expanded their centre city floor spaces rather than curtail them as has been the trend in urban U.S.A.

The celebrated English–French culture conflict in Canada has sometimes been compared with the white–black racial confrontation in America but the analogy is absurd. The French were never slaves. Instead they are and always have been a powerful, self-respecting charter group. Religion, language and traditions, together with other aspects of their cultural heritage, were never extinguished by a master race as were the fundamentals of Negro culture in America. The principal city of French Canada, Montreal, is the largest and most cosmopolitan of all Canadian cities and after Paris, the largest French-speaking city in the world. Its cultural mix and bi-lingualism make it socially one of the most exciting cities in North America. Only New York and Mexico City are in its class but neither of them is liveable in except for élites who, being able to afford the best anywhere, can retreat to plush enclaves protected by private security personnel.

The newer sections of downtown Montreal from Place Bonaventure to Place Ville Marie form a complex of underground specialty shops and above-ground office towers and hotels with metro connections to outlying districts. It is the best example of dovetailed private–public investment in integrated commercial development in all of North America. It is possible to spend one's entire day working, eating, shopping, socializing and relaxing in this complex multi-levelled city within a city without ever being obliged to go outside—an advantage in the frigid Canadian winter. A hundred acres of Old Montreal, which was the original walled city, have been restored in recent years in such a way as to preserve its historic streets and architecture. The older parts of Quebec City have also been preserved in similar fashion, partly by government subsidy and partly by ordinances which preclude the owners of private property from maximizing personal capital gains by alterations which would deface the historic aspect of the old city. No large city in the United States, not even New Orleans, has so conscientiously preserved its historic districts as have Montreal and Quebec City.

Automobiles and City

Montreal, however, has committed enormous blunders, American style.

It has demolished whole low-rent neighbourhoods that formerly housed the poor, to clear routes for vehicular freeways and to assemble land in prime sites for large-scale investors. Established neighbourhoods of the weak have been sacrificed to generate capital gains for the influential. Quartier Ste. Famille, which lies but a short walk from the office towers of Place Ville Marie, is a 25-acre area stocked with turn-of-the-century mansard-roofed town houses adorned with detailed cornices as well as wrought-iron rails and grilles. The districts is one of renters who are a mixture of professionals, McGill University students, hippies and poor people. In the past all of them have co-existed harmoniously. Now both land and people are threatened with 'urban renewal' by Concordia Estates Ltd., which has acquired 96% of the properties and wants to replace most of the old structures with new ones that will house three to four times as many people at higher rents. Present residents are irate. Their neighbourhood, they say, is not a slum but a desirable, if old, low-rent enclave that happens to exist in a location where escalating land values induce the owners to maximize their investments to the detriment of tenants, most of whom could not afford to stay on in newer, more expensive quarters.

In Vancouver, however, the advent of the freeway came later: so late that its citizens were fully aware of its disrupting consequences. When the City Council employed a firm of California consultants to design a freeway that would have bisected the urban core, the voters, feeling that their way of life was threatened, 'turned the rascals out'. After a series of angry protest meetings, some of them in City Hall itself, the citizens formed a new non-partisan political coalition that eventually elected a mayor and aldermen pledged to maintain the human qualities of their city. One of the new aldermen, V. Setty Pendakur, was elected one month after he had published a devastating book (Pendakur, 1972) which not only documented the history of civic protest against highway advocates and those business interests which had urged them on, but exposed the attitudes of the incumbent aldermen who were bent on promoting urban growth at all costs with only minimal public debate.

In the United States it has become difficult, if not impossible, to stem pressures for more urban freeways since Congress established a federal trust fund to finance an interstate highway system. This trust fund now draws an annual income of approximately $6 billion from gasoline taxes and pays up to 90% of the cost of urban freeways that are tied into the interstate system. Canada, to its good fortune, does not have such a fund to warp the minds and intimidate the wills of its politicians and regional planners. Time and again, American cities have attempted to get legislation passed by the Congress which would permit them to tap the federal highway trust fund for mass transit systems as alternatives to freeways. Except for minor experiments these efforts have been frustrated by vested-interest lobbies that are financed by automobile, steel, oil, rubber and cement manufacturers, together with allies in the construction industry and related labour unions. It is now abundantly clear that the urban environment of North America is being structured, wherever possible, not in response to basic human needs and desires but rather to accommodate and support the profits and economic growth targets of major corporations, regardless of resultant waste of natural resources or of the destruction of functioning social networks.

Even in the midst of the 1973–74 'energy crisis' no significant policy changes regarding the control of the automobile in urban areas were seriously entertained. The solution for America, as proposed by President Nixon, was to ration fuel and gasoline, tax them more, and to relax price controls so that increased costs might act as a deterrent to use. There has been no serious consideration in Washington of alternative systems such as legislation to curb highway construction and to use the highway trust fund liberally to subsidize efficient mass transit systems. Yet, clearly, the fuel shortage of early 1974 highlighted the wastefulness of energy use in the modern American city. Automobiles consume about one-quarter of U.S. oil needs, and over half of this is consumed in intra-urban driving. Even the oil company executives recognize this fact. Addressing a news conference at the University of Texas, the President of Gulf Oil predicted that the energy shortage will change the landscape and life style of America. 'People will probably go back to the European-type city', he noted, 'where the people actually live in the cores and are not strung out in the suburbs' (*New York Times*, 20 September 1973). However, federal monies and policies designed to make the centre city more attractive must come first.

But how likely is this scenario? Once an environment is structured for an accelerated consumption of finite resources because it is profitable for the influential then the evolution of that deliberately structured environment continues in a straight line as if responding to Isaac Newton's law of inertia. People's lives are locked into it and it cannot be changed without enormous risks to their security, which thus makes them captive to the system. Their inclination then is not to press for a new, more functional, but unfamiliar system. Rather it is to patch up whatever is known, despite its costs and inconveniences.

An Exception—Public Control of Urban Land

Ownership of private property holds a special, cherished place in the American psyche, therefore the idea of leasing land on which to construct private dwellings is not welcome, despite the fact that land values alone are increasing by around 10% annually in American cities, largely due to speculation on private property sales. The City of Saskatoon in Saskatchewan offers an alternative that others might envy. As it emerged from the depression of the 1930s and World War II, Saskatoon found itself, as did many other cities in North America, with a substantial amount of tax-delinquent vacant land on its hands. Instead of unloading it at give-away prices as was the common practice, Saskatoon went into the real estate business. Thereby it found itself able to determine beforehand where and how it would grow in the future. The city does not develop its own land but it does put convenants in its deeds so that private developers who purchase from its land bank must comply with its master plan. Since Saskatoon supplies municipal services only as development is consistent with its policies, competitive private speculation in undeveloped land is almost non-existent. Incompatible uses of land and consequent environmental decay are avoided. Precious open space is preserved for aesthetic and recreational purposes. To maintain its leverage the city now buys land on its periphery as fast as it sells. In that way it maintains a land reserve that is 20 years in advance of needs. The result is orderly, predictable and attractive development.

Saskatoon does not try to profit from this business although it does so incidentally because it can provide its public services more efficiently. In 1970 house lots with appropriate services installed sold for $3000 to $4000, although equivalent lots sold for $8500 in Edmonton (Richardson, 1972, p. 99). Both Stockholm and Canberra have been doing this sort of thing for years with similarly happy results. Considering the extent to which urban environments are victimized with chaotic results by private speculation in land it would seem that such examples would lead ultimately to making land a public utility in every urban community. Mathematicians have long known that it is not possible to maximize two variables simultaneously. If cities are to permit maximization of private profits in land speculation a maximization of environmental excellence for the general citizenry becomes impossible.

Epilogue

It is of more than passing relevance that Canada, together with the most progressive welfare states in Europe, is a constitutional monarchy. Although the crown is of only ceremonial significance on the political stage it remains symbolic of very real citizen rights which were formalized with the establishment of the feudal system in medieval times. As contemporary national economies become subordinate to the global goals of a national and socially indifferent corporate industry the citizens of each nation are put in positions analogous to those of pre-manorial communities. They are obliged to organize themselves internally to guarantee mutual security and welfare in the face of external threats.

The most important social reality of feudalism was its code of reciprocity or mutual dependence which bound both lord and tenant each to defend and respect the welfare of the other, despite wide differences in wealth and social status. While the lord controlled the land, tenants possessed well-defined rights to its use as long as they fulfilled their obligations. This reciprocal arrangement, guaranteeing employment and sustenance to the most obscure tenant, is the antecedent of the contemporary welfare state. It guaranteed survival and its means as a right.

To this day there prevails among the citizens of Europe and Canada a widespread assumption that it is the responsibility of the total society to guarantee to each member the right to well-being by means of education, medical care, employment, shelter and social security. It is as symbolic guarantors of such traditional manorial rights that the monarchs of the British Commonwealth, Scandinavia and the Netherlands maintain their present popularity. Were the crown but a symbol of authority without reciprocal obligation it would have been dispensed with long ago.

By contrast, the mystique of frontier-agrarian individualism which prevails in the conventional American mind runs counter to notions of collective social responsibility. Thus the American middle class for the past two decades has been abandoning its centre cities because it could thereby escape participation in more comprehensive social security programmes. Had such programmes been more fully developed and institutionalized they would now protect not only the poor but also members of the middle class itself from the kinds of environmental devastation and social conflict which prevail.

As natural resources become scarcer and more expensive the competition for them will increase unless there is a reverse trend toward the socially co-operative traditions of the medieval manor as translated into institutions of the modern welfare state. It is not an accident that the United States, among all advanced industrial states, is the least mature in its development of comprehensive social security systems. This, along with its racism inherited from a slave-holding past has thus far made it, despite its wealth, incapable of building and maintaining more liveable cities.

References

Canada Year Book, 1970–1971. Ottawa: Information Canada, Bureau of Statistics.

Greer, S., 1973. Toronto tries to put lid on growth. *The Vancouver Sun*, October 18, 1973, p. 4.

Kerner Commission, 1968. *Report of the National Advisory Commission on Civil Disorders*, New York: Bantam Books.

Levitt, K., 1970. *Silent Surrender: The Multinational Corporation in Canada*, Toronto: Macmillan.

Lorimer, J., 1970. *The Real World of City Politics*, Toronto: Lewis and Samuel.

Lorimer, J., 1972. *A Citizen's Guide to City Politics*, Toronto: Lewis and Samuel.

Pendakur, V. S., 1972. *Cities Citizens and Freeways*, Vancouver: University of British Columbia.

Richardson, B., 1972. *The Future of Canadian Cities*, Toronto: New Press.

Ryan, W., 1971. *Blaming the Victim*, New York: Pantheon Books.

Schultze, C. L. (Ed.), 1972. *Setting National Priorities: The 1973 Budget*, Washington, D.C.: The Brookings Institution.

Statistical Abstract, 1972. Washington, D.C.: U.S. Bureau of the Census, Government Printing Office.

Section III. Transportation and the American Environment

12
The Politics of Transportation

TIMOTHY O'RIORDAN

The two chapters that follow look at the effect of transportation networks upon North American settlement patterns. Davis treats this subject in the broad historical perspective, pointing out how the physical terrain presented both a challenge and an opportunity to early settlers, and emphasizing the impact of different transport modes on both the economic as well as the cultural landscape. On the other hand, Eliot Hurst analyses critically the motivations of the early railroad pioneers who were linked by common interests to financiers and other entrepreneurs and to government officials. While corruption was rife, the more important implication was the emergence of a capitalistic nexus of power which largely controlled the rate and spatial arrangement of economic investment. The pattern of railroad development was not accidental but in large part dictated by the motives and collective self-interest of a small number of very powerful men who opened up the American hinterland at a critical point in the economic development of the nation. While the physical legacy of this period is reflected on the modern North American landscape, the political power of transport interests has shifted to other transport modes, where it remains equally as effective.

For example, many believe that the canal-building era ended with the introduction of the railway. This is far from the case, for today many very expensive canals are still being constructed by the U.S. federal government through the Corps of Engineers. The Corps is a civilian arm of the Department of the Army, an élite organization detailed to construct mammoth public works projects such as dams, canals and harbours. The Corps is ultimately tied to the Congressional committees which authorize its funds, and to the vested interest groups which stand to benefit from its projects through an unusual lobbying organization known as the Rivers and Harbours Congress. The Chairman emeritus of the Congress is a key member of the Senate Public

Works Committee and its directors include the Chairman of the Public Works Appropriations Sub-committee, numerous other senators and congressmen and executives from large companies, dredging interests and the coal industry (Drew, 1970, p. 57).

An example of the lobbying power and political connections of the Congress can be seen in the way it successfully negotiated the completion of the Trinity River Barge Canal linking barge traffic to the Gulf Intracoastal Canal which runs from Florida to Texas. The canal lobbyist was Dale Miller, a Texan friend of President Lyndon Johnson, Chairman of his inauguration in 1965 and a close associate of senior Corps officials and Washington politicians and lawyers (Drew, 1970, p. 57). He represents the Gulf Intracoastal Canal Association, the Port of Corpus Christi, the Texas Gulf Sulphur Company and the Dallas Chamber of Commerce. The project, which was authorized along with numerous other Corps proposals as part of a package and hence is not subject to detailed congressional scrutiny, was originally estimated to cost $790 million in 1962, and cost over $1 billion when it was constructed in 1971.

The Corps have not been noted for their environmental awareness. 'With our country growing the way it is,' commented its Chief, 'we cannot simply sit back and let nature take its course.' 'This business of ecology,' noted his deputy, 'we're concerned, but, people don't know enough about it to give good advice. You have to stand still and study life cycles, and we don't have time. We have to develop before 1980 as much water resource development as has taken place in the whole history of the nation.' (Drew, 1970, pp. 51, 61–62.)

But these powerful attitudes of man–nature dominance and growth have held less sway in the wake of an environmentally aroused America in the 1970s. In 1971, President Nixon announced the abandonment of the cross-Florida Barge Canal project, even after $1 billion had been spent on it, and by 1972 the Corps were incorporating very detailed environmental analyses into all their projects, including the analyses of multiple goals such as regional development, environmental quality and social stability as well as economic growth.

The political nexus of power controls another sector of transportation in the U.S.: the automobile lobby. In 1956, the U.S. instituted a sales tax of 4 cents per gallon of gasoline, the revenue from which formed a fund known as the Highway Trust Fund. Since then the Fund has accumulated some $200 billion, growing by about $6 billion annually: close to $7 million per hour! The Highway Trust Fund has been used to finance much of 41,000 miles of interstate freeway that connects all the major metropolitan centres in the nation, provides fast efficient ring routes for many and bisects the vast majority, disrupting community life and encouraging the widespread use of the automobile for urban travel.

A combination of concern over deteriorating air quality, created in large measure from the inefficient combustion processes of the internal combustion engine, plus a growing concern over the adequacy of gasoline supplies (American cars account for 40% of the 16 million barrels of oil consumed daily in the nation) forced the U.S. Congress to consider improving mass transit systems in American cities. (In the past the existence of the freeway, the relatively inexpensive car and the seeming abundance of gasoline combined to reduce public use of mass transit systems from 42 billion passengers to 6 billion passengers per year over the past ten years. Consequently, most transit systems run with huge deficits and with improperly maintained and outdated equipment.)

In February 1973 the Senate passed a bill to divert some $800 million annually from the Highway Trust Fund to help finance mass transit: primarily a 10,000-mile priority system in key urban areas. The Fund is a coveted political prize, especially for small town and rural politicians, for it enables them to sponsor highway bills in their congressional districts. During March and April 1973 the principal lobbying arm of the automobile, trucking and construction interests, the Highway Users Federation for Safety and Mobility, financed a $3 million campaign and employed a staff of 50 to persuade key congressmen to block a similar bill before the House of Representatives. The David facing this Goliath was a public interest group called the Highway Action Coalition, consisting of three people and a budget of $22,000. They all but succeeded, for the House defeated the bill by a mere 25 votes. But the almost frightening political power of the highway held sway, for in subsequent Senate–House bargaining, to reach a compromise the House Rules Committee voted the whole matter out of Congress in March 1974.

In his 1974 State of the Union Message to Congress, President Nixon promised $2 billion in guarantees to the ailing American railroad and $19 billion for urban and rural mass transit systems over the next six years. This money will come out of general revenue, not the Highway Trust Fund, and compares unfavourably with the $500 billion U.S. oil companies will spend on oil exploration and production over the same period; much of this oil going to feed the enormous appetite of the private automobile. So the burden is going to fall upon state and local governments to finance these very necessary facilities, and since 1972 a number of communities, including Detroit and Los Angeles, have embarked upon very costly programmes of transit improvement. For example, Denver residents voted in a $4 million bond issue, Atlanta voters ratified a 1% increase in the city sales tax, and Bostonians are now faced with an additional $176 in household taxes per year, all to finance more extensive and more efficient transit systems and to permit a lowering of the fare.

Will the commuters be won back? It is difficult to answer this because so many Americans are designed for automotive transport and Americans simply cannot abandon their cars. In September 1971 the San Francisco Area Pollution Control District called its first smog alert and requested that commuters use public transport. Traffic counts revealed no diminution of traffic despite the hazardous atmosphere (Pirages and Ehrlich, 1974, p. 62). The smog situation is now so bad that Congress was forced by a very powerful and revealing book sponsored by Ralph Nader's Center for the Study of Responsive Law (Esposito, 1970) to upgrade its air quality standards in its

1971 Clean Air Act. Esposito showed how the automobile lobby conspired to delay the manufacture and installation of air pollution control devices for many years, and also produced evidence to prove that the automobile companies were not fulfilling their legal duties in meeting 1970 clean air standards and, indeed, were influencing the setting of such standards (Esposito, 1970, pp. 26–68). The Act called for a 90% reduction in all automobile emissions for all models built in 1975 or after and insisted that all metropolitan centres prepare comprehensive traffic control proposals to deal with air quality protection during periods of smog alert. By 1973, cities such as Boston and New York were holding public hearings on massive limitations of automobile commuting during pollution 'scare' periods and Los Angeles proposed a novel gasoline rationing system to reduce automobile use at such times. These are powerful proposals but their fate awaits the settling of the 'new mood' in America following the onset of the energy crisis in 1974. However, even if modified traffic control plans are adopted, these may encourage a return to transit patronage especially if the service becomes more comfortable and efficient. The implications for the future shape and density of American cities could be quite profound.

Airway transportation is also the subject of much attention by an America increasingly concerned about the quality of its surroundings. In April 1972 development of the supersonic transport plane was blocked by Congress and a year later all overhead flights of SST's were prohibited. This meant that the Anglo-French Concorde could not fly over the U.S. at supersonic speeds, a policy that prompted two U.S. airlines, Pan Am and TWA, to cancel their options on 13 Concorde aircraft. Currently, some half-million Americans travel by air daily, a figure that is expected to rise to 1·3 million daily by 1982. This means more planes and more airports. Chicago's O'Hare Airport, currently the busiest in the world, handles more than two planes per minute throughout most of the day. But airports mean large amounts of land and lots of noise, therefore few urban residents are willing to accept new airports or extensions to existing facilities without protest. Currently some $4 billion in suits are pending against Los Angeles International Airport, all four selected sites for a new New York airport are blocked by citizen oposition, and proposed airports in New Jersey, St. Louis, San Francisco, Minneapolis–St. Paul and Atlanta are similarly stalled. Land costs around many airports exceed $100,000 per acre so expansion is almost out of the question. Solutions being proposed include constructing an ocean airport off the New York–New Jersey coast (at a cost of $7 billion), the use of short take-off and landing (STOL) aircraft for short haul intercity use, and the siting of large airport complexes far out in the countryside linked to city centres by high-speed mass transit. Two examples of the last alternative have recently been completed, an 18-square-mile airport–industrial development outside Kansas City and an enormous 28-square-mile complex linking Fort Worth and Dallas, Texas.

Meanwhile, at the other end of the transport spectrum, environmental values and a desire for more exercise have produced a growing political lobby in favour of the bicycle. There are about 80 million bicycles in the U.S. today and sales have recently tripled to 15 million annually. A lobby group calling themselves the Friends for Bikecology want Congress to allocate $10 million dollars annually for a system of bikeways in all metropolitan areas. In view of long-term deficits of gasoline which appear likely, this call is being heeded as more and more Americans take to two wheels. Already a number of smaller communities have constructed a system of bikeways, and a proposal to link bikeways to rapid transit stations is now under serious consideration.

With changing American attitudes to the quality of life, many of the traditional transport modes are regarded as unsatisfactory mechanisms for moving people and goods. Either they must be environmentally cleaned up or they must be replaced by substitute modes that are more efficient, more conserving in their use of resources and more healthy. Whatever the result, the effect of all these demands upon the American landscape will be considerable.

References

Drew, E. B., 1970. Dam outrage: the story of the Army Engineers. *The Atlantic*, **114,** 57–62.

Esposito, J. C., 1970. *Vanishing Air: Ralph Nader Study Group Report on Air Pollution*, New York: Grossman.

Pirages, D. C. and Ehrlich, P. R., 1974. *Ark II: Social Response to Environmental Imperatives*, San Francisco: Freeman.

13
Transportation and American Settlement Patterns

JOHN DAVIS

The evidence of transportation upon the American landscape shows in manifold ways. Disused canals, portage routes and Indian trails are as much part of this as are the giant switching yards outside Chicago, the major airports and the nets of interstate highways and pipelines which weave across the country. The construction of routeways and terminals means physical change of landscape texture: the bankside vegetation of canals, the railroad ballast bed, the tarmac and concrete strips of major highways, the presence of ports, railroad depots, airports, car parks and garages. Much more widespread and important, however, are the landscape changes resulting from transportation links, for the economic and demographic maps of the United States are only really explicable in terms of transport, as indeed are the growth and location of most cities. Various forms of transport have had their effects in different ways at different times, some have been much more localized than others, reflecting both their period of construction and use and their relative efficiency.

Patterns within the landscape are often a reflection of routes. The rectangular survey adopted for much of the U.S. has been emphasized and stamped more indelibly on the modern landscape by urban and rural road patterns. The modern interstate roads, with their broad sweeps across country and their clover leaf intersections, provide a completely different pattern. At the same time the interstate highway, like the railroad, has been a divisive factor in urban development. In the future 'the wrong side of the tracks' as a derogatory term may be replaced by 'the wrong side of the freeway'.

Trailways and Canoeways

Routeways have always connected man and land. Before the white man came to North America the Indians used well-trodden trails which linked their camps to hunting grounds while the more sedentary tribes in the east and south-west developed more complex systems of trails leading from villages to fields and fishing areas. The Indian used snowshoes and canoes—two essential aids to movement, given the contemporary environmental conditions in North America. The immigrant European, with a more advanced technology and different concepts of life and economy, not only used and improved upon the Indian forms of transport, especially the canoe, but also introduced new techniques and methods. Thus Indian trails, canoe routes and portages were widely used by the fur trappers, voyageurs and traders with their Indian guides and consorts in the search for tradeable wealth. The Spanish, with their different concept of empire, also developed trails and introduced pack trains so as to explore territory, bring the Christian faith to the Indians and hopefully find, extract and export gold and silver. The 'road' linking many of the missions in California is a good though latish example of a trail developed under the Spanish–Mexican influence.

Despite the existence of these ready-made routeways, early European immigrants confined their settlements largely to the east coast. Inland movement was both difficult and expensive for goods and even people and so tide-water sites with navigable streams were much sought-after locations. Below the fall line a number of east-flowing rivers were navigable and to these areas came the first main settlements. It was not many decades before the variations in natural advantages became important influences on the distribution of towns. While religious, economic, political and other factors played their parts it is perhaps significant that the best natural harbours were along the New England and mid-Atlantic state coasts and that further south suitable harbours and places of refuge became sparse. Thus, since all the trade of the early colonies was with the British Isles and, to a lesser extent, Western Europe and since an early colonist had a vital dependence upon supplies from Europe for survival and economic growth, it is small wonder that the northern half of the United States eastern coast could boast of four towns of over 10,000 population by 1790.

Roadways

Travel in the interior was mainly by horse and mule along roads in and near the main settlements; there existed no connected system of routeways that were properly maintained and signposted. The first successful and well-built road, the Lancaster Turnpike, was completed from Philadelphia to Lancaster, Pa., in 1794, and its success spurred interest for more such turnpikes. By 1774 a system of post roads linked the main eastern towns from Portland via Boston, New York, Baltimore, Williamsburg, Charleston and Savannah to St. Augustine; a spur route northwards joined New York to Montreal via Albany. By 1800 this had been extended from a single north–south line to a much more complex net extending as far west as Nashville, St. Louis,

Detroit and Niagara with another road from Nashville via Natchez to New Orleans. In 1834 the system embraced the whole mid-west and south and had its western limits at Forts Smith, Snelling and Leavenworth. The direct effect of such roads upon the environment was small; with few notable exceptions the building of bridges and other aids to travel made little, other than purely local, impact; but such routes did help to draw settlements to specific places and help to develop specific patterns—inns, hostelries, stabling and post offices were sometimes drawn to the roads—especially once the route between two established points had been made. Once such habitations had come into existence others could more easily be drawn to the same areas as security was increased and loneliness decreased. For example, in 1887 Nebraska City was chosen as a military freight route terminal where some 138 lots were surveyed for houses, foundries and boarding establishments.

Riverways and Canalways

Due to the Appalachian barrier, the road system evolved in the settled part of the old colonies. The particular drainage characteristics of the Appalachian massif restricted the number of navigable rivers, especially those draining from the west of the divide to the Atlantic, therefore alternative routeways were sought for trade and commerce. The Louisiana Purchase of 1803 stimulated this search by transferring sovereignty of the Mississippi basin to the U.S., thus making the whole Mississippi–Missouri river system open not only to exploration but also to exploitation as a routeway. The Ohio and Mississippi became prominent as major transport arteries. The river routeway made it practicable for settlers and their belongings to be transported more easily, quickly and cheaply than by land, and bulk movement of freight became a practical possibility. The effects on the landscape were considerable: river towns sprang up at transhipment and portage points; some of these grew into flourishing commercial and industrial towns like Pittsburgh, Cincinnati and Louisville.

However, one link was missing, the link from the navigable tide waters of the Atlantic to the Mississippi river system. Many schemes were put forward for crossing the Appalachians by an artificial routeway, but only a very few of these were successful. The first, however, was the most successful: the 364-mile-long Erie Canal opened in 1825 after eight years in construction, at a cost of $8 million. The cost was worth it as freight costs were spectacularly reduced. Westwards along the canal route moved thousands of settlers while barges of wheat and other grains, iron, timber and other commodities were moved eastwards. The Erie Canal made use of the only really low-level route through the Appalachian block: the Hudson–Mohawk gap. Other canal routes to the north and south were less successful as they had to use flights of locks and in some cases railroads. The Pennsylvania main line used a railroad portage to overcome the problem of elevation.

Other canals appeared in the 1820s onwards, for example Champlain 1822,

Delaware and Hudson 1829, Chemung 1834, Chenango Valley 1837. The canals can be divided into four groups: those which attempted to link the seaboard with the Ohio–Mississippi system, those extending the navigable limits of eastward flowing streams, canals linking Lake Erie to the Ohio and finally feeder canals to existing canals, lakes and rivers. The actual choice of routeway, determined primarily by physical terrain conditions, made or broke existing settlements and often led to the growth of new and the great expansion of existing towns, e.g. Rochester and Syracuse. Akron, on the Erie and Ohio Canal, is an interesting example of a settlement which grew at the junction of the Pennsylvania and Ohio Canals initially as a resting place, barge repair and building centre and transhipment centre; only later did rubber manufacture, for which it is now famous, develop. Canals linking the Great Lakes (of which the Welland, 1833, was the first) were crucial to the Lakes becoming a major artery, confirming the importance of the Erie canal route to the interior and ensuring the relatively speedy peopling of the eastern mid-west. Industrial growth was also fostered by the canal era—for example, Pittsburgh benefited much from the early impetus to iron smelting created by the confluence of navigable rivers and canals.

But the construction of canals and roads by private enterprise was an expensive business; tolls could not cover running costs and capital investment, therefore eventually the United States government was prevailed upon to help. Reports to the Senate by Gallatin and others stressed the need of such aid and especially of the value of improved transport links. Canals were particularly favoured over roads since it was felt that their special advantages in moving freight were more necessary to the economic wealth and growth of the nation, than were roads that were only suitable for moving people. A cost-benefit analysis for such large public expenditure was of course unheard of. Fulton (1797), a contemporary chronicler, noted:

> 'The general utility of artificial roads and canals is at this time so universally admitted, as hardly to require any additional proofs. It is sufficiently evident, that whenever the annual expense of transportation on a certain route, in its natural state, exceeds the interest on the capital employed in improving the communications and the annual expense of transportation (exclusively of the tolls) by the improved route, the difference is an annual additional income to the nation.' (Report of the Secretary of the Treasury on the subject of the Public Roads and Canals made in pursuance of a resolution of the Senate on 2 March 1797, published by order of Senate, 15 January 1816.)

Railways

The canal era of the 1820s–1840s did not mean that roads were completely neglected. Corduroy (timbered) roads were built, and the Wilderness Road was constructed through the Cumberland Gap to link the Potomac to Harodsburg in Kentucky. The famous National (Cumberland) Road which extended from Cumberland (Va.) to Vandalia in what is now Illinois and which was macadamized as far as Springfield, Ohio, was intended to go further west still but, like the canals, was superseded by the form of transportation which did more to transform the American landscape and environment than any other until the automobile. This was the railroad. The earliest railroads in the U.S. were centred on Baltimore where the Baltimore and Ohio Railroad began in 1828, first with the rise of horse-drawn carriages and later steam engines (in 1831). The first scheduled steam railroad train ran on Christmas Day 1830 at Charleston S.C., and by 1833 South Carolina had 136 miles of railroad.

The railroad, being less dependent on water supplies and more tolerant of gradients than canals, more independent of weather conditions (especially winter freeze-up) than rivers, and able to carry a greater number of goods and people more quickly than roads and trails, was destined to alter the face of much of the U.S. The importance of the railroad and its actual and contemplated effects is indicated by its prominence in American folk music and literature.

As was the case with the initial post road and canal pattern, the initial railroad distribution was heavily concentrated east of the Appalachians and especially in the north-eastern part of the country. Southern New England was particularly quick to seize upon the railroad transport as its terrain was least suited to canals and river navigation. It was also in the north-eastern group of states that business and industrial enterprises were sufficiently developed and could obtain the necessary capital to finance construction. It is, however, interesting to note that the first railroad journal appeared in Tennessee on 4 July 1831. The *Railroad Advocate* was published by a group who had never seen a railroad but who came into being 'for the purpose of discussing ways and means to collect and disseminate information on the utility and practicability of constructing railroads and other improvements in the country'. (*Railroad Advocate*, 4 July 1831.) Images of prosperity stimulated by the railroad were most powerful: 'Can any man of common sense fail to perceive, that if east Tennessee possessed the same facilities for transportation which are, or soon will be, enjoyed by almost every other section of the Union, her situation would speedily undergo an almost incalculable change for the better?' The following year the *Railroad Journal* first appeared and argued strongly in the number for 6 October 1832 that costs of coal carriage in North Carolina would be appreciably less by rail than by turnpike. But, in spite of the far-seeing ideals of this group in Tennessee the south lagged far behind the north in railroad construction and did so until the closing years of the 19th century.

Like the turnpikes, the early railroads were characterized by a multiplicity of small companies. This led to practical problems of interchange, especially as at one time there were eleven different gauges. Gradually, however, gauges for all but mineral and some mountain lines were standardized at 4 ft 8·5 in, and by amalgamation, 'over-running' rights and other techniques the problem of interchange was reduced.

One outcome of this was the introduction of fast freight lines operated by successful independent companies such as Red, White, Sagenaw Valley, Empire Transportation, International and others. Gradually, connecting lines were also leased or purchased as the early companies grew and consolidated their holdings. As network extensions were enlarged only a few highly profitable companies survived the vigorous economic battle.

What were the effects on settlement patterns? The frontier was pushed further back as the railroad brought more dense agricultural settlement. Just as the stage coach and post routes induced the development of inns, relay stations and small farms to accommodate travellers and provide feed for horses and mules, so the railroads established their own depots and division towns. In the east such facilities tended to be in existing towns as the earlier railroads were mainly linking existing settlements. West of the Mississippi, and especially west of the Missouri, the railroads often had an open choice in the selection of suitable sites. Some effects on the landscape were transitory; the boisterous, noisy and sometimes colourful 'end of tracks' shanty towns described by many writers gradually moved on with the movement of the construction gangs. Behind them the cattle railhead towns like Ogallala, Abilene, Dodge City, Ellis and Newton enjoyed brief glory and importance before being either superseded by another town further west or made superfluous by railroads built into the range lands.

The coming of the railroad, either as continuous through lines or as short links from mines to towns or rivers and canals enabled the mining industry to develop to a hitherto undreamt-of scale. The railroads not only carried coal and iron but were large consumers themselves. The Pennsylvania anthracite and the main Appalachian coalfields may have owed their initial development to the canals, but before long mineral lines were playing an important role in feeding to the canals coal from hitherto inaccessible areas. Thus the landscape of eastern Pennsylvania became covered with tips, mines, adits, railtracks, small towns and other protuberances of mining and industry. Some railroads owned mines themselves; for example, the Philadelphia and Reading Railroad Company controlled canals, mines and blast furnaces.

With rail links connecting the mid-west to the Atlantic coast, and with the aid of the Great Lakes and major canal and river routeways, it became possible to harness the mineral, forest and agricultural wealth of the interior to supply the needs of both the east coast belt and overseas markets. Indeed, such overseas trade became essential as the supplies of raw materials outstripped eastern demand. Gradually, seaports became increasingly concerned with exports: first grain, later meat, mineral ores and coal. Even though today the St. Lawrence Seaway permits ocean-going vessels to reach the Great Lakes, a considerable amount of coal, grain, wood pulp and other commodities still make their way to the Atlantic and Pacific ports of North America predominantly by rail. Rightly, great importance has been laid upon the Great Lakes as a system of transport routes but it must be remembered that its function would be seriously limited if it were not for railroad access to them. Visually, this is apparent from the size of the railroad yards at Duluth-Superior, Toledo and Thunder Bay. It is the railroads which bring the materials from farm and mine areas to the ports.

The specific route of the railroad played a significant part in determining the distribution of crops and influencing settlement patterns. To be within economic waggon haul of the railroad depot meant that a farmer could grow crops for sale outside the area; to be too far away meant either subsistence farming or the need to rely on products which moved more easily: cattle and sheep. It was not so much the overenthusiasm of railroad mania but sound economic need which caused numerous spur and branch lines to be built in the agricultural regions, thus bringing areas within reach of the rail tracks. At the rail depots the need to provide storage for the farmers' produce led to the erection of silos and elevators, which today remain the dominant feature of the landscape in much of the Great Plains and the Prairies.

Technological improvements in train speed and in food preservation (refrigeration) opened the way for produce specialization in selected parts of the country, and for many years Florida, S. W. Texas and parts of Arizona and California were very dependent upon rail links to the distant market for the distribution of their citrus fruit and salad crops. Farming in the north-east also changed. With better and cheaper grain and meat coming from the mid-west and later from the Great Plains the farmers in the east lost their markets for such products and changed increasingly to the production of milk, poultry and some types of truck farming. Also, the opening up of the western lands drew people from the New England farm areas, especially the more marginal areas, and as a result the eastern farming area shrank as farm and land abandonment took place.

Possibly the greatest effect on the American landscape made by the railroad was the use of land grants. In order to induce railroad construction the federal government made large blocks of land available to the railroad companies under various conditions which were easily met. The first Federal land grant was made in 1850 to the Illinois Central Railroad and by the time of the last important land grant in 1871 grants totalling over 170 million acres had been given for railroad building. If one adds land grants for canals and other forms of transport, then some 200 million acres of land were made available by the federal and state governments to further transport development. This excludes the help received from the states in land and all the financial help made available at different levels of government both in the form of loans and grants.

Though much of this land was not directly useful to the railroad companies it did provide a most profitable source of investment. Some land was sold to settlers, usually in the customary lots of 160 acres, but in many cases the size and shape sold varied to suit the needs of particular purchasers. The railroads saw these sales as a source of steady income and were often prepared to offer land for sale on payment spread over a period of years. They would also provide advice on farming, move the purchaser's goods, chattels and family in at a very reduced rate and grant interest waivers in times of poor harvests. The argument was that once an area was peopled it would need more goods and would also provide produce for sale and thus the railroad would gain business. Partly for this reason the railroad companies were often keen to promote settlement and even tried to attract groups to particular areas by offering free plots of land for churches or schools, and sometimes even the free transport of materials needed for such buildings. By promoting settlement the railroad companies saw more orderly and rapid land occupancy and also the possibilities of small-scale industrial developments and certainly long-term business advantages. Behind the agricultural settlement came urban centres, for the railroads were also influential in establishing many a small town such as Hastings and Holdredge in Nebraska. A large part of the Great Plains area in particular was either indirectly or directly settled and farmed as a result of the railroad land grants.

The railroads, especially the logging lines, enabled the wealth of the forest lands of the northern Great Lakes states to be more fully exploited. Whereas timber had, before the railways, been floated to market, after being dragged to the riverside by horse, it was now possible to bring the timber from further afield and also to use rail all the way to the logging mills, ports or markets. While remnants of the old lines with the associated lumber camps stand abandoned, in some parts of Michigan, Wisconsin and Minnesota the old logging railroads continued in operation till very recently, sometimes as a type of tourist attraction, e.g. the Cadillac and Lake City line till 1971.

The railroads brought expansion, consolidation and innovation with them; they also all but obliterated the canal. Few canals were able to withstand railroad competition for very long; canals had not reached their full development potential when the railroad era began and while in 1841 the mileage of the two was equal, within ten years the railroad mileage had vastly outstripped that of canals. Some canals quickly passed out of use, except for very local traffic, others continued in operation till the 20th century but, with very few exceptions, the canals have disappeared as a means of transport. Thus placid stretches of canal route are left in many places in the east, some converted for recreational use, as in Illinois (at Morris just south of Interstate 80) others with such a use as a potential, still others standing derelict, stagnant and an eyesore, e.g. in the vicinity of Akron. Suggestions of the former use and importance of such canals can still be seen in the location of some industries though in most cases they now turn their backs on their original umbilical cord. The railroad also forced the decline of some towns, a few in the canal areas but most in the Mississippi river system. Here towns which flourished during the steamboat era were gradually, or in some cases suddenly, left 'high and dry' as railroads reduced the amount of general freight being carried from place to place by water, and as bridges spanned the rivers so ferry terminals became redundant. Mark Twain in 'Life on the Mississippi' remarks on the number of declined riverport towns to be seen as a result of the railroad development.

Looking at the detailed freight figures of today it is perhaps hard to believe that water transport has shown a fairly continuous rise in total tonnage moved over the past sixty years, but most of this is accounted for by long-distance haulage of a fairly narrow range of bulky raw materials, e.g. coal, oil and timber. Today the inland waterways, such as the Mississippi–Missouri system, the Great Lakes and a few canals, for example the Intracoastal and especially the Great Lakes Canals, Soo and Welland, account for more trade than at the height of the canal era, but their overall impact is much more localized and limited.

Motorways

While it was the steam engine, both on river boats and on railroads, which revolutionized transportation in the 19th century, in the 20th it was the internal combustion engine. The early 'horseless carriages' were neither successful or reliable but by 1910 the automobile had sufficiently advanced in design and reliability to have a firm place in the transport network, albeit mainly in limited areas. Innovation and the necessities of the 1917–18 war led to great improvements and this, together with the large-scale mass production methods pioneered by such people as Ford, led to a revolution in transport of great magnitude and significance on the American continent.

What have been the results? First, a greater mobility than could ever economically have been achieved by railroads or waterways; door-to-door travel and transport by one mode was now feasible where previously transhipment and often changes of mode had been necessary. Overall speeds were also often increased as time lost in transhipment or change was eliminated. Secondly, the speedier movement and greater hauling power meant that the old economic limit of distance from a railhead could now be lengthened, thus influencing land-use patterns. Thirdly, the new form of road transport called first for better roads and subsequently for more roads. In the horse-drawn age very few roads were macadamized, most were dirt and, except in some towns and on certain very major trunk roads, the road surface was extremely poor. The automobile demanded graded roads, the elimination of potholes and ruts, and bridges rather than fords. The obvious result in the landscape was to give the roads greater permanence and to make them very different from the surrounding countryside. By the 1930s the road traffic registrations covered nearly 23 million vehicles so that the demand for specially built roads became more vocal and the first of the modern turnpikes was put into construction. By 1960, with a registration of 74 million vehicles and the emergence of the articulated lorry, the political lobby of the highway users had become sufficiently influential to cause the federal government to introduce a $100 billion Interstate and Defence Highway System of freeways: some 42,000 miles of limited-access, toll-free highway. This is not the place to discuss the economics, technology or merits of this enormous undertaking, but it is relevant to note the implications of the turnpikes (tollroads), interstate highways and similar roads upon the landscape. First, their sheer size—at least two lanes in each direction with central reservation and often hard shoulders, rising to three lanes or more in busy areas and with considerable land usage at intersections and junctions. Secondly, the impact on cities, especially the amount of building destruction necessary to bring a freeway into a city, is considerable, particularly as it means bringing the road near to the centre of existing cities. Admittedly this reconstruction does provide an opportunity for the large-scale redevelopment of blighted urban areas, but sometimes these roadways, elevated, ground-level or in 'cuts', created their own local blight and just as the railroads in the past acted as sector boundary-producers in the city, so too does the interstate urban motorway system.

Another effect is to encourage mobility, especially the outward movement of people and services. Lack of transport in the past kept settlement compact. The railroad enabled some outward spread which increased with the growth of the street railway, street-cars and tramways in the latter part of the last century. Thus the growth of towns like Chicago, Boston, New York, Detroit and others was much dictated by the patterns of street-car routes. It was the street-car, not the automobile, which first created the suburb, separated the city core from residential areas, zoned the city into functional areas and encouraged the spatial separation of working-place and dwelling-place.

However, suburban growth, or perhaps more aptly 'sprawl', was encouraged and made possible on a much greater scale by the automobile and towns like Los Angeles spread out rapidly in the 1950s and 1960s as a result of this increased mobility. The interstate freeway system makes this outward movement even more diffuse, for while it is true that the interstate and similar highways permit access only at junctions, yet suburbs and housing sections can, could and do develop at these intersection points. Route 128, encircling Boston, and other ring roads as well as trunk arteries into cities have aided greater diffusion.

They and the automobile have also encouraged changes in the function and environment within the central city and the C.B.D. The blight, tax delinquency, smog, crime and sheer drabness are causing the middle-income groups and small businesses to move out from central areas to the peaceful affluent suburbs. The retail function of the urban core is also suffering as the more affluent prefer to shop in the spaciousness of suburban shopping malls. The influx of poorer, often non-white, people into the inner and frequently ghetto areas has speeded this movement to the suburbs. The movement of other service functions traditionally associated with the old city centres to the suburbs further encourages the decline of the city centre and inner city. Obviously, many factors are at work in this trend but the resulting environmental change is at least partially attributable to mode and network changes in transportation.

In general terms one can argue that America has become a motorized society, with all the implications this has for industry, urban planning and the environment. It means that, at least in the major cities, the motor car and the provision of its associated facilities takes up a very large part of the central, and therefore most expensive, areas: it has been estimated that in the case of central Los Angeles about half of the area is devoted to streets, parking areas, gas stations, repair centres and auto showrooms, etc., and similar if not quite such high percentages could be echoed for many cities apart from some of the oldest in New England and the north east. While Los Angeles is perhaps the most notorious for its smog—a mixture of gases and particulate matter created by automobile exhausts but made more toxic by the energy of sunlight—all large North American cities suffer this problem despite vigorous efforts by federal enforcement agencies to reduce noxious emissions.

Airways

The American environment has always presented problems and challenges to the transportation industry. Terrain and climate have affected all modes from the canoe through stage coach and waggon train to railroad and automobile; the effect of the sheer territorial size of the country together with its affluence are the factors which explain in large measure the importance of internal air communications. Very much a product of the 20th century and especially of the last thirty years, the airline system of the U.S. is the most comprehensive, dense and heavily used of any of the major countries of the world. The airline has provided the final answer, to date, for the speedy movement of people and freight, at a price. The map of travel time relative to size of country has shrunk very rapidly since the inception of regular air services across the country. Admittedly air freight is still in its infancy, due to cost, but it has shown a very great rise over the past two decades. Already some specialist machinery and electrical and electronics parts are flown to markets, as are some drugs and such foods as fresh strawberries from California, and flowers. It is, however, in air passenger transport that the main developments have been made. All cities of over 100,000 population are linked by regular air services, and many of the smaller ones too. Increasingly, privately owned aircraft, company owned and private charter companies are making the airline network even larger and more complex. Its landscape effects are, first, that virtually every settlement of any size has its own airstrips and most towns have regular airports. These airports consume land and affect surrounding land use; especially in the case of the international airports and major city airports which have grown enormously in size to cater for the increased size of the aircraft using them and thus the need for increased length of runways, etc. A really large airfield, for example, O'Hare Airport at Chicago, can cover ten square miles, and when one remembers that Chicago has three and New York four airports, the land consumption is considerable. Also, in the case of airports in particular, noise pollution has its impact environmentally. This, together with the need for adequate flat land at a tolerable price level, is why airports are increasingly being located in estuarine areas, or close to the shores of lakes, etc., for example, New York (Kennedy), San Francisco and Boston.

Air transport and the automobile compete with other modes of transport with respect to freight movement (Table 13.1), but this is not the case for passenger movement (Table 13.2). While total passenger miles have increased, modes have fared very differently, the most spectacular being the advance of

Table 13.1. Changes in volume of intercity freight traffic in billions of ton miles

		Rail		Motor vehicle		Inland waterways		Oil pipelines		Air	
	Total	volume	%	volume	%	volume	%	volume	%	volume	%
1940	651	412	63	62	10	118	18	59	9	—	0·002
1950	1094	628	57	173	16	163	15	129	12	0·3	0·03
1960	1330	595	45	285	21	220	17	229	17	0·8	0·06
1970	1921	768	40	412	21	307	16	431	22	3·0	0·2
1971	1975	791	40	430	22	307	15	444	23	3·4	0·2

Source: *U.S. Book of Facts and Statistics* 1974, page 538 and I.C.C.

Table 13.2. Changes in volume of intercity passenger traffic in billions of passenger miles

		Private car		Air		Bus		Rail		Inland waterways	
	Total	volume	%	volume	%	volume	%	volume	%	volume	%
1950	508	438	86	10	2	26	5	32	6½	1	0·2
1960	784	706	90	34	4	19	2½	22	3	3	0·3
1970	1185	1026	87	119	10	25	2	11	1	4	0·3

Source: *U.S. Book of Facts and Statistics* 1970, page 536 and I.C.C.

air and road at the expense of rail. The advantages on the one hand of door-to-door transport as provided by the car and, on the other, speed over long distance as provided by air, means that the railroad has lost both long-distance and shorter-distance travel. Today, apart from a few long-distance lines, most of the rail passenger traffic is relatively medium-short distance or commuter travel. The impact of the new metroliner service between New York and Washington, which on a city-centre to city-centre schedule provides a comparable service in time with the airlines and faster than the road, is awaited with interest. Should it be successful, then such a system would be developed for other corridors such as Chicago–Detroit–Cleveland and also San Diego–Los Angeles–San Francisco. Nowadays, in many parts of the country the old railroad depot/station either stands derelict or has been converted to another use; increasing mileages of railroad travel have been abandoned through amalgamations and the associated reduction in duplicated services, and changing economic needs, such as forest or mineral depletions. Today, most lines which do remain are used for freight only, not many carry more than a very few passenger trains.

Pipeways

One form of transportation used exclusively for freight remains to be considered. The first pipeline was constructed and opened in 1865, six miles from Pithole Oilfield in Pennsylvania to Miller's Station on Oil Creek River. Today there are some 129,000 miles of trunk pipeline and 47,000 of feeder lines. In the main these are oil, oil-products or gas pipelines, though recently experiments have been made with moving crushed coal in water suspension and also solid products by pipeline. Many such lines are buried underground and only appear at the surface to cross rivers, ravines, etc., so that once they are laid and the land put back to former usage little remains except inspection/pumping stations and guider markers. However, near or at the terminals the impact can be different. The so-called 'Spaghetti City', Houston, gets its name from the multiplicity of pipelines across and alongside the Ship Canal in the petrochemical industrial area, and many inland oil refineries and refineries on the Great Lakes and their associated industries owe their *raison d'etre* to the pipeline as an efficient means of transport.

Conclusion

Looking at the United States in the 1970s, the impact on its landscape of the road, waterway, airway and pipeline is far greater than men like Gallatin, Jefferson and others ever foresaw. Certainly they were right in believing that routeway was a key to prosperity but while transportation certainly helped to weld American unity no-one could have predicted some of the consequences of the evolving communication networks. Apart from the rise and fall of communities, the division and blight of the modern city and the pervasive smog, the development of transportation has also tended to reduce regional dissimilarity and dampen creative self-expression. Individuality in early building styles, fences and even urban layout has given way today to regularity and standardization. From the air or from the ground, almost any middle-sized American city looks and feels the same. Airports are similar, ferry terminals look alike and even the modern American railroad, Amtrack, is now standardized.

While this sameness is in part a reflection of modern mass production methods it also points up a declining interest in individual expression, a reluctance to be unorthodox and a ready acceptance of the totally expected. The impact of repetition and routinization on the landscape is a tendency to dampen any sense of stimulation or interest in surroundings. Landscape change may all too easily lead to landscape similarity, for while transportation networks provide avenues for the flow of people, goods and ideas, reception and diffusion of these commodities will in part depend upon the collective aspirations of men.

References

Davies, R. E. G., 1972. *Airlines of the United States since 1974*, London: Putnam 1963.

Locklin, D. P., 1966. *The Economics of Transportation*, Homewood, Illinois: Richard D. Irwin.

Stover, J. F., *The Life and Decline of the American Railroad*, New York: Oxford University Press.

Taylor, G. R., 1951. *The Transportation Revolution, 1815–1860*, New York: Harper.

Tunnard, C. and Pushkarev, B., 1963. *Man Made America: Chaos or Control? An inquiry into selected problems of design in the urban landscape*, New Haven: Yale.

Williams E. W. (Ed.), 1972. *The Future of American Transportation*, Englewood Cliffs: Prentice Hall.

Winther, O. O., 1964. *The Transportation Frontier, 1865–1890*, New York: Holt, Rinehart and Winston.

14
The Railway Epoch and the North American Landscape

MICHAEL E. ELIOT HURST

Baran and Sweezy (1966, p. 219) have identified three epoch-making innovations, each of which, in their own words, '. . . produced a radical alteration of economic geography with attendant internal migrations and the building of whole new communities'. Two of the three are new transportation modes, the railway and the automobile, the third a generator of power perquisite to one of the other two, the steam engine.

Geographers and other social scientists have assessed the impact on the landscape of transportation innovation largely in idiographic terms, but while one should neither deny the worth of such analysis nor underestimate the landscape change involved, closer analysis reveals the role of individuals and groups formulating policy within the confines of a developing capitalist system. By the 1880s, in both the United States and Canada, the railway pre-dated the industrial corporation of the 1890s and 1900s, in terms of both size (in 1874 the Pennsylvania Railroad had assets of about $400 million) and fortunes made (the Durants of the U.S. and the Strathconas of Canada). The 20th century impact of the automobile is equally strong; the process of suburbanization, and the ever-attendant freeway—images which probably loom largest in today's society, with all their contingent developments of residential, commercial and highway construction—have all along been propelled by the mass production line and corporate fortunes.

Prolegomenon

Transportation geographers and historical geographers alike have become enmeshed in a procrustean bed of their own making. The 'process/stage' view of the world put forward in the recent past (NAS-NRS, Ad Hoc Committee, 1965) has a uni-dimensionality about it which is troublesome to both sub-disciplines. Thoman (1965) has criticized transportation geographers who, in attempts to ensure respectability, have analysed the evolution of Canadian railroad networks within narrow doctrinal notions of 'location theory' or 'network analysis'. More recent efforts to reiterate this concern (Eliot Hurst, 1973, 1974) have received some support; Borgstrom (1973) for example notes,

> '... we can either play with artificial notions about transportation systems as we used to play with electric trains on a plywood reality, or we can see through these paradigms to the decision-making processes which create the systems and strive to understand (and ultimately, to affect) the forces behind, and the spatial relevance of those decisions ... Until we infuse our analyses with the sensitivity of humanitarian concern, our exercises in network geometry and inventory control, however rigorous, will continue to be as superficial and self-serving as they are marginally useful and negligibly relevant'.

'Decision-making processes' did not create the railway systems; with the development and expansion of the capitalist system the need for a more rapid mode of transportation became imperative. As technological time collapsed spatial relationships, world markets expanded and capitalism had to break down spatial, national and ethnic barriers. Thus, when developed, the railway was quickly adopted to achieve those ends, as was the steamship and, later, the aeroplane. Therefore it is not surprising that there is such a close correspondence between the rise of capitalism (especially industrial capitalism), and technological innovations and their adoption, for above all else it is capitalism that shaped Canadian and American landscapes.

Similarly in historical geography, a debate continues on the role of the 'hunt-the-technique syndrome' as a kind of 'cul de sac of numerate dilettantism' (Harley, 1973). Models, content analysis and naive positivism become the procrustean limits of such past landscape images as those that might have been expressed in Captain John Palliser's records of his 1857–60 expedition to the Great Lakes and westward in the newly unfolding prairiescapes of British North America (Koroscil, 1971a, b; 1973). Baker (1972, Chapter 1) joins Harley in stressing the counter view that such approaches may contribute little to our understanding of past landscapes, although they may reveal a great deal about the naivety and immaturity of the practitioners.

This is not to label all historical geography as mere reconstruction of 'imagined views' of a past world. Such imagined views may well be as misleading as the content analysis of Palliser's journals. Bowden (1969), in reviewing images of the western interior of the United States held by the educated élite of New England in the 1830s, found their view of the 'desert landscape' to be shared by few, and that the myth of the Great American Desert before the Civil War was itself fictitious! 'Official mythologies are common to all countries. All countries cherish one or two particular periods of their histories which they enable and embellish to justify and give meaning to their present and to give a purpose to their future. This habit ... is extremely important to recent and ramshackle [nations]' (Barzini, 1972, p. 16). Those words were about 'a great, old, and solid nation', Italy, but are more applicable to North America.

Far from reflecting a romantic 'national dream', the impact of transportation on the North American landscape is the product of hard-nosed attitudes by entrepreneurs and others in the context of a burgeoning and profitable capitalist economy. For while values and feelings influence economic behaviour as much as those of culture and romance, human behaviour is not random or unstructured, but '. . . is subject to shared responses, [and] common situations' (Vance, 1970, p. 140). Nevertheless, within a commonly shared value system there are 'exceptional acts' whereby the individual entrepreneur or firm seeks to shape the organization of business that will garner greater individual, as opposed to generic, profits. 'Such is the rationale of capitalism, that the standard theory of spatial organization [e.g. "central place theory"] that we as geographers apply . . .' does not help us to understand those business practices which apparently do not fit into the perceived landscape regularities (Vance, 1970).

The very notion of the entrepreneur carries with it the values of capitalism, including the degree to which an 'exceptional' act in terms of risk will be made in order to gain profit and success. In other words, the reconstruction of past and present geographies must step outside the mechanistic regularities of mainstream transportation and historical geography, and assess the role of a Cornelius Vanderbilt or a William Vanderbilt and their values of 'The public be damned. I am working for my stockholders'. In such terms it is possible to see for example the 'railway epoch' in Canada not so much as the infilling of the interstices of some Christallerian latticework, but as the deliberate tying of a hinterland Western Canada to a metropolitan core around the St. Lawrence. In the context of commercial capital, individual fortunes and political gain, the railway's stimulus as a moulder of economic landscapes takes on a more powerful dimension than that sketched earlier by geographers.

Moreover, the list of financiers and railway men who were active during the formative years of the Canadian landscape is also a veritable 'Who's who' of the political milieu. The linkage runs from merchant capitalism to finance, transportation and land speculation. In the words of the Canadian Prime Minister in 1854, Sir Alan McNab, 'railways are my politics', although today it might well have been better expressed as 'politics are my railways'! This means that in the area of transportation we are less likely to find sufficient public authority over private enterprise to ensure that some sort of social utility is constantly being kept in view or to see business and government working hand-in-hand to solve a wide range of economic geographic problems. These problems included the control over fierce and debilitating competition between railway companies as the economic landscape grew in complexity and geographic scope; the creation of more orderly management arrangement through which special interest users of railway services, such as industrialists and farmers, could exercise some influence over decisions which vitally affected them; and the infusion of massive publicly donated capital to ensure the growth of the rail networks and continued technological change. All of these issues were confronted in a context of the shared concerns and values of capitalism itself.

A systematic examination of these interwoven political and economic forces casts a completely new light on landscape evolution. Regulation of destructive competition, for example, was not the result of a public will being imposed on recalcitrant entrepreneurs (Kolko, 1965). Rather it was necessary to assume a 'rational' allocation of power between conflicting business and agricultural interests. To some extent there was disagreement amongst powerful economic interest groups as to how their goals were to be achieved. But they shared the notions of 'progress, efficiency and political reality'. Each tried to identify itself with the 'public interest', and each tried to argue for efficiency in terms of an orderly system of social control of the railways (Kerr, 1968).

But there was simply no public interest, rather a host of private interests contending to outbid one another in the political context. Economic interests made their claims, whether for the 'public good' or not, through the institutions of government, who tended to share similar values (Adams and Gray, 1955, p. 21). Thus the government masked their pronouncements on 'public interest' in an elaborate rhetoric so that those who did lose would find it easier to accept. Should Pittsburgh be allowed freight rates to enable its entrepreneurs to develop flourmilling at the expense of Buffalo interests? Should farmers be forced to pay a higher percentage of their income for shipping grain or livestock to market? (Kerr, 1970). Were pooling arrangements, such as the New England quintuple alliance, in the best interests of the economic geography of that area? (Darr, 1970). Or were the Burlington and Sante Fé bids to link Chicago and Kansas City in the interests of the orderly growth of the railway networks? (Smith, 1970).

These are difficult questions to resolve, and even a common identity between railwaymen, businessmen and government does not necessarily make their solution any easier, except when the realities of capitalism are understood. Public interest philosophy may well influence the *language* of regulatory policy, but it fails to alter substantially a regulatory *practice* which allowed economic interests to use the resources of governments to seek their own ends. Without such a realization of the *realpolitic* of the growth and change of transportation modes and networks in the North American landscape, no understanding of the dynamics of geographic transformation would be possible; at the basis of the economic geography of that landscape lies capitalism, its values and its identities.

The Railway Epoch

An important function of governments in the 18th and 19th centuries was the creation of transportation networks as the corequisites of a commercial private-enterprise system. The building of these networks directly absorbed enormous

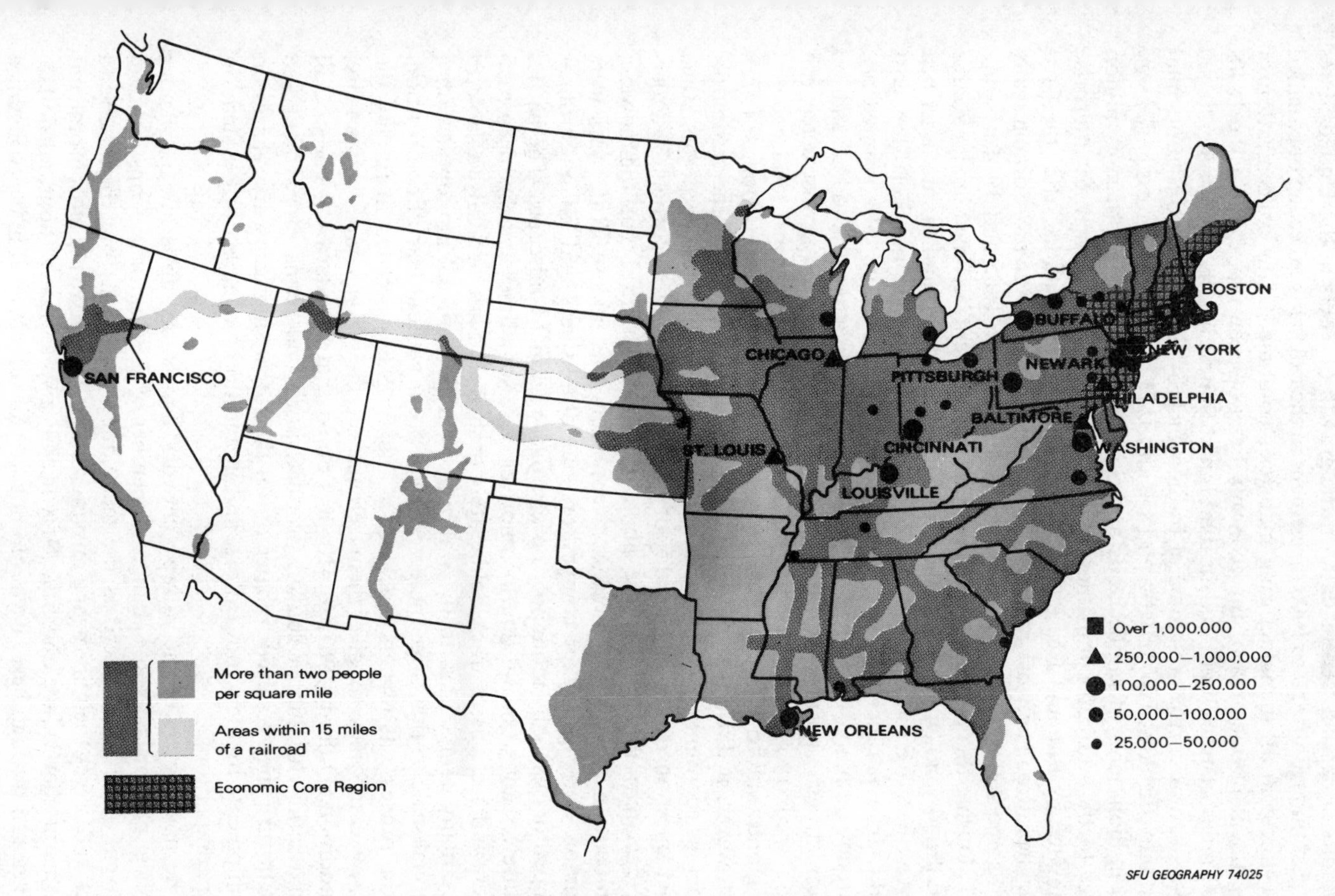

Figure 14.1. Development of the American railway system. From *Cities and Immigrants: A Geography of Change in Nineteenth Century America* by David Ward. Copyright © by Oxford University Press, Inc. Reprinted by permission

amounts of capital. Census data on the growth of assets in the United States suggest that from 1850 to 1900 investment in the railroads exceeded or equalled investment in all manufacturing industries combined. This concentration of investment in one area of the economy has not been repeated. In the 1840s 7000 miles of track were laid in the United States, but by 1860 some 30,000 miles of track carried a volume of traffic which by then equalled that of canal shipping (Fishlow, 1967). Between 1850 and 1871 the U.S. Congress gave away 130 million acres of public lands and State governments another 49 million acres to the railroad companies, a gift from taxpayers to private enterprise estimated by Fite and Reese (1965) to be worth $489 million. After the Civil War the railroad network in the United States increased even more dramatically with some 200,000 miles of track being laid between 1865 and 1914. (Figure 14.1) During this period the U.S. Federal government became a major source of railroad financing since, in addition to the massive land grants described above, the government granted the railroads $65 million in special low-cost credit. In a parallel gesture the Canadian government in 1881 granted *one* railroad company (the Canadian Pacific) 25 million acres of land, $25 million in direct cash subsidy, and another $34,041,082 in already constructed rail lines; ancillary grants at the federal, provincial and municipal level brought the grand total to $106,280,334 and 43,962,546 acres of land (Chodos, 1973).

However, it did not stop there. The rail network provided the setting for ferocious economic competition and the emergence of a new corporate ethic. When questioned about the legality of some of his activities, Cornelius Vanderbilt declared 'What do I care about the law? Hain't I got the power?' Competition among the railway entrepreneurs was intense. Rate wars were common, forcing weaker competitors out of business and giving stronger competitors virtually monopoly powers over regional landscapes. The battles themselves got so brutal that locomotives were sometimes smashed into each other, and sections of track deliberately destroyed. Standard Oil was able to force many competitors to the wall and to achieve virtual monopolies on a regional scale so that price increases could occur without fear of competition. In the process, its owners Rockefeller and Flagler succeeded in forcing the railway companies to give them rebates on their *competitors*' shipments! In addition, Dillard (1967, p. 410) has noted that Standard Oil got '. . . all data relating to shipper, buyer, product, price and terms of payment', so that in turn they received 'rebates on all their own shipments, by rebates on all shipments by their competitors, and in addition a complete spy system on their competitors'.

The opening up of the railway network had some subtle and pernicious implications for landscape change. A member of the California Constitutional Convention of 1878 described the technique: 'They start out their railroad track and survey their line near a thriving village. They go to the most prominent citizens of that village and say, "If you will give so many thousand dollars we will run through here; if you do not we will run by. And in every instance where the subsidy was not granted, this course was taken and the effect was just as they said, "to kill off the little town"' (quoted in Josephson, 1962, pp. 84–85). According to the same report, the railroad companies, 'blackmailed Los Angeles County for $230,000 as a condition of doing that which the law compelled them to do'.

By 1887 the United States federal government was forced to act against high profits, graft, corruption and discriminatory practice by the railway companies. The Interstate Commerce Act of that year established the Interstate Commerce Commission (ICC) supposedly to regulate the railways in the public interest. But reality and federal action are frequently two different things. A few years after the passage of the Act, the then U.S. Attorney General Olney wrote a letter to a railway president, that read in part: 'The Commission . . . is, or can be made of great use to the railroads. It satisfies the popular clamour for a government supervision of railroads, at the same time that supervision is almost nominal. Further, the older such a commission gets to be, the more inclined it will be found to take the business and railroad view of things . . . !' (quoted in McConnell, 1970, p. 197). That prediction has been borne out by the facts; subsequent regulatory agencies, the FCC (Federal Communications Commission), CAB (Civil Aeronautics Board) and others, as well as the ICC, '. . . have in general become promoters and protectors of the industries they have been established to regulate' (McConnell, 1970, p. 199). The history of Canada's Royal Commissions and the Canadian Transport Commission are no exceptions. Thus the economic landscape became the battlefield for a competitive capitalism and private sector–government cooperation.

Landscape Colonization and the Northern Pacific Railroad

There is a common image of 'frontierism' in North American geography; the dynamic frontier is seen as drawing pioneers and settlers, and then an infrastructure, ever westwards. Frontiers, many historians maintain, are simply hinterlands to a metropolitan East, an illusion of a reality whereby the satellitic pioneer Wests were converted into developed colonies of the East. The keystone of eastern backing was a desire for financial stability and steady but controlled growth, for on these the reputation of the eastern capitalists as financiers and the fate of their enterprises finally depended. Transportation links facilitate this colonization, but instead of a straight line progression from East to West, the infrastructural lattice is formed in a series of quantum jumps. Thus, at times, developed metropole and developing satellites were separated by gaps even when connected by the slender linkage of the railroad. Into this void stepped the subsidiaries of transportation companies: the colonization organization. The Northern Pacific Railroad was one of the first in the early 1870s to develop such an organization to infill its railroad land great areas (Peterson, 1929).

The Northern Pacific, chartered in 1864 and begun in 1870, had 229 miles of track running through Minnesota by 1871; its land grant was 12,800 acres/

mile of track through the states, and twice that, once it had reached Lake Superior for its Pacific route. The Northern Pacific, when the competition for settlers was keen in Minnesota, became the first to develop an organization to carry out its colonization plans. This organization, 'to promote . . . Immigration by colonies, so that neighbours in the Fatherland may be neighbours in the New West', included a land department designed to sell the land as quickly as possible, and an emigration department, with a European agency, to 'facilitate and render certain the rapid sale and settlement of its lands, and to promote the early development of the entire belt of Northwestern States and Territories tributory to the company'. Some of those attracted from New England were not as happy as the railroad company had been at what greeted them; they complained '. . . that the climate was like New England in November, that snow was everywhere; that the soil was only a foot deep instead of three . . . [as described in the advertising]; that the timber was good for firewood only; and that the land was well watered as advertised, but so well watered that a fifth of the township was swamp land!' (Peterson, 1929, p. 134). That winter was a severe one for the colonists, and the Northern Pacific actively suppressed adverse reports leaking back to the East. But the illusory landscape of the Northern Pacific's territory could not be shattered, for there seems to have been no abatement of interest amongst easterners to emigrate in 1873 and 1874.

A similar colonization attempt occurred where the Northern Pacific crossed the St. Paul Railway (Figure 14.2) about 40 miles west of the Detroit Lake colony. Here the scheme was in the hands of the Northwestern Land Agency of Duluth, who were to receive a 10% commission on all sales. Similar inducements were offered by the railroad company: reduced fares, reception houses, aid in draining the Red River flats, etc. Glyndon, as this second colony was called, was extensively advertised via a broadsheet, the *Red River Gazette*, which was widely distributed, and through advertising space in newspapers in New York, Boston, Chicago, Cincinnati, Hartford, Toledo, St. Louis and Philadelphia. In addition, small advertisements were placed in 850 country newspapers and two Swedish ones. The Northern Pacific underwrote the cost of a local Glyndon newspaper, since it contained news of settlement along the line. Some $10,000 was spent on advertising, and sales of land were around $25,000 by the summer of 1872. Settlement was slow, however, due to adverse publicity concerning the winter weather and the degree of help available from the railroad company.

A third type of colony along the railway route was Yeovil; this colony was the brainchild of the Rev. George Rodgers of Stalbridge in south-western England, and was to be located just east of Glyndon. In April 1872 this idea was proposed to the Northern Pacific's London agent. Rodgers was to be paid a moderate salary by the Railroad company, and the expenses for a trip to the townsite. Rodgers reported very favourably on conditions as he found them in Minnesota, a fact which was picked up by both U.S. and British newspapers. On his return to England he marshalled together two groups to

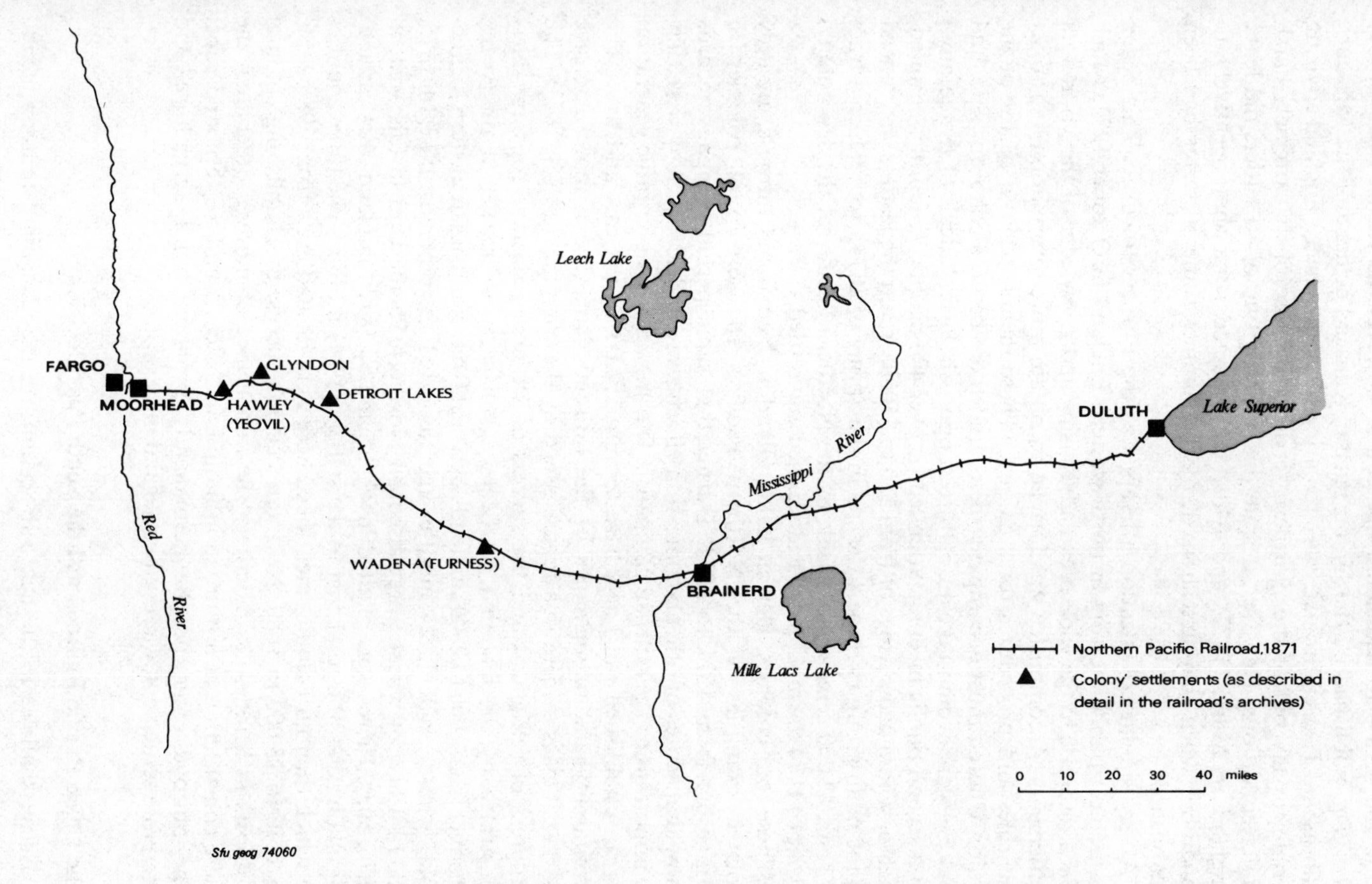

Figure 14.2. Colonization spurred by the St. Paul and Northern Pacific Railways

settle the area. He had some difficulties, however, as *Advice from an Old Yeovilian*, published by the Union Pacific, Kansas Pacific, and the Missouri, Kansas and Texas railroads, was circulating in south-western England advising people to stay out of the Northern territory because of its extremely harsh winter conditions. However, despite this, a first group of 80 settlers did leave England in March 1873, and while they reacted somewhat negatively to conditions when they first arrived, by the summer some 50 farms had been established.

A very similar colonization attempt was the Furness townsite. This began its life at a public meeting in north-western England in October 1872, which was a response to the widespread advertising and knowledge of the Minnesota settlements. A committee was formed to obtain more information and to set out rules and procedures for a colony. (These included that a pastor of the Union Church must accompany them, that there be no alcohol for sale, and that everyone would take an oath to support the moral ideals of the colony.) At a second public meeting, attended by over 400 people, 73 families enrolled, each of whom had savings of from $50 to $5000. Again the leaders went ahead of the main group to inspect the site in the Spring of 1873, from which a tract of some 42,000 acres was set aside by the Northern Pacific for the new colony. By April 1874 the main party had reached and settled at Furness.

These four colonies, Detroit Lake, Glyndon, Yeovil and Furness, are only isolated examples of the Northern Pacific's sponsorship of colonies in Minnesota from 1871 to 1874. During that same period company records show that scores of other colonization efforts were under way. Why were they so important to the railroad company? The answer is simply traffic generation for an established rail line whose termini lay far to the east and west. The profitability and maintenance of the railroad demanded a more extensive market. This is evident in a report issued by the company itself in 1873: 'The progress of settlement and the success of settlers in raising crops are fairly illustrated by the fact that . . . [we] . . . will carry to market of this year's product from 1000 to 2000 car-loads of wheat from countries in western Minnesota, whose residents 24 months ago imported their breadstuffs.' In 1875, 5000 bushels of wheat were raised on Northern Pacific land in this area, in 1876, 50,000 acres were planted in wheat, and by 1877 total land sales came to a million acres at a total cost to the settlers of $4½ million! The three counties in which the four colonies mentioned here were located grew from 398 people in 1870 to 13,000 in 1880. Northern Pacific's success in establishing colonial outposts in Minnesota underlines the importance of motives other than the mere construction of a railroad in moulding the early mid-western landscape. The railroad could not be permanently profitable until agriculture, and therefore settlement, was established and prosperous.

The Union Pacific Railroad and the South Pass Decision

As we shall see in the next sub-section, on the Canadian transcontinental rail link, the U.S. equivalent was a political necessity, but because of its length, and the lack of settled areas through which it could pass, it was considered by entrepreneurs as showing little promise of profitability. The State therefore had to intervene if it were to get its intercontinental link. Between 1862 and 1864 the U.S. Congress passed a series of Acts incorporating a Union Pacific Railroad Company, lent the enterprise public credit, granted public lands and generally made the deal attractive to private entrepreneurs. Loan money was thus generated and the project started, but in its eagerness to see the railroad under way Congress gave little attention to its operation, other than setting minimum engineering standards. The Acts provided for the termini but did not lay out service areas or specify a particular route. Capitalist government was at this point more concerned with reaching the West coast to bring the hinterland under more direct control than it was with dampening down speculation, chicanery and graft.

Seizing on weaknesses in the legislation was Thomas C. Durant, Vice-President of the new Union Railroad Company, who remained in effective control of route location until after the completion of the line. Because this same Durant was also active in companies organized to construct the railroad line itself, the link between operation and construction was forged.

> 'As officers of the Union Pacific, Durant and his group had control of funds of which they had supplied only a small fraction; and while the company had access to money for its building, it showed little likelihood of operating at a profit. On the other hand, there was a clear chance for great profits in constructing the railroad. As officials of the Union Pacific Railroad Company, and with little financial risk, Durant and his group were able to make decisions that favored themselves as the railroad's contractors rather than furthering the future utility of the line. This they did.' (Grey, 1970, p. 52)

How did the link between operator and constructor and the maximization of the latter's profits work? Simply because a route requiring little grading left a maximum of profit from the fixed loans available for each mile. But there were other profitable inducements: competition and public demand for a Central Pacific Railroad generated enormous pressure for rapid completion of the line. In an effort to win as much as possible of the limited subsidized mileage set out by Congress, and thus to increase the chance for greater total profits, the company chose its route to facilitate rapid and direct passage westwards. At some points speed overrode cheapness of construction, but overall a route with minimum grading could satisfy both requirements as long as engineering standards were met. Grey (1970) notes that in company reports and private letters it is made quite clear that little or no attempt was made to relate the line to the economic potential of the area it served. A cheaply constructed line was to have pre-eminence over existing commercial centres in the region—hence centres like Denver and Salt Lake City were by-passed, whilst either smaller towns were linked up or new ones were generated (Cheyenne, Laramie, Ogden). In the process the cultural landscape of Wyoming was remoulded from the older pioneering landscape. Even the land grants

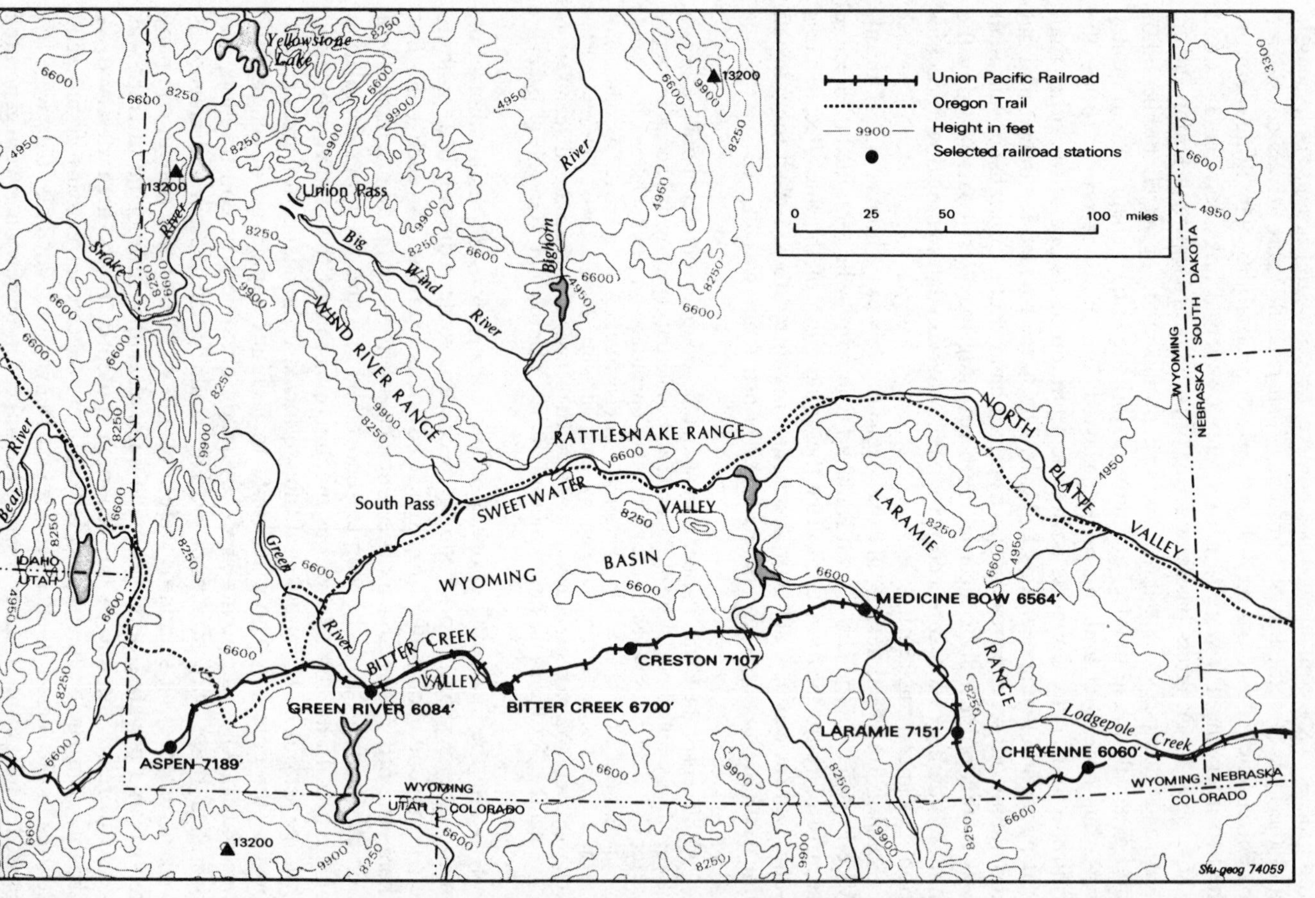

Figure 14.3 The South Pass Route of the Union Pacific Railway

were ignored as traffic generators (in contrast to the Northern Pacific case mentioned above), the land simply being accepted as a bonus which might supply some timber and ballast for construction, but which was of little use in meeting construction costs or raising construction profits. Again in Grey's words (1970, p. 53) '. . . the owner-contractors of the Union Pacific subordinated economic and social geography to physical geography, for the physical nature of the country determined the amount of necessary grading and possible construction profits'.

These motivations influenced the choice of a route through the eastern Rockies, for it was widely believed at the time that the railroad would run up the North Platte River and over the South Pass (Figure 14.3). But the South Pass was rejected in favour of the 'Gangplank' and Evans Pass which enjoyed very easy gradients. The South Pass, although recommended by earlier surveys, including that of the U.S. Army, was ignored because of its clash with the aims of the company's owners. For though it had a lower summit, it was longer, hence increasing time and construction costs. Because the Evans Pass route maximized company profits any other attractive route was excluded from serious consideration. This interpretation, backed by company records and letters, is somewhat at variance with an earlier interpretation (by Vance, 1970) of the avoidance of the South Pass, which saw the Evans Pass choice as one which would enhance the commercial operation of the railroad; i.e. the South Pass was uneconomic to the operator, not, as has been shown here, uneconomic to the constructor!

To a transportation geographer such a sequence of decision-making within the capitalist economic milieu is extremely interesting. Not only does the cultural landscape of southern Wyoming bear the imprint of the Union Pacific's presence, but it shows that route choice is frequently based not on existing or potential population concentrations or on the potential traffic generation but on profitability, in this case to the construction companies linked via Durant and others to the very railroad company itself. The non-choice of the South Pass made Cheyenne, Laramie and Ogden important centres, but it also affected the rate of development of Wyoming, for the North Platte and South Pass routes were in fact better watered and hence agriculturally a better choice. But such a choice did not maximize profits for Durant and his co-directors.

Railways and the Canadian Landscape

Canada was discovered, explored and developed as part of the French, and later the British, mercantile system. It grew to independence and 'nationhood' in a relatively brief historical time period, during which goods, people and capital were moved out from a metropolitan core (Toronto–Montreal) toward a developing hinterland. However, until 1867 the name Canada belonged only to what are now the Provinces of Quebec and Ontario; in 1867 the Dominion of Canada was created by confederating three provinces: Nova Scotia, New Brunswick and old Canada, the latter becoming divided from that time into

Quebec and Ontario. Newfoundland, Prince Edward Island, Manitoba and British Columbia remained outside the Confederation at its initial establishment.

The powerful anticipations of the Canadian establishment in Ontario and Quebec are reflected in a budget speech of the 1870s:

> 'Coming further east still, let us but have our Canal system completed, our connection with the Pacific Railway at the head of Lake Superior, the North West becoming rapidly settled, the exports of the settlers passing through our canals and the whole system of the Ontario railways complete, and the result will be that the trade of the city of Toronto which has doubled in five years will be quadrupled, and the case will be the same with Hamilton, London and other cities in ... [Western Ontario] ...' (quoted in Mackintosh, 1939, p. 16.)

Political investment in transportation systems to link metropolitan core and expanding hinterland was recognized as early as 1839 in Lord Durham's report on British North America:

> 'The defence of an important fortress, or the maintenance of a sufficient army or navy in exposed spots, is not more a matter of common concern to the European, than is the construction of the great communications to the American settler; and the State, very naturally, takes on itself the making of the works, which are a matter of concern to all alike.' (Durham, 1839)

But political support of the railway collusion between political and financial spheres was particularly prevalent in transportation several decades before the transcontinental links were funded. To keep the St. Lawrence trade route competitive with routes developing to the south, new transportation links were studied. The new railway technologies soon seemed superior to the canal, so by the 1830s and '40s short railway links were being completed, connecting trading centres with their immediate forest/farming hinterlands. The more important of these links had the additional purpose of tapping the traffic of American railways. The Great Western, for example, was built across the Ontario peninsula between Buffalo and Detroit to provide an efficient short route for American mid-western commerce across Canadian territory. (The reverse process also occurred. Thus Canadian goods were allowed to flow in bond across American territory: the St. Lawrence–Atlantic Railway was built from Montreal to Portland, Maine, to provide the St. Lawrence metropolitan core with a winter ice-free outlet.)

This interest in railway construction interlinked politics and financial capital. In a developing economy as yet hardly distinguished by large private accumulations of capital such collusion was essential. Almost to a person, the promoters went into public life: charters had to be bargained for, routes selected, money and credit raised, elections financed and cabinet ministers bribed. The men who received government-guaranteed loans, and eventually direct cash/land grants, local bond subscriptions and heavy investment by

British and local financial groups, were either members of the government or extremely close to it. As one Hamilton entrepreneur working on behalf of the Great Western commented: '. . . if it were known that members of our legislature would oppose an Act of Parliament *because* they could not effect a pecuniary arrangement with the company, it would shake the confidence of capitalists and deter them from investing their means in the great enterprises of this country.' Sir Alan McNab, with his railway politics, from that same Hamilton base, was president of the Great Western! 'Few of the leading or minor politicians of the day were not mixed up with private railways or contracting companies, as well as with their official duties in directing railway construction as a public policy.' (Glazebrook, (1964), Vol. 1, p. 169.)

Myers (1914/1972) has catalogued the chicanery of the period, as Members of Parliament granted charters to themselves and each other, and met their special interests as landowners, lawyers, contractors and so on by judicious 'log rolling'. For example, the St. Lawrence and Atlantic Railway intertwined A. T. Galt and Peter McGill as representatives in legislatures and of the British American Land Company and the Bank of Montreal. The New Brunswick–Nova Scotia Railway, chartered in 1847 with a capital of $2 million, reads as though it were largely a roster of Parliament itself. 'Heading the procession of incorporators was Sir Alan N. McNab; there were five members of the Legislative Council, including the active John Ross . . . a number of trading and sundry other men of capital—Sir George Simpson, Governor of the Hudson's Bay Company and Paul Fraser of the same Company . . . some seigneurs and various other individuals of note either in politics or trade.' (Myers, 1914/1972, pp. 157–8) The Grand Trunk Railway brought together Galt and McGill again, as well as George Etienne Cartier (solicitor general as well as company solicitor!) and various legislative members who were cofounders of the Canada Life Assurance Company, other railway companies, and sundry banks and manufacturers.

Politics was in fact a business: the Canadian parliament was crowded with men who were there to initiate, expand or conserve class or personal interests. Of the 206 Members of the 1876 Parliament, 86 were merchants, 15 were lawyers and 12 were gentlemen of leisure; the rest included manufacturers, insurance company presidents, shipbuilding and lumber capitalists, contractors, and a few journalists, physicians and farmers (Myers, 1914/1972). Sir Edmund Hornby (1928, p. 90), who came to Canada as a lobbyist for the Grand Trunk, has fortunately left us some of his impressions:

> '. . . my work was almost exclusively "lobbying" to get a Grand Trunk Bill through the House . . . The Canadian Ministers were willing enough but weak—the majority of doubtful quality, and although up to the last moment I felt there was a chance of getting the Bill through, I was always doubtful, since it was clear that some twenty-five members, contractors, etc., were simply waiting to be squared away either by promise of contracts or money, and as I had no authority to bribe they simply abstained from voting and the Bill was thrown out. Twenty-five thousand pounds would have bought the lot . . . I confess I was annoyed at my ill-success and had half a mind to split upon some dozen members who had been a little indiscreet in their proposals

> to me. As usual it was a Psalm-Singing Protestant Dissenter who, holding seven or eight votes in the palm of his hand, volunteered to do the greasing process for a consideration. Upon my word I do not think there was much to be said in favour of the Canadians over the Turks when contracts, places, free tickets on railways, or even cash was in question.'

The Grand Trunk was finally incorporated in 1853, with a government guarantee of $3000/mile, and for certain branches a land grant of one million acres. Under the terms of a Guarantee Act passed in 1849 under the steermanship of the finance minister, Francis Hincks (later a director of the Grand Trunk and a participant in the negotiations over the Canadian Pacific Railway) the government had institutionalized capital assistance to the railways; by 1867 the provincial government had incurred a debt of $25·6 million on behalf of the Grand Trunk and a lesser $4·9 million for the Great Western. Laissez-faire capitalism was obviously not the way to develop the Canadian economic landscape.

The Canadian Pacific Railway

All this simply laid the foundations for the most spectacular example of landscape transformation by public–private capital: the Canadian Pacific Railway (CPR). This mammoth undertaking, which has been eulogized as the realization of a 'National Dream' by Canadians (Berton, 1970), was in part an attempt to maintain Canadian political and economic sovereignty over the wheat hinterland of the prairies and the forest and mineral resources of the Pacific Coast (Figure 14.4). For this purpose, as we shall see, the State acted as an agent of private capital to maintain the hegemony of Laurentian and British capital over that of the American. In retrospect it is somewhat ironic that such an enterprise should pave the way for the eventual American domination of the Canadian economic landscape. (For an account of American capital control and the demise of the British and subsequently the Canadian capitalist class, see Levitt, 1970; Lumsden, 1970; Teeple, 1972 and Laxer, 1973.)

In 1871 George Etienne Cartier moved in the Federal House, seconded by A. T. Galt, that a resolution be adopted to build a Pacific Railway by private enterprise but with liberal public aid in grants of land and money (*Hansard*, 1871, p. 1028). The push for a charter for this transcontinental link came from David MacPherson, a Senator, and director of Gzowski and Company, one of the Grand Trunk's principal contractors. No company was named in the Act that followed but, as one official historian of the CPR noted, the upcoming link was 'a new property which cried out to have money spent on it', and that money was spelt out, $30 million and 50 million acres of land. The editor of a contemporary newspaper, *The Globe*, saw the links: he complained that the government could come to terms now with 'any company who may be disposed in return to give their political support, or may become subservient followers, ready enough to grab the land and the money; but in the

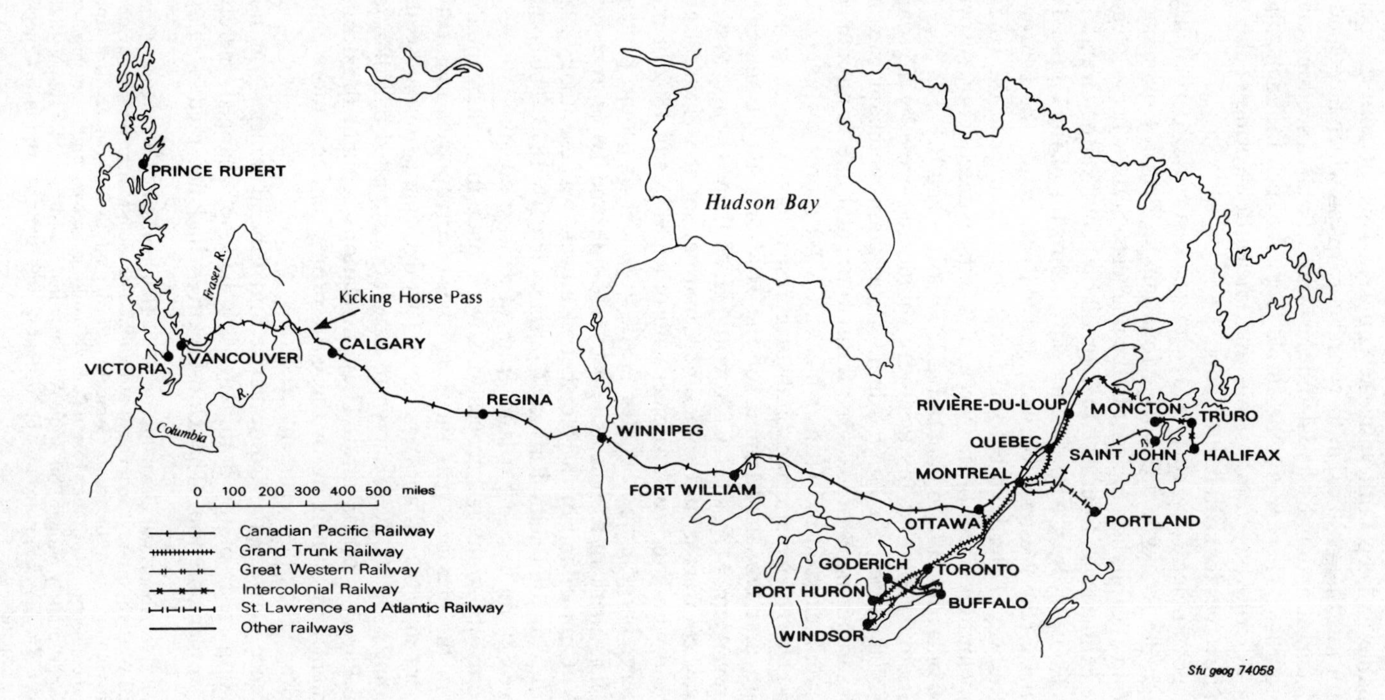

Figure 14.4 The Canadian Pacific Railway

end quite unable to keep faith with the country. This, of course, opens at once a door for just that political jobbery and corruption which have made the Intercolonial a great national scandal and the Grand Trunk a seething mass of political immorality'. There was to be ample scope for such thoughts, and for what he later described as 'a monstrous and unclean off-spring of the secret and immoral intrigues that call it into being'.

While the government pushed ahead with track surveys, particularly for passes through the Rocky and Selkirk Mountains in British Columbia, they awaited a group of capitalists to assist in the financing of the scheme. Negotiating for Prime Minister Sir John A. MacDonald on behalf of the Grand Trunk Railway was C. J. Brydges, who was not only a director of the Grand Trunk but also of the Great Western Railway, and a Receiver for the Detroit and Milwaukee Railway. Later he was to be Lord Commissioner of the Hudson's Bay Company and President of the Mechanics Bank (subsequently suspended with $168,000 bills in circulation but only $2500 cash in hand).

But the Grand Trunk and Brydges were shouldered aside by a group led by a Montreal financier, Sir Hugh Allan. Allan, grain merchant, steamship owner and insurance investor, saw the transcontinental line as an immediate construction, unlike the Grand Trunk which saw it as a more slowly staged one. Along with a group of American capitalists representing the Northern Pacific, Donald Smith and Donald McInnes of the Hudson's Bay Company, his brother Andrew Allan, M.P.'s J. J. C. Abbott,* H. Nathan, Thomas McGreevy (also a railway contractor and sometime Chairman of the House Railway Committee, and who later was a subject with other M.P.'s in 1892 of proven election bribery charges), and Senator A. B. Foster. This group, principally by shedding its American associates and by adding some Toronto capitalists such as F. W. Cumberland of the Northern Railway, was eventually granted a charter as the Canada Pacific Railway Company in early 1873. The company was to receive $30 million and a land grant of at least 50 million acres. But before the company could act, accusations were made in Parliament that in the election in the previous year there had been 'corrupt relations between the government and the Allan "ring"!' A Royal Commission subsequently found that some $350,000 were shown to have been supplied by Allan to the Conservative Party's election fund. These revelations brought the demise of the Government and a shelving of the CPR project under a more cautious railway-building Liberal government in 1873.

The difference between the Conservative and Liberal policies was minor; the pecuniary interests of the dominant class had not changed, merely the pace of its implementation. The Conservatives had accepted a relatively

*Just as an example, Abbott was also a director of the Hamilton and S. W. Railway Company, the Montreal and Western Railway, an incorporator of the Canadian Marine Insurance Company, was on the Boards of the Citizens Insurance Company, the Merchant's Bank, the Bank of Montreal, and had financial connections with the Molson family's banking, brewing and steamship interests. He succeeded, appropriately enough, Sir John A. MacDonald, as Prime Minister in 1871. The British newspaper, *The Speaker*, noted of Abbott, that he was 'the man who in 1872 negotiated the great bribery scheme'.

short time limit for the whole line through Canadian territory; the Liberals accepted the rail link as the ultimate goal but believed its pace of construction should be much slower. In both cases the merchant capitalists contrived to enjoy wealth and power. In the Liberal interregnum railway building continued, although largely of the nature of feeder links or portages to and between waterways and the United States. The dominant financing remained public and the relationship between government and private entrepreneurs persisted. The rather more niggardly Liberals did, however, lower the 'cost' on railbuilding to $10,000 and 20,000 acres of land per mile!

A start was made on the most easterly part of the proposed intercontinental link or Pacific Railway as it now tended to be called; the plan was for a rail link from Fort William to Winnipeg. From the latter a line was to be built southwards to connect to the American railways, including the equally ill-famed St. Paul and Pacific Railway which was to be a vital crux in the second CPR design. (The St. Paul and Pacific was caught up in scandal and swindle for many years, eventually being turned over to Donald A. Smith of the Hudson's Bay Company and George Stephen, President of the Bank of Montreal, both leaders of the new CPR. The St. Paul branchline was very important to the CPR both as an American link and as an ample barrier against the invasion of rivals (see Myers, 1914/1972, pp. 244–263).)

By 1878, however, only a little more than 100 miles of track had been laid from the western shores of Superior, much to the chagrin of the British Columbian mercantilists waiting to profit from their new transportation link. The Liberal interregnum saw in fact, marked, lengthy and sometimes bitter negotiations with B.C. until its demise in 1878 and the return to power of Sir John A. MacDonald. Despite the earlier scandal, Sir John's willingness to grant a high tariff to Canadian manufacturers earned him a large campaign fund, newspaper influence, and the election itself (Cartwright, 1912, pp. 189–91). What MacDonald did was simply to attach to national government the interests of the most ambitious entrepreneurial business groups. A Liberal historian, Frank Underhill (1960) noted, '. . . the interests of the . . . speculative . . . groups, who aimed to make money out of the new national community or to install themselves in strategic positions of power within it . . . required the lavish expenditures of taxpayers' money in public capital investment if their ambitions were to be realised. In return their support was needed to keep the Conservative government in office'. This re-established the triple alliance: Confederation Government, Conservative Party and mercantile capital.

But this was only part of the alliance strategy; Confederation was based on a perceived need for agricultural expansion to the West. The rising demand for foodstuffs for the growing industrial British cities was expected sooner or later to provide a market for Canadian prairie wheat. The opening and settlement of the West needed a transcontinental railway which would have to span 1000 miles of uninhabitable muskeg between the metropolitan core and the new Red River integument. The railway would eventually pay for itself by hauling wheat to Montreal for transhipment to Britain, and by hauling the manufactures of the core to the new western settlement. As a fundamental step in that policy the socio-economic system was partially closed against the importation of American goods by a high tariff imposed in 1878 and the stage was set for a new contract to be let for an intercontinental railway.

The new CPR company headed by George Stephen (President of the Bank of Montreal), and including in the background Donald Smith of the Hudson's Bay Company, was granted 'extraordinary privileges, and powers, immunities and rights with which to transform the western landscapes. The transformation was to be as risk free as possible.' The 1880 contract granted the company: (i) a cash grant of $25 million to aid construction; (ii) a land grant of 25 million acres; (iii) a grant of an additional 1·6 million acres for road beds, stations, workshops, yards, etc. for the main and branch lines; (iv) the admission, free of customs duties, of any materials to be used in constructing the railroad; (v) land granted under (iii) to be free of any municipal, provincial or federal taxes, and that granted under (ii) to be free of similar taxes for a finite 20-year period; (vi) a guarantee of a 20-year railway monopoly; (vii) the transference to the company of those parts of the line already constructed by the government: the 700 miles so constructed had in fact already cost the various governments $37,785,000 (Myers, 1914/1972, p. 267; Chodos, 1973, p. 156). Myers in fact estimates that with the extinguishing of Indian titles, etc., the capitalization was about $30 million more than noted here and that the direct land grant even 'before the great inrush of settlers' was worth at least $79,500,000. In addition '. . . it was distinctly provided that in making the land grant only land fairly suitable for settlement was to be allotted to the contractors. In other words, they were to have the pick of the finest lands, for the contract further specified that they were to have large powers of selections of the land. They were also to control the land sales'. (Myers, 1914/1972) This Canadian largesse was even more generous than the landscape powers granted American railways: the land grants in the United States nearly always consisted of arbitrary attenuating sections, so that the railway company had to take its chance as to whether it was granted good agricultural land, forest resources, mineral resources or barren land. The CPR received attenuating sections but it had powers of selection over those sections. Not counting the tariff and tax concessions, the Canadian government underwrote the private CPR company by about $140 million in cash and land equivalents.

While the American frontier landscape was initially settled by pioneer settlers, behind whom came the transport links and state–private investment, in Western Canada the railroad essentially determined settlement policies and landscape change. As Naylor (1972, p. 17) has noted '. . . the state structure—that is to say, the CPR and its military arm, the Canadian Pacific Mounted Police—went first, expropriating the Indians and Metis and allocating new lands to the settlers who followed. The horizons and to a remarkable extent the personnel of the federal government and the CPR were often inseparable'. At the time of the Riel Rebellion (for self-determination of the Metis in Saskatchewan, in 1885), the then CPR President, William Van Horne, 'offered to transport militia units West, provided he was given undivided control of the whole movement'. After all, the CPR had to have physical control of its 25 million acres, the Metis held some of the best, and the government could not have political usurpers on its hands. 'The rebellion was put down, and with the railway having thus proved its value to the nation, the Government guaranteed another CPR bond issue [$5 million plus permission to sell an additional $15 million worth of bonds on the London, England market] so that construction could continue. It was the finest moment of the government–CPR partnership in action . . . the partnership's mastery of the North West stood unchallenged.' (Chodos, 1973, p. 24.) As one critic in the House of Commons of the time, G. E. Casey, noted:

> '. . . the gentlemen who sit opposite are merely the political department of the Canadian Pacific Railway. It is really the CPR which governs. This is a conclusive proof that these honorable gentlemen are mere trustees for that railway of the political power of the country, as other gentlemen may be trustees of their bonds or land grants. It is a waste of time to argue with them as to whether they should obey the orders of the masters or not. They must carry out the behests of the Company' (quoted in Myers, 1914/1972 edn., p. 300).

Thus the CPR can be seen as a mercantile, state-chartered monopoly: George Stephen representing financial interests through the Bank of Montreal, Donald Smith and his links through the Hudson's Bay Company representing landed interests, and William Van Horne, with his wide experience in American railway construction, representing productive rather than political influences. As the CPR line followed the old fur trade routes the lands round the fur trading posts still in the hands of the Hudson's Bay Company (alias Donald Smith) became potential urban centres with increased real estate values! The central Canadian capitalists who gained from the protection of Canada's manufactures were the same set of people who benefited from government assistance to the railways! Canadian taxes flowed freely into railway enterprises and from there into the financial sector and into manufacturing industries, including those supplying steel and rolling stock to the railway, fertilizer and farm equipment to the Western farmer. Such was the design; with a few setbacks the design succeeded beyond expectations, for today the CPR is more than just a railway company—it owns large tracts of enormously valuable urban land (including prime waterfront land in Vancouver and Toronto), it plays a major role in telecommunications, it is a shareholder in a variety of mining and real estate interests and of course it owns hotels, shipping lines and airlines.

Conclusion

There is only one world; although one may or may not perceive it accurately. Our task as geographers is to understand that world; even though Karl Marx's prediction that 'time would annihilate space' is making that task more complex rather than easier. This chapter has taken part of that world (the Canadian landscape) and has attempted to understand its development by showing first the close interconnections between the interests of various governments and the interests of various railway companies. Secondly, this chapter has shown that railway technologies in Canada were built largely to service the merchant-capitalists who were frantically plundering natural resources and agricultural products from the hinterland to the metropolitan Toronto–Montreal core. Hence decision-making surrounding railway construction was a direct response to the exigencies of merchant capital, and the fact that one man could be grain merchant, lumber merchant, railway promoter, banker and politician should come as no surprise. In the United States the situation was a little different: (and later in Canada, the parallels become closer) industrial capital (capital engaged directly in the production of surplus value) made use of, and some cases expanded upon, the existing railway facilities. Traditional models in transportation and historical geography frequently miss this perspective yet it offers an integral insight into the values and motivations of the people who were so influential in moulding the North American landscape.

References

Adams, W. and Gray, H. M., 1955. *Monopoly in America: The Government as Promoter*, New York: Macmillan.

Baker, A. R. H. (Ed.), 1972. *Progress in Historical Geography*, New York: Wiley-Interscience.

Baran, P. A. and Sweezy, P. M., 1966. *Monopoly Capital*, New York: Monthly Review Press.

Barzini, L., 1972. Romance and the risorgimento, *New York Review of Books*, October 5, p. 16.

Berton, P., 1970. *The National Dream*, Toronto: McClelland and Stewart.

Borgstrom, E. E., 1973. Letter to the editor. *Transportation Geography Newsletter*, No. 3, 20–21.

Bowden, M. J., 1969. The perception of the western interior of the United States, 1800–1870: a problem in historical geography. *Proceedings, Association of American Geographers*, **1,** 16–21.

Cartwright, R., 1912. *Reminiscences*, publisher unknown.

Chodos, R., 1973. *The CPR: a Century of Corporate Welfare*, Toronto: James, Lewis and Samuel.

Darr, R. K., 1970. The quintuple alliance: an early pooling agreement. *Kansas Quarterly*, **2,** 24–29.

Dillard, D., 1967. *Economic Development of the North American Community*, Englewood Cliffs, N. J.: Prentice-Hall.

Durham, Lord, 1839. *Report on the Affairs of British North America* (Ed. Craig, G. M.), Toronto: McClelland and Stewart.

Easterbrook, W. T. and Warkins, M. H., 1967. *Approaches to Canadian Economic History*, Toronto: McClelland and Stewart.

Eliot Hurst, M. E., 1973. Transportation and the societal framework, *Economic Geography*, **49**, 163–180.

Eliot Hurst, M. E., 1974. *Transportation Geography: Comments and Readings*, New York: McGraw Hill.

Fishlow, A., 1967. *Railroads and the Transformation of the Ante-Bellum Economy*, Cambridge: Harvard University Press.

Fite, G. C. and Reese, J. E., 1965. *An Economic History of the United States*, Boston: Houghton-Mifflin.

Fogel, R. W., 1962. A quantitative approach to the study of railroads in American economic growth. *Journal of Economic History*, **22,** 163–197.

Glazebrook, G. P. de T., 1964. *A History of Transportation in Canada*, Toronto: McClelland and Stewart (2 volumes).

Grey, A., 1970. The Union Pacific Railroad and South Pass. *Kansas Quarterly*, **2,** 46–57.

Harley, J. B., 1973. Change in historical geography: a qualitative impression of quantitative methods. *Area*, 69–74.

Hornby, E., 1928. *Autobiography*, Boston: Houghton-Mifflin.

Josephson, M., 1962. *The Robber Barons*, New York: Harcourt Brace Javanovich/Harvest Books.

Kerr, K. A., 1968. *American Railroad Politics*, 1914–1920: *Rates, Wages and Efficiency*, Pittsburgh: University of Pittsburgh Press.

Kerr, K. A., 1970. A new view of government regulations: the case of the railroads. *Kansas Quarterly*, **2,** 30–38.

Kolko, G., 1965. *Railroads and Regulation*, 1877–1916, Princeton: Princeton University Press.

Koroscil, P. M., 1971a. Historical geography: a resurrection. *Journal of Geography*, 415–420.

Koroscil, P. M., 1971b. The behavioural environment approach. *Area*, **3,** No. 2, 96–99.

Koroscil, P. M., 1973. Behavioural analysis in historical geography. Paper delivered at the Institute of British Geographers meeting, Birmingham, England.

Laxer, J. M., 1973. *The Political Economy of Dependency*, Toronto: McClelland and Stewart.

Levitt, K., 1970. *Silent Surrender: The Multinational Corporation in Canada*, Toronto: Macmillan.

Lumsden, I., 1970. *Close the 49th Parallel*, Toronto: University of Toronto Press.

Mackinstosh, W. A., 1939. *The Economic Background of Dominion-Provincial Relations*, Ottawa. (Reprinted Toronto, McClelland and Stewart, 1964.)

McConnell, S., 1970. Self-Regulation, the politics of business. In *Economics: Mainstream Readings and Radical Critiques* (Ed. Mermelstein, D.) New York: Random House.

Morton, A. S. and Martin, C., 1938. *History of Prairie Settlement and 'Dominion Lands' Policy*, Toronto: Macmillan.

Myers, G., 1914/1972. *A History of Canadian Wealth*, Toronto: James, Lewis and Samuel.

NAS-NRC, Ad Hoc Committee, 1965. *The Science of Geography*, Washington, D. C.: NAS/NRC Publ. No. 1277.

Naylor, R. T., 1972. The rise and fall of the third commercial empire of the St. Lawrence. In *Capitalism and the National Question in Canada* (Ed. Teeple, G.), Toronto: University of Toronto Press, pp. 1–41.

Peterson, H. F., 1929. Some colonization projects of the Northern Pacific Railroad. *Minnesota History*, **10,** 127–144.

Ransom, R. L., 1964. Canals and development: a discussion of issues. *Papers and Procs., American Economic Association*, **54,** 365–376.

Smith, V. C., 1970. The Burlington and Kansas City: a chapter in a railway strategy. *Kansas Quarterly*, **2,** 78–87.

Teeple, G., 1972. *Capitalism and The National Question in Canada*, Toronto: University of Toronto Press.

Thoman, R., 1965. Some comments on the Science of Geography, *Professional Geographer*, **17,** 8–10.
Underhill, F. H., 1960. *In search of Canadian Liberalism*, Toronto: Macmillan.
Vance, J. E., 1970. *The Merchant's World: the Geography of Wholesaling*, Englewood Cliffs, N. J.: Prentice-Hall.

Section IV. Land Use: Changing Perceptions and Policies

15

Land Use Management in the United States: The Problem of Mixing Uses and Protecting Environmental Quality

TIMOTHY O'RIORDAN

The following three chapters by Friday and Allee, Stankey, and O'Riordan and Davis illustrate in part the problems facing the use of land in modern America. American agriculture is currently passing through a most interesting phase. For a long time it was plagued by an inefficient combination of price supports which encouraged overproduction and acreage restrictions and/or output quotas which left highly valuable agricultural land lying idle. Consequently, while agricultural incomes appeared to rise, these were more than offset by the rapid escalation in the price of land, increased raw material costs and very expensive capital investment for the machinery and storage facilities to ensure the necessary high productivity. Even at a time of world food shortages, American agriculture has much idle capacity. The current picture is influenced by the decision by President Nixon in 1973 to reduce substantially agricultural price supports so that the farmer can more readily create his own market opportunities. The trend is towards larger farms run more as businesses than family enterprises while the small-time farmer is gradually disappearing (Clawson, 1970; Johnston and Quance, 1973).

Therefore, the problem facing American agriculture is not so much a lack of capacity to produce the necessary food as that of ensuring that production costs created by factors beyond the farmer's control are stabilized and that the environmental damage created by overproduction is controlled (Clawson and Held, 1965; Morgan, 1966). Allee and Friday take a closer look at this latter aspect, noting that agriculture is a substantial contributor of environmental pollution in the United States, producing quite a significant impact on the quality of the rural landscape. The problem is that as costs of production rise, the incentives to absorb the external effects of agricultural practices correspondingly weaken.

A basic issue surrounding land management policies in the U.S. is that of catering to an increasing variety of demands made upon the land resource. Stankey deals with this question quite explicitly with relation to current forest management controversies. American forestry management is facing an era of conflicting goals and shifting public demands. For over half a century, following its inception in 1905, the U.S. Forest Service was more of a custodian than a supplier of timber. Nearly 95% of U.S. timber needs were met from private forest land. Since 1940, however, timber demands have outstripped the capacity of commercial operations to meet national needs: the demand for wood products has increased 70%, lumber 49%, pulp products 235% and veneer and plywood 475% (U.S. Forest Service, 1972). Consequently the big U.S. timber firms have turned to the nation's National Forests to meet the supply shortages and Congress has supported this trend by appropriating large budgets to encourage the production of timber from such lands at the expense of other legitimate uses of the National Forests, including recreation, wildlife management and $13 million for watershed protection (Hill, 1971, p. 1).

Combined with the fact that the State of Washington alone is currently exporting 1·7 billion board feet of timber to Japan (total U.S. output is around 40 billion board feet annually) a major public controversy has erupted over what precisely should be the objectives of the U.S. Forest Service. The centre of this controversy is the issue of clearcutting, i.e. the removal of all standing timber in large patches, which now takes place over about half a million acres of the 1·5 million acres of National Forests that are harvested. But, in terms of volume, clearcutting accounts for about 60% annually of the wood removed from the National Forests (U.S. Library of Congress, 1972). It is becoming increasingly questionable whether such 'dominant use' practices with their attendant environmental side-effects are permissible, in view of public demands for an integratively managed forest resource. Two interesting publications document this controversy more thoroughly and both present extensive recommendations for reform in the Forest Service. These books are *The Last Stand*, an analysis by Ralph Nader's Center for the Study of Responsive Law (Barney, 1972) and an earlier report by the Stanford Environmental Law Society (Pendergraft, 1972).

The U.S. Forest Service has moved to confront these criticisms. As Stankey documents, it has shifted the centre of gravity of its professional staff away from foresters towards multiple professional teams including social scientists, landscape architects, fish and wildlife biologists, and other 'environmental' professionals. It has also made a firm commitment to incorporate public opinion at early stages of management decision-making and is currently reassessing its very basic premises with respect to such practices as multiple use, sustained yield, clearcutting and even-age management. Good discussions on the multiple-use controversy can be found in Martin, 1969; McArdle,

1970; Stirling, 1970; Whaley, 1970; Alston, 1972. The Public Land Law Review Commission (1970) grappled with this question, but much subsequent criticism has surrounded its controversial 'dominant use' recommendation (Hagenstein, 1972).

With growing demands for a variety of recreational uses of the National Forests (as documented by O'Riordan and Davis) the American forest today is regarded as much as a heritage as a productive timber resource; a far cry from the highly utilitarian notions propounded by early U.S. Foresters such as Gifford Pinchot.

The revolution in life-styles that has accompanied the emergence of post-industrial American society has resulted in enormous demands for the use of both public and private lands and waters for recreational use. As O'Riordan and Davis discuss, outdoor recreation is more than a pastime in America, it is big business and a fundamental way of life. The growing prosperity of post-war society has brought with it disposable time and disposable income in great quantities, with the result that Americans can now get to the open spaces for which they have always yearned. The effects of this revolution are quite dramatic in terms of the management of public and private open space. The growth of recreational use has far outstripped the supply of available land and water so managers must now face the difficult task of regulating use without recourse to the use of pricing—a traditional economic method for controlling demand. Because most public recreational space is available at either zero or nominal charge, the only mechanism available to managers to limit numbers is to restrict personal freedoms either by rationing use, by zoning different activities in space and/or time or by discouraging access. In addition, the right of the private individual to enjoy the free use of his own property is increasingly challenged as the growth of private recreational property threatens to curtail the availability of public open space.

Americans have always cherished the illusion of freedom of action and minimum intervention by the state, but today they are being challenged by a reality which is steadily reducing those freedoms in order to ensure that the public interest is safeguarded. Therefore once again in the conflicts that must inevitably develop, we see how the theme of illusion and reality links the forces that shape the development of the American landscape.

References

Alston, R. M., 1972. Forest—Goals and Decision making in the Forest Service. Ogden, Utah: U.S.D.A. Forest Service Research Paper, INT-128.

Barney, D. R., 1972. *The Last Stand*, Washington, D.C.: Centre for the Study of Responsive Law.

Clawson, M., 1970. A new policy direction for American agriculture. *Journal of Soil and Water Conservation*, **25,** 3–7.

Clawson, M. and Held, R. B., 1965. *Soil Conservation in Perspective*. Baltimore: Johns Hopkins University Press.

Hagenstein, P. R., 1972. 'One Third of a Nation's Land'—evaluation of a policy recommendation. *Natural Resources Journal*, **12,** 56–75.

Hill, G., 1971. National Forests: timbermen vs conservationists. *New York Times*, November 15, pp. 1 and 48.

Johnston, G. L. and Quance, C. L. (Eds.), 1973. *The Overproduction Trap in U.S. Agriculture*, Baltimore: Johns Hopkins University Press.

Martin, P., 1969. Conflict resolution through the multiple use concept in Forest Service decision making. *Natural Resources Journal*, **9,** 228–236.

McArdle, R. E. (Ed.), 1970. An introduction—why we needed the Multiple Use Bill. *American Forests*, **76,** 10–59.

Morgan, R. J., 1966. *Governing Soil Conservation: Thirty Years of the New Decentralisation*, Baltimore: Johns Hopkins University Press.

Pendergraft, J., 1972. *Reforming Timber Management*, Stanford: Stanford Environmental Law Society.

Public Land Law Review Commission, 1970. *One Third of the Nation's Land*, Washington: Government Printing Office.

Stirling, E. M., 1970. The myth of multiple use. *American Forests*, **76,** 25–27.

U.S. Forest Service, 1972. *Outlook for Meeting Future Timber Demands*, Washington, D.C.: U.S.D.A. Forest Service.

U.S. Library of Congress, 1972. *Report on Clearcutting on Federal Timberlands*. Washington, D.C.: Environmental Policy Division, Congressional Research Service.

Whaley, R.S., 1970. Multiple use decision making—where do we go from here? *Natural Resources Journal*, **10,** 557–565.

16
The Environmental Impact of American Agriculture

RICHARD E. FRIDAY AND DAVID J. ALLEE

The central question in the United States is no longer 'Can we produce enough food?' but 'What are the environmental consequences of attempting to do so?' (Brown, 1970). While it is still all too true that a large share of the world's growing population goes to sleep hungry each night, bountiful resources and a productive agriculture have spared North Americans from that prospect so that they may afford the luxury of worrying about the impacts of food production activities on their surroundings. Figure 16.1 shows that, in the long run, North American farmers have had little problem in providing food

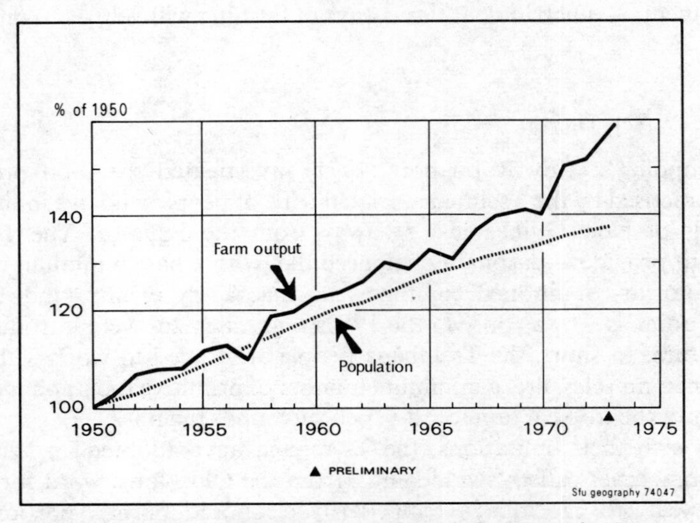

Figure 16.1. Farm output and U.S. population growth. Source: U.S. Department of Agriculture, 1973a, p. 18

for a growing domestic population. In fact, until the early 1970s, most federal agricultural programmes were orientated toward dealing with large surpluses.

Americans in 1973 spent less than 16% of their collective disposable (after tax) income on all types of food (if food eaten away from the home is excluded, food costs accounted for only 12·3% of disposable income). This was in spite of the soaring food prices during the early 1970s caused largely by an unexpectedly large export market for American farm products. Yet with growing affluence and a more efficient agriculture Americans on average have been spending less and less on food (Table 16.1).

Table 16.1. Per cent of disposable personal income spent on food

Year	% spent on food at home	% spent on total food
1960	16·2	20·0
1965	14·6	18·1
1970	12·7	16·2
1971	12·4	15·7
1972	12·3	15·7
1973	12·3	15·8

Source: U. S. Department of Agriculture (1974).

Farming and Environmental Impact: Three Examples

It is not so much the fact that American farmers produce more but *how* they produce it that has resulted in increasingly serious impacts on the American Environment. A quick look at three ways of farming will help us to understand why.

Example 1: New Guinea, 1963

Anthropologist Roy Rappaport (1971) investigated the food production system practised by the Tsembaga, a local tribe of people who live in the central highlands of New Guinea just 5° away from the Equator. The Tsembaga manage to provide a reasonably balanced diet with a bare minimum of capital imputs: no air-conditioned combines, no machinery of any kind, not even draught animals. It was only in the 1950s that machetes were introduced into their culture. In short, the Tsembaga people are forced to work with the aid of only their muscles, fire, a minimum number of primitive tools and the natural resources of the tropical rain forest which surrounds them.

Faced with their limitations, the Tsembaga have adopted an agricultural system sometimes called 'swiddening' (from the Old Norse word for 'singe'). Once a year, the Tsembaga temporarily clear and burn a portion of the secondary forest which had been farmed perhaps 20 years before. Crops are planted and harvested as they mature. About 18 months after a particular field is first cleared, it is allowed, even actively encouraged, to return to the status of the secondary-growth rain forest which surrounds it. Meanwhile, a new field has been reclaimed from the forest and is already yielding its first harvests.

If we were to fly over the land of the Tsembaga, we would see scores of small circular fields carved out of the forest, but most would be in some stage of reafforestation. At any one time, the Tsembaga actively cultivate less than 10% of their arable land. Standing on the ground in a cultivated field, we would not find the apparent organization that we have come to expect from a well-kept garden. Instead, there would be a dishevelled array of plants consisting of a ground-level mat of sweet potato vines, with a taro leaf reaching through here, a clump of sugar cane projecting still higher there and the wide leaves of the tall banana occupying the upper reaches. There would also be tree seedlings sprouting up from the ground everywhere, but not from any lack of attention to weeding. In fact, the Tsembaga farmer is almost as careful to protect the invading seedlings as he is to nurture his food plants.

Example 2: Georgia, 1825

Trees are not allowed to grow in fields here. Our host points with obvious pride to the cleanly cultivated field of short, bushy cotton plants in the hot Georgia sun. It is not an easy job to keep it that way, a fact attested to by the appearance of both our host and his four slaves hunched over their hoes in the field. Mechanical aids are scarce, consisting of basic tools and wagons, most of them being made from common local materials such as wood and leather. The exception to this is the cotton gin, invented in 1793 and improved many times since then, which permits one hand to gin (separate the cotton fibres from seeds) the same amount of cotton that used to take many hands. Draught animals provide almost all of the non-human power.

Our host will probably complain vehemently about the growing scarcity and cost of good labour, relating how prime field hands were selling for up to $900 on the Atlanta slave market last week, since the importation of new slaves has been illegal since 1808. However, on the whole, our host is optimistic. The export market for cotton has been growing by leaps and bounds and cotton production has increased threefold in the past two decades. In fact, it was the booming export market for such labour-intensive crops as rice and cotton, combined with the scarcity of free labour in the American colonies, that made slavery such an irresistible alternative in the first place.

We are in a good position to observe the impact of growing commercial markets on land use. The soil we are standing on was wilderness only a few short years ago. The area is rapidly filling up and King Cotton is still marching on, just now reaching the fertile lowlands of the lower Mississippi River Valley. But it is not just a case of an expanding area coming under cultivation in response to a growing market. Our host, like many others, has moved his family, his slaves and his equipment to this place from the already infertile soils of Virginia. Of course, chemical fertility as a substitute for natural fertility is, as yet, unheard of. Although the issue is by no means yet settled, this westward expansion will continue for decades to come. Ever since the early days of the republic, honest men have disagreed over the question of federal land policy. These debates will continue, but a gradual evolutionary process has already begun in the disposition of publicly owned lands whereby they are increasingly sold at lower prices and in smaller minimum-sized units.

The lure of a growing commercial export market for cotton has increased the degree of specialization over what had existed before, but this region is still not a sea of cotton plants. Chickens and pigs and cows are a common sight around almost every house. A large share of the fields is used to grow food crops for the slaves. However, this will be changing shortly with the coming of steam-powered transportation on the rivers, which will open up trade with other areas more suited to commercial food production. Especially in the most productive cotton regions, and particularly in those years when the highly variable cotton prices reach an irrestible peak, American cotton farmers will increasingly decide to buy their food staples elsewhere and concentrate on growing cotton.

Example 3: Iowa, 1973

We have come a long way from the 'swiddening' form of agriculture in New Guinea. We are surrounded for practically as far as the eye can see by corn. Not only are trees not usually allowed to grow in these fields, but neither is any type of weed, alfalfa, oats or anything else.

This corn is nothing like the maize that was borrowed from the Indians. Yields are much greater: well-managed, responsive corn land today can produce well over 100 bushels of corn per acre, yet as recently as 1945 the average was only 34 bushels per acre (Pimentel *et al.*, 1973, p. 444). (The average U.S. corn yield in 1973 was 91 bushels per acre.) Today's corn fields are made up of the uniform progeny of some very carefully selected parents, all lined up in perfect formation as if they were marching in an endless parade. It is difficult to overstate how carefully these varieties have been selected: for the timing of their maturity, for disease resistance, for nutritional content, but especially for yield and the ability to respond to the barrage of modern crop-management techniques. This last quality is especially important because of the vastly improved growing environment that man is able to provide today. For example, an average acre of corn in the United States receives about 115 lb of chemical fertilizer, with some highly responsive areas receiving 150 lb or more. In 1950, the average application rate was only 15 lb per acre (Mayer and Hargrove, 1971, p. 5).

Farmers are still complaining about the scarcity and high costs of good labour. In fact, an average crop acre in the U.S. today receives less than nine hours of human labour as compared with 18 hours in 1945 (Pimentel *et al.*,

1973, p. 445). But capital inputs are not nearly as scarce as they once were. Even a moderate size grain farm has any number of machines lumbering across it during the year to plough, to plant, to fertilize and to harvest. As a result, the amount of horsepower per farm worker in the United States has risen from 10 in 1950 to 47 in 1971 (Pimented *et al.*, 1973).

The Examples Compared

Faced with definite limitations on what is possible, the Tsembaga have evolved an agricultural strategy which imitates and takes advantage of the characteristics of the tropical rain forest which surrounds them. The tropical rain forest has evolved into a mature and highly stable ecosystem. For one thing, it very effectively recycles nutrients before they are leached from the thin layer of fertile topsoil and stores them as part of the vegetative cover. Swiddening only temporarily replaces this ecosystem with a group of food crops. Cutting and burning the natural vegetative cover releases a valuable storehouse of plant nutrients. Permitting rapid reafforestation, even to the point of encouraging the growth of tree seedlings in the middle of the garden, protects the fragile topsoil from erosion and renews the recycling of nutrients before they are all leached away. The Tsembaga garden also imitates the stability of the rain forest in other ways. A wide variety of food plants are deliberately intermingled in one plot of land just as many different species inhabit the natural forest. Plant-specific insects are controlled by dispersing each type of plant. Even if one crop fails completely, the Tsembaga farmer is left with an essentially balanced diet from his one garden.

But swiddening farming is a far cry from the more ‘advanced’ forms of agriculture where man seeks to modify the environment by permanently replacing the natural vegetative cover of large areas with a few species of food plants of his own choosing. The resulting monoculture is essentially a high-risk enterprise. Highly sophisticated, capital-intensive monoculture farming exists in the corn belt, wheat belt and numerous other ‘belts’ where entire regions are planted to a few carefully chosen varieties of a single species. These highly simplified systems of modern agriculture are best compared to an immature natural ecosystem. They are made up of a small number of species chosen especially for their usefulness to man, but not usually for their ability to survive under variable conditions. As Rappaport (1971, p. 130) succinctly put it, ‘In man’s quest for higher plant yields, he has devised some of the most delicate and unstable ecosystems ever to have appeared on the face of the earth.’

Thus, in seeking to control crop production exactly to suit his needs, the North American farmer has created an artificial ecosystem, the maintenance of which requires the continuous input of vast amounts of energy. Corn provides a good example of a monoculture system and the kinds of energy inputs required to maintain that system. Because of the lack of a rotation with a nitrogen-fixing legume crop, increased amounts of chemical fertilizers, which are very energy-demanding to produce, are used. A way must be found to control the invasion of highly adaptable weeds in the richly fertilized soil. Herbicides, chemicals produced with heavy energy inputs, have become the most common control mechanism. Ironically, the use of chemical weed control makes continuous corn growing almost necessary because the residual effects interfere with the growth of other rotational crops. Next comes the problem of controlling diseases and insects damaging to corn. In the absence of such biological controls as rotations and strip cropping, pesticides, again energy-demanding to produce, are used in large quantities. Of course, all of these energy-intensive inputs are applied with machines powered by fossil fuels.

One recent investigation concluded that in 1970 the production of an average acre of corn in the United States used the equivalent of 80 gallons of gasoline (or 2·9 million kilocalories) of non-solar energy (Table 16.2; Pimentel *et al.*, 1973, p. 446). This was more than three times the energy required in 1945. Practically all of the increase in energy inputs has come from fossil fuel sources. While the amount of corn produced per acre has increased greatly during the same time, the use of non-solar energy has increased even more rapidly so that the ratio of the calories in the corn output to the non-solar input calories has declined by nearly 25%. Nevertheless, to keep things in perspective it should be noted that this total amount of man-applied energy still represents only about 11% of the solar energy which is converted into corn biomass in a one-acre field.

Comparison of the energy ratio of U.S. agricultural production with that of a more primitive type of farming, such as swiddening, is also highly instructive. In 1963, it was estimated that American agriculture as a whole used 1·15 units of energy, practically all of it non-human, to produce one energy unit of agricultural output (Hirst, 1974, p. 137). In the same year, the Tsembaga farmers had to rely on their human energy alone as input into their gardening, but they harvested about 16 units of energy in the form of food for every one unit they put in as human labour (Rappaport, 1971, p. 124). It is no wonder, then, that various agricultural experts have concluded that the application of energy resource inputs in American corn farming has account-

Table 16.2. Energy use in corn production[a]

	kcal/acre			
	1945	1950	1959	1970
Total inputs	925,500	1,206,400	1,889,200	2,896,800
Corn yield (energy output)[b]	3,427,200	3,830,400	5,443,200	8,164,800
Kcal return/input kcal	3·70	3·18	2·88	2·82

[a]Reproduced with permission from D. Pimentel *et al.*, *Science*, **182,** 443–449 (1973). Copyright 1973 by the American Association for the Advancement of Science.
[b]Each lb of corn was assumed to contain 1800 kcal and a bushel of corn was considered to be 56 lb.

ed for 60 to 80% of the increased corn yields since the 1940s (Pimentel *et al.*, 1973, p. 445).

Land Labour and Capital in American Agriculture

To reconstruct the development of agricultural technology in the United States, we need only go back to the relative availability of land, labour and capital inputs in 1825. From the very beginning in the United States, labour has been in short supply relative to the amount of natural resources. First, indentured servitude and, later, slavery, became accepted institutions in the American colonies largely because of the strong pressures exerted by a scarce supply of free labour. Even in the early days of American farming, when 'unused' land was abundant and rapidly coming under the plough, the substitution of capital for labour was important. As early American inventions and improvements in such things as the cotton gin, the plough and the reaper illustrate, American innovators have long been at the forefront in the area of labour-saving technology. The process has vastly accelerated, however, with the end of the easy expansion of productive farmland and the rise in the opportunities for non-farm jobs. Therefore, with land and capital (raw materials, inventiveness) regarded as abundant, American farmers have simply substituted an energy-intensive and materials-demanding technology for labour and careful husbandry (Figure 16.2).

This whole evolution of achievement in American agriculture presents us with a paradox. The growing use of capital inputs to take the place of traditional labour and land explains much of the success story in terms of yields and food surpluses. Yet these inputs include in part such items as chemi-

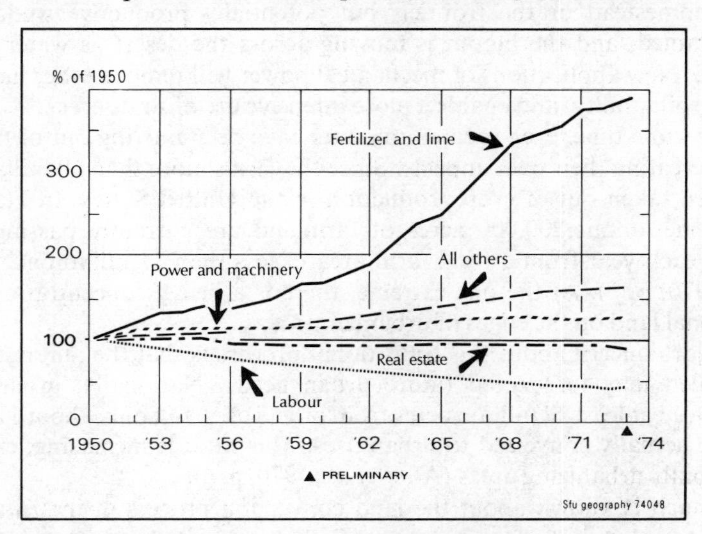

Figure 16.2. Relative quantities of selected farm inputs in the U.S.
Source: U.S. Department of Agriculture, 1973a, p. 12

cally produced fertilizers, insecticides, fungicides and herbicides now recognized as being potentially harmful to human welfare through their side-effects, direct and indirect, on man's environment.

The Impact of Agriculture upon the American Environment

The capital-intensive form of American agriculture has resulted in a number of environmental effects of which land-use changes, pesticide residues, fertilizer discharge and animal waste pollution, plus erosion and sedimentation, are probably the most serious. The environmental consequences of each of these factors will be discussed below, and throughout there will be an analysis of the institutional factors which help to create these problems.

Land-use Changes

The long-regarded relative abundance of land as a natural resource, combined with the rising importance of capital inputs in the food-producing process, has resulted in important, often conflicting, land-use changes in the United States. On the one hand, the possibility of the large-scale application of man-made improvements to land has led to the continual development of new farmland that possesses some unique characteristic, such as high fertility or favourable climate. The day of 'new' land development is far from being at an end. Over the past three decades, during most of which time there also existed large-scale government-financed programmes to divert and retire 'surplus' cropland, more than 30 million acres have been reclaimed for crop use, often of a highly intensive nature. The new lands may not have the character of the homestead or the frontier, but potentially productive wetlands are being drained, and the bloom is moving across the desert as water is being provided. New applications of mechanical power will remove other limitations (such as soil salinity) and enable a more intensive use of some areas.

At the same time, even greater amounts have been moving out of farm use, thereby creating their own impacts. Since the 1940s more than 60 million acres have been taken out of crop production in the United States. In New York State alone, about 200,000 acres of farmland are currently passing out of farming each year from a total farm area of less than 11 million acres (Allee *et al.*, 1970, p. 5). At the one extreme, there is a heavy concentration of this transitional land on the edges of urbanizing areas.

A major concern about this transitional process is that the amount of land being idled may far exceed future urban needs. Nationally, urban sprawl has removed at least 17 million acres from others uses, but only about 11 million acres are actually converted to urban uses. This ratio is increasing, especially in the rapidly urbanizing areas (Allee *et al.*, 1970, p. 6).

Not much is known about the land conversion process in the rural urban fringe, but what is known has been well documented by Clawson (1971). In the highly speculative climate that exists where rural meets urban, landowners perceive it to be potentially more profitable simply to hold land rather than to cultivate it in the hope that it will be bought at highly inflated prices. In reality, only a portion of the total area affected by such speculative fever may be needed for true urban uses in the foreseeable future. Yet the stakes are so high in the form of possible capital gains that there is no shortage of gamblers. In the interim, capital investments may be allowed to deteriorate permanently. Many farming investments are highly specialized and have little value in other uses. (Investment in a new dairy barn, for example, may actually have a negative value to an urban buyer if he has to pay to tear it down.) Yet continual investment is essential for the competitive survival of the farm enterprise. In conjunction with the problems of rising tax assessments, problems with new neighbours, with labour and with vandals, the offer of a speculator who is not able to keep the land in production may be all that is required to change productive land into idle land.

Consequently, the rural-urban fringe is an eyesore and a source of much friction between incompatible uses. In addition, much of the idle land is unusable as open space even when there is a desperate shortage of open space in the vicinity of most American cities. For it is privately owned (and hence not suitable for public use without overcoming considerable legal problems), scattered into small parcels and liable to be unsightly as well. The challenge facing land-use planners is to find ways to encourage those mixes of land use which stress complementary rather than conflicting activities and which provide the necessary common amenities that private individuals cannot readily supply.

Although the transfer of land to urban uses has serious long-run effects on the urbanizing areas where it is concentrated, the vast majority of land that is no longer farmed today is still located in more rural areas. Since 1950, more than half of the counties of the United States, practically all of them rural, have declined in absolute population. Even in a highly urbanized state such as New York, out of the 200,000 acres of land that leave farming each year, no more than 15,000 acres move directly into urban uses. At least that many more acres are prematurely made idle by the urbanization process. The rest have reached what is known as the 'obsolescence edge' of today's farming. Probably more acres of farmland have been displaced by this process than by erosion or poor conservation practices.

The story of how farmland reaches the obsolescence edge involves both technology and the more indirect effects of urbanization. One important factor on the technological side is that all land is not equally responsive to the new agricultural technology. In general, the advance of agricultural technology has been able to substitute for some natural shortcomings of the land while it enhances the importance of other good qualities such as drainage, topography and responsiveness to fertilizer. Many lands have become so disadvantaged by changing technology that they have been pushed out of farming by technological obsolescence alone. They fail to provide a satisfactory income even if the land has no alternative use or value.

During the past three decades, various 'pull' factors have helped to speed up this process greatly. Improved roads, the private car and mobile industries in search of labour have brought almost all farmers in the north east, and many farmers elsewhere, within commuting distance of alternative jobs. Also, general trends toward increased urbanization, higher incomes and greater leisure time have helped to create a new, expanding market for those farms that are close to the point of technological obsolescence. Marginal farms are increasingly being transformed into open country homes, vacation homes and private recreational lands. It takes a surprisingly small number of otherwise urban-orientated people who have an interest in owning a part of the countryside to provide an active market. For example, if the 200,000 acres of farmland that are released annually in New York State were sold in parcels of 40 acres each, it would take only 5000 non-farm families (out of a state population of 18 million) entering the market each year to transfer it all (Allee, 1966, p. 1298).

Unique displacement problems and policy issues have resulted for those farming regions which find themselves moving closer to the obsolescence edge. As the surrounding year-around farm population dwindles, communities often find themselves with a shrinking base for their community services and organization. On the other hand the 'successful' communities which attract people and diversity of employment may lose the open-country environment that attracted development in the first place. Also, a local economy based on commercial recreation may be even more seasonal than farming and may provide incomes only slightly more desirable. But, of course, for many of these communities, there may be no clearly superior alternatives.

No matter what the final use of the farmland on the obsolescence edge may be, it is likely to return with time to some type of forest cover. The opportunity would seem to exist of adding millions of acres to our already strained supply of harvestable forest land. But what results is a 'suburban forest' (Conklin, 1966; Waggoner and Ovington, 1962), a forest owned by many people in small hard-to-log tracts, that is expected to provide many other things apart from forestry income. Various surveys have shown that the major ownership values tend to be recreation, aesthetics, a future housing site and just the satisfaction of owning some land (Sizemore, 1973). A large share of this new breed of owners is at present openly hostile to the very idea of harvesting trees from their property. If the commercial logging industry is ever to gain access to this forest, it will have to learn a lot about these new owners and their objectives. In particular, it will have to go out of its way to cater to those values while still harvesting the timber.

Pesticides

Pest control is necessary every time that man upsets the delicate balance of nature in order to produce food and fibre, which is to say ever since the beginnings of agricultural production. The Tsembaga, who can exert only enough control over nature to change the balance slightly and temporarily, are able to 'control' pest build-ups by promoting reafforestation and interspersing different types of plants. Modern pest control methods can be thought of merely as an extension of this continuous struggle to maintain an imbalance of nature for man's benefit. Pesticides have played a most important part in the rapid advance in this century of the attempt to control the food-producing environment. Yet, even with the aid of these chemical tools, the U.S. Department of Agriculture estimates that more than 30% of the potential crop production in the United States is lost to insects, diseases and weeds.

The term pesticide obscures a mixture of several hundred chemicals which is in constant flux. For example, the broad class of herbicides has increased in usage at a faster rate than either insecticides or fungicides. DDT, which some people considered to be the insecticide to end all insecticides 25 years ago, had declined in domestic usage by more than 60% from its peak in 1959 by the time government restrictions were placed on its use, but in its place have come numerous other synthetic compounds.

The use of chemical pest controls has often seemed to accelerate in the manner of a vicious circle. For example, the adoption of synthetic pesticides has made it possible for U.S. consumers to demand cosmetically flawless produce, such as scab-free apples and worm-free sweet corn. It has made it possible to provide the complex food marketing structure with the uniform and long-lasting ingredients necessary for today's highly processed foods. In addition, the advent of effective synthetic organic pesticides also helped to make possible further concentration and specialization in food production. For example, herbicides have helped to make it possible to raise corn year after year in the same field; but it turns out that growing corn continuously greatly intensifies the potential damage from the corn rootworm, a major pest in corn production. Thus, this is a classic case of where the use of a pesticide has led to intensified monoculture which has resulted in the need to apply still more chemicals in order to maintain that unstable balance.

There are so many potential alternatives to the present methods of chemical pest control that choosing from among the alternatives in any rational manner is going to be an extremely complex undertaking. For example, it has been estimated that there are more than 750 basic pesticidal chemicals in existence, of which less than 30 at present account for the majority of pesticide tonnage actually manufactured in this country (Nicholson, 1970, p. 184). Add to this the fact that each commercial grade pesticide is not just a single chemical, but a mixture of somewhat different chemicals. Furthermore, each individual chemical characteristically metabolizes into at least several products (DDT alone has at least five recognized major breakdown products). Each of these metabolites may be more or less toxic, more or less stable, than the original chemical.

One basic choice facing farmers and policy makers is whether to use chemicals which are highly toxic but which very quickly metabolize into less harmful forms, or whether to use less toxic products which are effective because they are non-biodegradable. The former group carries some danger to the applicator and any life in the immediate vicinity of the application, while the latter group can create an extensive chain of damage in quite an unexpected manner by accumulating through the food chain into highly lethal doses.

A second set of choices involves non-chemical controls. Cultural controls (such as the removal of crop residues for insect breeding control or tillage for weed control) were in use long before the chemical pesticide revolution. But cultural practices affect many other aspects of farm production, such as erosion and the need for fertilizers, so that it is not usually feasible to use cultural controls for maximum pest control. In addition, under present incentives, the farmer often finds it cheaper and easier to control pests with chemicals than to prevent them by cultural means.

Other solutions, no matter how apparently promising, all have their drawbacks. Breeding of plants resistant to a particular disease or insect is costly, time-consuming and very difficult, making it feasible to guard against *some* diseases and pests. Insect sterilization programmes have been tried (and were very successful in the case of the screw worm in the south-eastern states), but again these are expensive and must be carried out on an extensive scale to be effective. The use of insect parasites and the controlled application of disease to kill off noxious insects have also been tried. This has a special common sense appeal because it involves finding and adapting nature's own controls to do the job, but there are problems. The introduction of a parasite into a new area is a tricky business and is likely to generate its own side-effects. Also, it is by no means certain that insects will be any less successful in developing resistances to diseases than they have been with insecticides. Needless to say this method is costly, in time, capital and manpower, perhaps twice or three times as costly to the individual farmer as automated application methods of conventional insecticides. In the current agricultural cost benefit calculus such an investment for so little short-term gain in productivity is rarely regarded as feasible.

Of course, it is possible to tolerate a certain amount of pests and put up with scabby apples, wormy corn and lower yields. Also, by mixing crops a little, the crop environment would become less suitable for pest population growth. However, for producers, the cost and return incentives have made the trend all the other way. For consumers, any of these changes could imply higher cost food and a shift in diet towards a less interesting mix of foods.

Fertilizers

The high productivity of American farming has always depended on the availability of large amounts of plant nutrients, although not necessarily from chemical fertilizers. In the beginning, the newly cultivated North American farmland was often rich in accumulated organic matter which, upon cultivation, began to break down to supply large amounts of mineralized nitrogen and

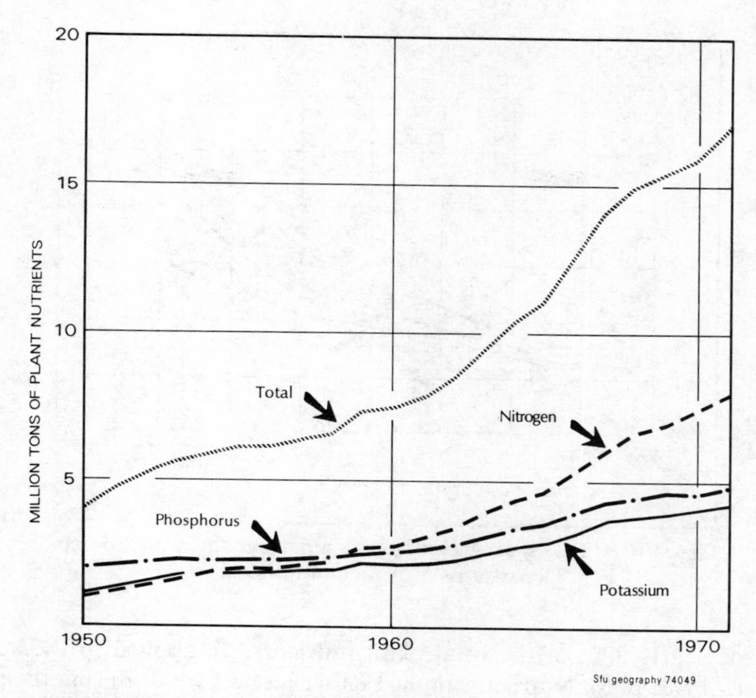

Figure 16.3. Increase of fertilizer use in U.S. farms, 1950–1970. Source: Harre, 1971, p. 5

other plant nutrients. This was a highly inefficient way to fertilize crops because of the lack of control over the mineralization process. In fact, nitrogen losses to the environment probably were much greater during that era than they are today.

The use of chemical fertilizers has increased dramatically in the United States during the past few decades (Figure 16.3). In 1972, American farmers applied more than eight times as much nitrogen in chemical fertilizers as they did in 1950. During the same period, the use of phosphorus has more than doubled. A major incentive was the relatively low cost of fertilizer in relation to other factors of production. Figure 16.4 shows that, for many years, everything the American farmer used was going up in price except fertilizer. It was largely the advent of relatively cheap, plentiful chemical fertilizers which was primarily responsible for the maintenance of a chronic overproduction potential in U.S. agriculture during the 1950s and 1960s.

The relatively declining cost of chemical fertilizers has affected the way in which farmers produce crops in many ways. For example, it influences the choice of crop varieties. The newer hybrid corn varieties do not yield especially well with little or no fertilizer, but they are amazingly responsive to heavy fertilizer applications. The relatively low cost of fertilizer also influences the timing as well as the amount of application. It has been estimated that, in

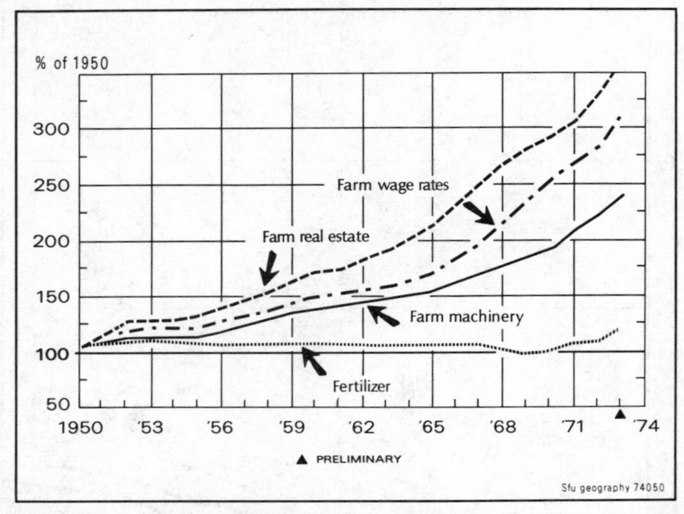

Figure 16.4. Prices of selected farm inputs in the U.S. Source: U.S. Department of Agriculture, 1973a, p. 13

the U.S., nearly 30% of the total plant nutrients are applied in the autumn (Hargett, 1973, p. 6). Nutrients applied either just before or during the period of the most rapid plant growth maximize the efficiency of fertilizer use by plants. But applying the fertilizer in the autumn or as a single pre-plant application in the spring is both less costly and less troublesome.

Plant nutrient waste elements can create a variety of potential environmental problems. The effect most related to human health is the possible contamination of drinking water with nitrates. Amongst other health-related effects, this can result in excess nitrate formation in infants, thereby causing methemoglobinaemia (blue babies). Excess nitrates can also cause health problems to ruminant livestock. The U.S. Public Health Service has established an upper limit on the nitrate concentration of water used for human consumption of 45 mg per litre, a figure that is increasingly exceeded in many rural water supply sources.

Other environmental effects of fertilizers are due to the simple fact that plant nutrients do what they are supposed to do. They make living things grow faster and longer, but not always where man intends them to. As nutrients accumulate in water bodies their concentrations eventually pass the critical thresholds limiting plant growth. The result is the proliferation of water weeds and algae which can deplete life-giving dissolved oxygen when they decay and contribute to problems of noxious taste and odour, turbidity, clogged filters, toxic chemicals and aesthetics. In the end, fish, recreationists, industrial and domestic water-users all may suffer.

Nitrogen is the most abundant element in the atmosphere, but plants utilize nitrogen almost entirely in the form of nitrate or ammonium. Agricultural nitrogen has long been under suspicion as a cause of deteriorating water quality because it is highly soluble, and therefore highly mobile, in the nitrate form. Under the right conditions, even nitrogen applied as ammonium is rapidly transformed to soluble nitrate. As was shown in Figure 16.3, the use of nitrogen in chemical fertilizers has increased at a faster pace than any other plant nutrient.

Phosphorus did not command nearly as much attention as nitrogen as an agricultural pollutant until recent years because it is applied in much smaller amounts and it does not remain in a water-soluble form for an appreciable length of time. Soil is a very effective phosphorus filter, effectively immobilizing the vast majority of fertilizer phosphorus applied in a soluble form. Even when soluble phosphorus reaches a body of water, the vast majority is 'scrubbed' out by the suspended and bottom sediment.

However, it is precisely because phosphorus does not occur in large concentrations in natural waters that it can be environmentally hazardous. The ratio of phosphorus to nitrogen tends to be much lower in natural waters than the optimum ratio for rapid plant growth. The effects of even small amounts of phosphorus enrichment can be dramatic. Experiments have shown that as little as 10 parts per billion of soluble phosphorus will accelerate the growth of blue-green algae and 50 parts per billion may result in profuse algae blooms (see Black, 1970, p. 76; Sawyer, 1947; USDA, 1968, p. 17). The rapid nutrient enrichment and eutrophication of western Lake Eire during the past several decades was certainly aided by a 480% increase in the soluble phosphorus concentration (from 7·5 to 36 parts per billion) in the lake, while nitrogen concentrations increased by only 30% (Verduin, 1970, p. 64).

To what degree agriculture's use of chemical fertilizers actually contributes to environmental damage is still hotly debated. There is fragmentary, indirect evidence on both sides. Table 16.3 represents one judgment of the relative national contributions of nutrients from various sources. In this case, run-off from agricultural land looms as a large contributor. Researchers in Illinois have reported a substantial increase in the nitrate concentration of surface waters supplied by surface run-off from agricultural lands (Wang and Evans, 1964). Using very indirect methods of measurement, some researchers have argued that agricultural lands contribute about one-third of the phosphorus found in the rivers of the Mid-west (Verduin, 1970, p. 67).

On the other hand, many researchers are concluding that, for various reasons, agricultural fertilizers are not as grave a threat to the environment as was first feared. For example, the Tennessee Valley Authority has conducted experiments on a small watershed in North Carolina where they are able to monitor all of the run-off water. Fertilization did increase the amount of nutrients removed in the discharge waters, but by very small amounts (Coffman, 1971, p. 21). Another study in New York State compared the nutrient contributions of three tributaries to a lake, two draining agricultural areas and one serving as a waste discharge outlet for a local municipality. The researchers concluded that the vast bulk of soluble nutrients (particularly phosphorus)

Table 16.3. Estimate of nutrient contributions from various sources[a]

	Nitrogen		Phosphorus	
	Pounds/ year (millions)	Usual concentration in discharge (mg/1)	Pounds/ year (millions)	Usual concentration in discharge (mg/1)
Domestic waste	1100–1600	18–20	200–500[b]	3·5–9·0
Industrial waste	> 1000	0–10,000	[b]	[b]
Rural run-off: Agricultural land	1500–15,000	1–70	120–1200	0·05–1·1
Non-agricultural land	400–1900	0·1–0·5	150–750	0·04–0·2
Farm animal waste	> 1000	[b]	[b]	[b]
Urban run-off	110–1100	1–10	11–170	0·1–1·5
Rainfall[c]	30–590	0·1–2·0	3–9	0·01–0·03

[a]Reprinted from *Journal American Water Works Association*, Volume **59,** by permission of the Association. Copyrighted 1967 by the American Water Works Association, Inc., 6666 West Quincy Avenue, Denver, Colorado 80235.
[b]Insufficient data available to make estimate.
[c]Considers rainfall contributed directly to water surface.

added to the lake water originated with sewage wastes and not agricultural wastes (Oglesby and Mills, 1973).

The evidence is inconclusive, partly because the source of nutrients is not always easy to determine and partly because some substances enter and leave the ecosystems in various forms. A method to overcome the latter problem is to calculate all additions and subtractions of nutrient material in a balance sheet fashion. An interesting nitrogen accounting system recently attempted (Stanford *et al.*, 1970) provides some evidence that the combined loss of nitrogen from erosion and leaching may actually have decreased in the United States between 1930 and 1969. No attempt was made to estimate direct losses from the leaching of fertilizer nitrogen, but the combined losses from erosion and leaching of soil nitrogen declined dramatically from 9 million tons to 5 million tons. This is despite the fact that total nitrogen inputs doubled and the use of fertilizer nitrogen increased from 0·3 million tons to 7 million tons annually during the same period. Of course, a comparison with a year other than one during the dust bowl era, with its large erosion losses, would be more indicative.

In the light of evidence such as this, many scientists are today becoming less and less apprehensive about the relative contribution of chemical fertilizers to environmental deterioration. Yet, undoubtedly, the widespread use of agricultural chemicals does give rise to concern. Assuming for the moment that the present food production system does create significant problems for the safety of drinking water and the stability of aquatic environments, what can be done to prevent this from happening? Obviously, one way would be to eliminate the problem at its source by using less of the chemical fertilizers. This could be achieved by a variety of complementary measures: by forcing farmers not to use fertilizers through quotas or taxes, by rationing mechanisms or by enticing them to use substitutes through incentives.

No one enjoys the prospect of going hungry for the sake of healthy fish or even unborn human beings. Fortunately, at least in most developed countries, this choice does not have to be made. Other inputs can be substituted for chemical fertilizers just as fertilizers have replaced other production factors. In fact, much of the effect of restricting the use of chemical fertilizers would be to reverse some of the most important trends in food production during the past several decades. For example, organic sources of plant nutrients, such as cover crops and animal wastes, could be better utilized. The word 'rotation' would again become an important concept in the agriculturalist's vocabulary. Even the idea of using farmland as a 'living filter' for the disposal and purification of urban sewage is gaining respectability.

Other types of substitutions would further reduce our dependence on chemical fertilizers. For example, the use of land and human labour could be increased, reversing the trends of declining use of both these inputs. In fact the recent increases in energy prices, shortages of fertilizer and growing export demand have resulted in sizeable additions to U.S. farm land in 1973–74 (some 60 million acres). These changes may well prove to be temporary, but, for the first time in nearly six decades, the farm labour force is not declining at an appreciable rate. Again, the mix of crops could be altered toward crops that are less demanding of nutrients. For example, corn is a highly desirable source

Table 16.4. Estimated balance sheet of nitrogen (N) in the United States for 1930 and 1969

	Amounts of nitrogen 1930	1969
	Millions of tons	
Inputs of nitrogen from:		
1. Fertilizer N	0·3	6·8
2. N fixed by legumes	1·7	2·0
3. N fixed (non-symbiotic)	1·0	1·0
4. Manure	1·9	1·0
5. Unharvested portions of crops	1·1	2·5
6. Rainfall	0·8	1·5
Total	6·8	14·8
Removals of nitrogen by:		
7. Harvested crops	4·6	9·5
8. Erosion	5·0	3·0
9. Leaching of soil N	4·0	2·0
10. Leaching of fertilizer N	0	?
11. Denitrification	?	?
Total	13·6	14·5

Source: Stanford *et al.* (1970).

of feed for livestock and, with dramatic improvements in technology, has been increasingly grown in place of other grains during the past few decades. But corn also requires larger inputs of nutrients for optimum growth. Other, less demanding, grain crops could be grown in place of corn in many parts of the country. Even within a particular crop, it is possible to change the choice of existing varieties and the nature of the breeding research efforts in order to stress those varieties which perform well with smaller inputs of fertilizer. Up to this point, we have been increasingly using and developing those newer varieties which produce at their best only with high levels of fertilization. Changes in timing of fertilizer applications and cultural practices could greatly reduce fertilizer losses.

The problem of fertilizer pollution will probably ultimately be controlled by either one or a combination of the above methods, but it is also at least theoretically possible to concentrate on reducing the damages caused by plant nutrients instead of the stray nutrients themselves. For example, agricultural run-off and drainage water could be treated to remove nutrients. However, this is an expensive solution. Agricultural run-off is very diffused and diluted compared with municipal sewage. Aquatic weeds could themselves be used as fertilizer or even a form of food. Which alternative or combination of solutions is finally adopted will depend on a variety of difficult-to-predict social, political and economic factors.

Animal Wastes

Animal waste 'pollution' is certainly not a new phenomenon. Until recently, even the wealthy were rarely further away from their livestock and its attendant manure than their doorsteps. There is a late 19th-century photograph which shows a herd of sheep contentedly grazing on the ellipse behind the White House. Early pioneers travelling across the Western frontier often wrote in their logbooks of finding streams and rivers yellow and putrid from the excrements of wild buffalo.

Raising livestock may have been man's very first agricultural activity, but the waste-disposal problems of today's concentrated livestock practices are new. This problem has been exacerbated by the steady increase in the demand for animal products in recent years—a matter in part encouraged by rising affluence and in part promoted by new technologies in the poultry and cattle industries. Figure 16.5 shows the sizeable increases in per capita consumption of selected livestock products in the U.S. On the supply side, the concentration of animals in extremely large production units has improved the efficiency of the operation though to the detriment of the animals themselves and to the environment generally. Modern poultry operations handle 100 thousand or more birds at any one time. In the cattle feedlot beef animals spend the last three to five months of their lives under a concentrated and confined fattening programme. Today, more than 60% of the cattle marketed in this country come from feedlots with a capacity of 1000 head or more (U.S. Department of

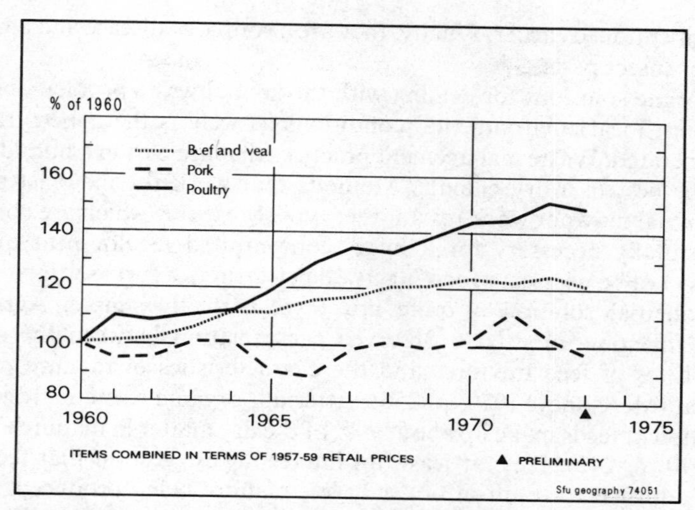

Figure 16.5. Increase in the per capita consumption of selected livestock products. Source: U.S. Department of Agriculture, 1973a, p. 45

Agriculture, 1973b, p. 76). Due to rapid technological advancements in the production of high food-value crops, such as corn and soybeans, it is now better business (i.e. cheaper) to fatten the livestock in confined quarters with harvested and commercially blended feed preparations than to allow the animals to rummage for their own food in a pasture. Highly automated feeding systems have also made this possible. Finally, competition has made the larger scale irresistible to many producers.

The magnitude of the animal-waste disposal problem has been compounded by developments in the manufacture of chemical fertilizers because formerly almost all animal manure was spread on the land as a source of valuable plant nutrient. But the introduction of relatively cheap and highly efficient inorganic fertilizers has spelt the demise of animal wastes as a useful product. Now these create a disposal problem despite their inherent value. In the absence of sanctions, most livestock producers improperly treat these wastes before discharging them into nearby watercourses, though an increasing amount will be spread on neighbouring cropland, if fertilizer prices continue their recent climb.

Animal manure, feedlot run-off and even the effluent from treated wastes are very rich in organic wastes and possess up to 100 times the biochemical oxygen demand of normal municipal sewage. If these wastes enter water bodies they can cause serious oxygen depletion, as well as resulting in putrification and obnoxious odours. The odour problem is particularly acute where animal wastes are stored near residential areas in warm humid climatic areas. Animal wastes also contain significant concentrations of nutrients, up to 3% nitrogen and 1% phosphorus, and therefore add to many of the water pollution problems mentioned earlier. Finally, they are a source of disease and a breeding ground for insect pests.

Appropriate solutions for dealing with the animal waste problem must take into account local environmental conditions as well as the concentration of wastes generated. Waste management practices that are best in a humid climate will not be best in a dry climate. Methods that work for the wastes of one type of animal may not work for another type. Measures which are considered environmentally necessary for a huge, concentrated feedlot near an urban centre may not be necessary for a 50-cow dairy farm in a rural setting.

One potential solution is quite novel yet perfectly simple. Agricultural scientists for some time have observed a correlation between the chemical characteristics of feed mixtures and the characteristics of manure produced from them (McKinney, 1970, p. 256). After all, organic waste materials from the digestion of feeds make up about 90% of the dry matter in manure (McCalla *et al.*, 1970, p. 242). It is, at least, an interesting suggestion that feed ratios could be adjusted to result in not only less manure being produced, but also in manures that pollute less.

Other alternatives involve reversing two of the dominant trends in recent livestock production. First, feedlot operators could be given incentives to disperse their activities across the countryside. This would decrease the problem of excessive concentration of waste discharges and should ensure sufficient nearby land for disposal and treatment. Of course, such a policy, like most of these alternatives, would raise livestock costs and prices by nullifying many of the cost-reducing innovations which have made it feasible to employ concentrated feedlot methods. Secondly, animal manure could again become useful as a plant nutrient. The trend has been away from this because of the economic pressures that have evolved: falling prices for chemical fertilizers versus the rapidly rising transportation and labour costs of spreading manure on cropland located far from the concentrated feedlots. This structure of incentives could presumably be altered either by the use of the economic carrot in the form of subsidies paid for spreading manure on cropland or by the stick in the form of a tax on the use of chemical fertilizers.

Many livestock and soil experts consider cropland-spreading to be the best method of disposing of animal wastes with minimal environmental hazards. Soil possesses remarkable qualities as a living filter for decomposing and mineralizing animal manure, in addition to immobilizing plant nutrients, especially phosphorus. It should be borne in mind, however, that animal manure spread over frozen ground in the winter or just before heavy rainfall in the spring may be subjected to heavy leaching before it can be safely incorporated into the soil. Despite this, in general, treatment by soil filtration is both relatively cheap and environmentally sound. To treat animal wastes to the level of municipal sewage wastes would be enormously expensive as animal wastes are so concentrated. It has been calculated that such a policy would cost more than $200 per dairy cow per year (Moore, 1970, p. 290).

The conversion of animal wastes to useful products is one of those extremely appealing technical solutions to the animal waste problem. Everybody likes to be paid to have his pollution carried away. There has been considerable talk and a few experiments have been made concerning the recycling of animal wastes back into components of animal feeds. This makes some biological sense because animals generally convert far less than half of their feed into animal tissue. The manure itself contains about 60 to 75% digestible materials (McCalla *et al.*, p. 242). In spite of this, it is unlikely that the recycling of feeds will move from being technically feasible to being economically attractive to livestock producers. The monetary incentives just are not present for the individual, competition-orientated producer.

A typically American solution is simply to separate livestock production from other non-compatible uses by zoning. Some observers wryly note that former city dwellers often no sooner get their homes built out in the countryside for the express purpose of enjoying the rural atmosphere than they start to complain about some of the rural atmosphere that they are experiencing (Wadleigh, 1971, p. 63). Theoretically, zoning areas for the exclusive use of livestock production sounds good but, realistically, one must face up to the dismal record of local zoning in controlling land use in rural areas.

Sediment

Sediment is often hailed as the largest single water pollutant in the U.S. because it is equivalent by weight to more than 700 times the solid load from sewage (U.S. Department of Agriculture, 1968). Of course, weight or volume is not a satisfactory scale on which to assess the effects of different pollutants. The fact still remains however, that at least 4 billion tons of sediment move into the tributary streams of the United States each year (Wadleigh, 1968). Because farming is a major land use in this country (cropland and rangeland combined account for nearly 45% of the 2266 million acres in the United States), it is to be expected that more than one-half of the total sediment yield originates on agricultural lands.

The detrimental effects of sediment on the environment are three-fold. First, the erosion of productive topsoil is in itself a direct loss which takes thousands of years to replace. This is the type of damage that a farmer is most likely to consider in deciding whether or not to adopt soil conservation practices. Secondly, sediment can have complicated and distant effects on the physical characteristics of the environment. As the early Sumerian farmers of the Tigris–Euphrates valley learned the hard way some 5000 years ago, removing sediment from irrigation channels, as well as from streams and harbours, is an expensive undertaking. In fact, some archaeologists claim that the subsequent decline in the sophisticated urban society of the Sumerians can be traced to their inability to control sediment in their vast irrigation system (Jacobson and Adams, 1958). Sediment is also a problem in that it occupies storage space in reservoirs and ponds, it creates turbidity which reduces the recreational uses of water, it degrades aquatic habitats, and it makes it more costly to treat water for industrial and consumptive uses.

But sediment is not always dysfunctional. As noted above, it can act as a chemical scavenger by removing phosphate and other undesirable nutrient ions from solution. Therefore some scientists are now suggesting that a certain level of sediment in water is desirable in order to prevent eutrophication by controlling the level of dissolved plant nutrients. On the other hand, it may well be that these unwanted chemicals absorbed on sediment particles gradually increase in concentration to the point where they partially redissolve and establish a new equilibrium in the water body. At any rate, this interaction goes to illustrate the point that the ideal solution to agricultural pollution problems will not be easy, if indeed it is humanly obtainable, because of many complicated types of interactions.

Sediment can be reduced but never eliminated as any naturally flowing body of water will produce sediment. However, improvements to existing agricultural practices can help to reduce the problem though not all obvious solutions are appropriate. For example, terracing is rapidly declining as a soil conservation practice, in part because of the cost of construction and maintenance but mainly because it is becoming less and less compatible with efficient, low-cost farming practices, such as using machinery to plant, till and harvest a larger number of rows at one time.

The more that is learned about the process of soil erosion, the more it is realized how relatively simple changes in the way in which farmlands are planted, tilled and managed can significantly reduce soil erosion. It has long been recognized that such practices as contour planting and strip-cropping can substantially reduce soil losses from hillsides. More recently, the method of tillage has been shown to have a significant impact on the amount of sediment produced. In general, land which is tilled in such a way as to leave a rough surface with large soil particles and cropland where the plant residues are not completely ploughed under or removed suffer less soil erosion than other lands. Even some rotations of row crops are more susceptible to erosion than others.

Policy Options

The successful resolution of the environmental problems emanating from agriculture is not strictly a technological question. There is little doubt that Americans know how to produce food with fewer environmental effects. There is already a wide range of known alternatives with a great variety of different costs and effects. The choice is never one of completely doing without. There is nearly always a substitute that is almost as good as present practices in terms of individual production efficiency.

The alternatives in the control of fertilizer pollution provide one example. On the surface, this does not appear to be an extremely difficult task. Many of the possible controls seem rather simple: substituting land, labour or rotations for chemical fertilizers, taking more pains to place the fertilizer closer to actively growing plants in time and space, and so on. It is a difficult task, of course, because of the perverse incentives. Farmers are paid for producing a dependable and appealing source of food at the lowest cost possible. They are not paid for providing safe drinking water or a healthy environment for sport fish. Most types of farming are sufficiently decentralized so that there is little question about the lack of a monopoly market power. A farmer can be as environment-conscious as any conservationist and continue to befoul the environment as much as any power-wielding industrialist precisely because he does exist in a highly competitive business world. In order to earn his living, he is forced to the same actions as his 'environment-be-damned' neighbour. Given the trends in monoculture, in technology and in long-term fertilizer prices during the post-war period, it should come as no surprise to us that farmers have found irresistible the pressures from first, higher profits and second, competitive pressures to keep costs low.

The first and easiest recourse to deal with these environmental problems is the application of more technology; but unless technology provides costless solutions, knowing how will not be enough. Furthermore, just because a project is considered to be technically feasible, the applied technology will not necessarily be developed and utilized. Under current incentives, the type of technology that would reduce pesticide pollution is not likely to be fully explored by private industry. Given the opportunity of developing and selling the standard chemical pesticides at a lower cost, the chemical industry is likely to continue to find the environmentally safer methods to be 'not economically feasible'. Farmers would simply not use the more expensive pest control techniques, given the existing incentive structure. As long as commodity prices are determined in part by those who use the least-cost pest control methods, the individual farmer will have little choice but to go along with the existing types of practices. The pressure of competition has made it mandatory for the individual to adopt the innovations that have proved successful, even though those changes have resulted in externality problems for others. Nitrates in the ground water, algae bloom in the lakes, pesticides in the fish and silt in the reservoirs are all examples of the resulting externality problems.

How then, can successful incentives be introduced? The history of soil conservation programmes in the United States is instructive here. First, where soil conservation practices have been successfully sold to farmers, it has not been on the basis of the benefits to the distant recreational or municipal user of the water supply. Soil conservation programmes have been sold primarily on the basis of their direct benefits: that they reduce soil losses on an individual farm or within a limited area. Furthermore, it has been shown that farmers have to be convinced that conservation practices will return those direct benefits within what they consider to be a reasonable length of time. Like the rest of us, farmers tend to discount the future heavily.

Another reason why soil conservation practices have been so widely adopted by American farmers in this century is that the government has intervened to change the incentives. The reasoning has been that, if the benefits are widespread among society, then society should help to bear the financial burden. Therefore the federal government adopted a strategy of promoting soil conservation practices by providing evidence of community approval, technical assistance, cost-sharing and cropland diversion.

The problem therefore lies in determining how much public intervention and at what level. For example, imagine the government were to impose certain restrictions on productivity by demanding new standards of pesticide control or fertilizer application. If the restrictions apply to only one region or crop, consumers are free to react by turning to the sources where there are no restrictions. The affected farmers would have to try to compete with producers in regions or crops where it was still business-as-usual. In the ensuing competition, a farm region or crop would be severely disadvantaged by restrictions that are quite reasonable and necessary on the whole. The highly disruptive effects of localized controls on the local economy make it unlikely that stringent environmental controls will be adopted except in areas where farming and agro-business are an insignificant share of the total economy. Furthermore, *local* controls, adopted on a *local* scale, would not improve environmental conditions nationwide. Local improvements would be offset by further deterioration in non-restricted areas from the increased use of polluting chemicals and practices.

How much would additional environmental protection cost if applied nationwide? A detailed economic study undertaken at Iowa State University predicted that if fertilizer usage on six major field crops were restricted nationwide to approximately 50% of the 1969 level, per acre yields would decrease substantially and per unit production costs would rise. Retail prices would increase so sharply that consumers would end up spending 23% more for their food. Yet the most surprising result is that, at least until farmers bid up the price of land sufficiently, net income per farm would rise by more than 50% (Mayer and Hargrove, 1971).

There exist a myriad of potential trade-offs in simultaneously controlling different forms of agricultural pollution. Some of the more obvious links are between fertilizer pollution, animal wastes and soil erosion. For example, restrictions on the use of chemical fertilizers may encourage the winter spreading of animal manures over frozen ground in the northern climates, a practice that is known to lead to especially high rates of nutrient losses. On the other hand, the substitution of animals wastes for chemical fertilizers during the warmer months may avoid the substantial nutrient discharges into water supplies from the wastes of confined animals.

Another example involves nutrient pollution and erosion control. It may be highly desirable to control erosion if the total phosphorus entering the water is the main concern, but if just the soluble phosphorus is the problem, fertilizer restrictions and erosion control may be at odds. Some scientists have concluded that a bit of well-planned erosion may cleanse water bodies by scrubbing out of the water the soluble phosphorus and placing it in a sedimentary form, largely unavailable for aquatic growth.

But pollution control policy cannot exist in a vacuum: potential trade-offs with various other types of government policies also exist. Historically, for example, the recent concern for fertilizer pollution would surely not be as great if it were not for the various federal policies of handling the overproduction potential by retiring and diverting cropland. One of the effects of the mammoth land diversion programmes of the 1950s and 1960s was to encourage farmers to use the remaining land more intensively than they otherwise would have. They did this by substituting more of the purchased capital inputs, including pesticides and fertilizers, for the land that they could not otherwise use.

The market will not solve the environmental problems of agriculture without some help. Nor will education and new technology be enough. Competition will continue to force farmers to ignore many of the side-effects of their production. Many of the substitutes for existing practices will only be feasible if all producers have to adopt them. Public action will be needed to change the incentives in order to reflect the externalities that are worth avoiding. Public intervention into the private activities of farmers is hardly new, but the complexity of the public choices ahead may be far greater than those made in the past.

References

Allee, D. J., 1966. Changing use of rural resources. *Journal of Farm Economics*, **48,** 1297–1305.

Allee, D. J., 1967. American agriculture—its resource issues for the coming years. *Daedalus*, **96,** 1071–81.

Allee, D. J., Hunt, C. S., Smith, M. A., Lawson, B. R. and Hinman, R. C., 1970. *Toward Year 1985: The Conversion of Land to Urban Use in New York State.*, Ithaca, N. Y.: Special Cornell Series No. 8, Cornell University.

American Water Works Association, 1967. Task Group 2610-P. Source of Nitrogen and Phosphorus in Water Supplies, *American Water Works Association Journal*, **59,** 344–366.

Black, C. A., 1970. Behavior of soil and fertilizer phosphorus in relation to water quality. In *Agricultural Practices and Water Quality* (Eds. Willrich, E. L. and Smith, G. E., Ames, Iowa: Iowa State University Press, pp. 72–93.

Brown, L. R., 1970. Human food production as a process in the biosphere. *Scientific American*, **223,** 160–170.

Clawson, M., 1971. *The Suburban Land Conversion Process in the United States: An Economic and Governmental Process*, Baltimore: Johns Hopkins Press.

Coffman, R., 1971. Newest findings on fertilizer runoff. *Farm Journal*, **95,** 21–23.

Conklin, H. E., 1966. The new forests of New York. *Land Economics*, **42,** 203–208.

Goldberg, M. C., 1970. Sources of nitrogen in water supplies. In *Agricultural Practices and Water Quality* (Eds. Willrich, E. L. and Smith, G. E.), Ames, Iowa: Iowa State University Press, pp. 94–125.

Hargett, N. L., 1973. *1972 Fertilizer Summary Data*. Muscle Shoals, Alabama: National Fertilizer Development Central Bulletin Y-53, Tennessee Valley Authority.

Hirst, E., 1974. Food-related energy requirements. *Science*, **183,** 134–138.

Jacobsen, T. and Adams, R. M., 1958. Salt and silt in ancient Mesopotamian agriculture. *Science*, **128,** 1251–1258.

Mayer, L. V. and Hargrove, S. H., 1971. *Food Costs, Farm Incomes, and Crop Yields With Restrictions on Fertilizer Use*. Ames, Iowa: CAED Report No. 38, Center for Agricultural and Economic Development, Iowa State University.

McCalla, T. M., Frederick, L. R. and Palmer, G. L., 1970. Manure decomposition and fate of breakdown products in soil. In *Agricultural Practices and Water Quality* (Eds. Willrich, E. L. and Smith, G. E.), Ames, Iowa: Iowa State University Press, pp. 241–255.

McKinney, R. E., 1970. Manure transformations and fate of decomposition products in water. In *Agricultural Practices and Water Quality* (Eds. Willrich, E. L. and Smith, G. E.), Ames, Iowa: Iowa State University Press, pp. 256–265.

Moore, J. A., 1970. Animal waste management to minimize pollution. In *Agricultural Practices and Water Quality*, (Eds. Willrich, E. L. and Smith, G. E.), Ames, Iowa: Iowa State University Press, pp. 286–297.

Nicholson, H. P., 1970. The pesticide burden in water and its significance. In *Agricultural Practices and Water Quality* (Eds. Willrich, E. L. and Smith, G. E.), Ames, Iowa: Iowa State University Press, pp. 183–193.

Oglesby, R. T. and Mills, E. L., 1973. *Owasco Lake and Its Watershed*. Ithaca, N. Y.: A Report to the Cayuga County (N.Y.) Planning Board and Cayuga County Environmental Management Council.

Peters, D.C., 1970. Pesticides and pest management for maximum production and minimum pollution. In *Agricultural Practices and Water Quality* (Eds. Willrich, E. L. and Smith, G. E.), Ames, Iowa: Iowa State University Press, pp. 209–223.

Pimentel, D., Hurd, L. E., Bellotti, A. C., Forster, M. J., Oka, I. N., Sholes, O. D. and Whitman, R. J., 1973. Food production and the energy crisis. *Science*, **182,** 443–449.

Rappaport, R. A., 1971. The flow of energy in an agricultural society. *Scientific American*, **225,** 116–132.

Roelofs, W. L. and Arn, H., 1968. Red-banded leaf roller sex attractant characterized. *New York's Food and Life Sciences Quarterly*, **1,** 13.

Sawyer, C. N., 1947. Fertilization of lakes by agricultural and urban drainage. *Journal of New England Water Works Association*, **61,** 109–127.

Sizemore, W. R., 1973. Improving the productivity of non-industrial private woodlands. In *Report of the President's Advisory Panel on Timber and the Environment*. Washington, D.C.: Government Printing Office, 234–295.

Stanford, G., England, C. B. and Taylor, A. W., 1970. *Fertilizer Use and Water Quality*. Washington, D.C.: U.S. Department of Agriculture, ARS 41–168.

U.S. Department of Agriculture, 1968. *A National Program of Research for Environmental Quality*. Washington, D.C.: Joint Task Force Report of the USDA and the State Universities and Land Grant Colleges.

U.S. Department of Agriculture, 1973a. *Handbook of Agricultural Charts*. Washington, D.C.: Agriculture Handbook No. 4, 55.

U.S. Department of Agriculture, 1973b. *Changes in Farm Production and Efficiency*. Washington, D.C.: Statistical Bulletin No. 233, Economic Research Service.

U.S. Department of Agriculture, 1974. *National Food Situation*. Washington, D.C.: U.S. Department of Agriculture.

Verduin, J., 1970. Significance of phosphorus in water supplies. In *Agricultural Practices and Water Quality* (Eds. Willrich, E. L. and Smith G. E.), Ames, Iowa: Iowa State University Press, pp. 63–71.

Wadleigh, C. H., 1968. *Wastes in Relation to Agriculture and Forestry*. Washington, D.C.: U.S. Department of Agriculture, Misc. Publication No. 1065.

Wadleigh, C. H., 1971. A primer on agricultural pollution: Summary. *Journal of Soil and Water Conservation*, **26,** 63–65.

Waggoner, P. E., and Ovington, J. D., 1962. *Proceedings of the Lockwood Conference on the Suburban Forest and Ecology*. Hartford, Conn.: Connecticut Agricultural Experiment Station Bulletin 652.

Wang, W. C. and Evans, R. L., 1964. Dynamics of nutrient concentrations in the Illinois River. *Journal of Water Pollution Control Federation*, **42,** 2117–2123.

17

Forest Management Policy: Its Evolution and Response to Changing Public Values

GEORGE H. STANKEY

Forestry today finds itself surrounded by controversy. The United States Congress has held special hearings on forest land practices (Subcommittee on Public Lands, 1971) and proposals for legislation revamping forest land management policies and programmes have been made. There has been increasing reliance upon the judicial branch by environmental groups to redress administrative practices perceived as inconsistent with good land management.

Clearcut logging (the practice of removing all standing trees from an area) has received much attention. There is concern that the practice results in the destruction of the land's productive capability (Wood, 1971). Other management programmes have also come under fire, particularly road construction, watershed protection and the allocation of forest lands to recreational, scenic and wilderness purposes.

However, the issue is not simply one of pitting persons opposing logging against forest managers. The United States Forest Service is also under increasing pressure from industrial interests to expand timber production, increase road systems, and generally 'develop' the National Forests considerably beyond their present level. For instance, the recent report of the President's Timber Supply Advisory Committee (1973) called for a 50 to 100% increase in the annual harvest of old-growth timber and the designation of all non-withdrawn commercial forest land for timber production. As a result, the question of how to resolve the many competing demands placed on the forest resource has often become lost in a simplified and polarized concept of resource management featuring 'developers' against 'preservationists'.

In some ways, the debate concerning forest land management is surprising. Professional forest management was a recognized discipline prior to the turn of the century, emigrating to the United States from Europe in the late 1800s. Today, nearly 40 universities and colleges offer accredited programmes in forestry. Forestry is a well-developed profession and shares some common characteristics with other professions, such as law and medicine, in its establishment of a professional body (the Society of American Foresters in the United States), the development of a code of ethics, a programme of professional licensing and the publication of professional journals.

Moreover, forestry is buttressed by a considerable body of legislation. Literally thousands of pieces of legislation govern the various aspects of forest management. Finally, the minutiae of detail affecting day-to-day affairs for the U.S. Forest Service are covered in the 22 volumes of the *Forest Service Manual*.

However, despite a cadre of professional resource managers, backed by a well-established educational system and abundant legislation and administrative regulations, much disapproval of forest land practices persists.

Is there, in fact, substance to the charges of forest land mismanagement? If so, what are some of the specific issues that have prompted dissatisfaction and what, if anything, has been done to redress public grievances? To provide some insight into these and other questions, I will review the results of an investigation of one recent controversy outlined by Burk (1970), focusing on the factors that led to the rise of the problem, the findings of the investigation, and also attempting to specify some underlying causes of what has been called 'the clearcut crisis'.

Forest Management in the Wyoming Forests: a Case Study

In response to public concern over forest management on four National Forests in north-western Wyoming (Bighorn, Bridger, Shoshone and Teton National Forests) the U.S. Forest Service assigned a team of investigators in 1970 to review existing management practices, particularly timber harvesting (Wyoming Forest Study Team, 1971).

The study team was instructed: (i) to explore fully the concerns and apprehensions of individuals and organizations in Wyoming; (ii) to conduct the study in accordance with the National Environmental Policy Act and (iii) to report its findings objectively and candidly. Their final report provided 61 general and specific recommendations for alterations and refinements in current management practices.

Logging had been conducted in north-west Wyoming since the late 1880s, primarily for fuel wood, railroad ties and mine timbers. Forest Service management began in the early 1900s. Until the 1950s, forest protection (against insects and fire) was the main activity, pending improved market conditions to make harvest feasible.

Improved technology, coupled with increasing demands for timber in the 1950s, led to the conversion of the old forests to young, vigorous stands through cutting. The accelerated harvest brought increased vitality to the local economy. However, it also led to changes in the landscape, as the patchwork pattern of clearcut blocks spread, some covering up to 1000 acres. With the increased rate of harvesting came growing public concern about possible detrimental effects on wildlife, water quality, recreational resources and scenery. Local agitation grew through the 1960s, with charges that production of saw logs had come to threaten the unique environmental qualities of the region.

The overriding issue in both the public criticism that led to the investigation and in the investigation itself was that timber harvesting was the dominant value considered in management plans and that, as a result, other significant values had been neglected and often lost. Wildlife represents an excellent example. The elk herds of north-western Wyoming have a national reputation. Hunting results in a harvest of more than 11,000 animals annually. However, despite the significant economic, aesthetic and recreational role elk and other big game played, the study team concluded that past management activities had been largely insensitive to these values; e.g. normal migration routes had been severely disrupted by the presence of large clearcuts. Road construction associated with timber harvest was conducted with little regard to potentially disruptive effects on wildlife. Recreational values had been similarly neglected. For instance, trail heads providing access into some wildernesses had often been obliterated by clearcuts or logging residues, or had been converted to roads.

In addition to recreation and wildlife, much concern has centred on the effect of timber harvesting on water quality. Of particular concern here are the effects of road construction. Research in a variety of locations across the United States has documented roads, rather than logging, as the primary source of reduced water quality (Packer, 1967). Roads have been, however, part and parcel of logging activities and, again, the study team found that road construction had often been carried out with little regard for its impact on other resource values.

In summary, the study team observed that, although management efforts were improving, timber harvesting nevertheless continued to hold the dominant position in the scheme of things. Efforts to meet the allowable cut often precluded the adequate consideration of other important resource values. (Allowable cut refers to the amount of forest produce that may be harvested, annually or periodically, from a specified area, over a stated period. It is a biologically-based *constraint* on cutting, but has often been metamorphosed into a *goal.*)

The case of the Wyoming forest controversy is merely a microcosm of a broader conflict that has reached across the United States. In the Bitterroot National Forest in Montana, for example, forest management practices were investigated by a U.S. Forest Service study team (1970a) and by a University of Montana Select Committee (1970). Both studies were precipitated by local concern that timber production and specific management practices associated with timber production (e.g. terracing hillsides for machine planting of trees) had led to the neglect of other values and that harvests exceeded growth. The Forest Service study reported that, at least implicitly, resource production goals held priority over other land management considerations and that decision-making suffered from a lack of directions. The University committee was more direct: 'multiple use management, in fact, does not exist as the governing principle on the Bitterroot National Forest'. Moreover, they concluded that consideration of values such as recreation, watershed and wildlife appeared only as an afterthought.

Similar controversies have occurred in the eastern United States. Starting in 1964, public objections to the practice of even-age management on the Monongahela National Forest in West Virginia were recorded. (Even-age management is based on the principle of developing timber stands in which the commercial timber is of one age. Throughout a larger management unit, various age classes might be represented; however, in any one unit, age distribution is even. Clearcutting is a principal method of developing even-age stands.) Again, a Forest Service study team (1970b) was commissioned to examine the charges and to prepare a public report. The West Virginia state legislature created a 14-member Forest Management Practices Commission and, in addition, requested a moratorium on all clearcutting in West Virginia National Forests, pending completion of the Commission's report. Strong concern was expressed in both reports that inadequate consideration had been devoted to forest values other than timber.

The list of areas where resource managers and citizens have clashed continues to grow. Although different areas are involved, there are many points of commonality; the feeling of citizen-exclusion from the decision-making process, a concern that present directions will have long-run disastrous consequences for both the resource base and the economy, and so forth. But a common catalyst in all cases has been the sincere conviction among many that timber harvesting has become *the* principal forest use and that, as a result, other important values have been given short shrift. Many see timber harvesting as the focal use around which other uses, such as recreation or wildlife management, are fitted in, to the extent they can be accommodated without detracting from timber production. The accommodation of these other values, rather than being the result of a conscious planning effort, is more accurately described as one of incidental and secondary adaptation. Moreover, many of the efforts to bring about a greater balance in forest management programmes have focused on reducing the negative impact of timber harvesting on other values. Although this is a desirable direction, it seems necessary to give more positive attention to aggressively *producing* these other values, rather than just simply trying to minimize adverse impacts on them.

The Historical Foundations of American Forestry Policy

To gain an understanding of the current controversy between forest managers and public groups, it is useful to review something of the historical settings in which American forestry policy evolved. Economic, political and social conditions that prevailed prior to about 1900 had a profound impact on both the image of America's forest land held by the public and on the response of a fledgeling forestry profession.

The conquest of the wilderness was a primary activity for nearly two and a half centuries. Seemingly hostile, providing a hiding place for Indians, and blocking westward movement, the forests of North America were perceived primarily as a barrier to the civilization of a new country. Many settlers were farmers and, to pursue their livelihood, forests had to be removed. Gates (1968) estimates that upwards of 600 billion board feet of timber were removed to make way for crops, livestock and cities (about 12 billion board feet were cut in the U.S. in 1972). Trespass on public and private lands was common. An estimated 450 million feet of lumber was stolen from public lands in Michigan alone. Moreover, efforts to prevent trespass or to prosecute trespassers were not overly successful, due to insufficient funding for enforcement by a generally unsympathetic Congress.

At the same time that exploitation of forest lands was widespread, a series of legislative measures was adopted that greatly impeded efforts to protect timberlands. There was strong Congressional pressure to distribute forest lands to the private sector. For instance, the Timber and Stone Act of 1878 provided that a settler could purchase up to 160 acres of unreserved surveyed public land in California, Oregon, Nevada and Washington territory that was unfit for agriculture for \$2·50 an acre. Persons caught taking timber from public lands were provided with loopholes that permitted them to relieve themselves of liability by purchasing the land for \$2·50 an acre. Thus, at the turn of the century, the legislative branch promoted entry on to public timberland with little or no attention given to the establishment of a positive forest land management programme.

At the same time as forest exploitation was running almost unchecked, concern was being expressed from various corners as to the future of the country's forest lands. For instance, in 1867, the idea of reserving 'through all time' those lands producing quality timber was set forth (Ise, 1920). Proposals to protect timberlands, to ensure more economic appraisals of timber for the purpose of sale and to provide for soil protection, began to appear. President Harrison set aside over 13 million acres as forest reserves in the early 1890s. In 1896, the National Academy of Sciences appointed a commission to report on what measures would be necessary to make the forest-reservations programme successful. The commission's report pinpointed many crucial problems facing the forest reservations (uncontrolled fires, excessive grazing, erosion, etc.). It recognized the inappropriateness of reserving lands better suited to agriculture. It also argued that the reserves were *public*, not the exclusive property of any one group. As a result of the commission's findings, President Cleveland set aside 13 new reserves covering over 21 million acres (Gates, 1968).

At the turn of the century, professional forestry was virtually non-existent. There were fewer than 10 professional foresters in the country and an effort to publish a *Journal of Forestry* in the 1880s failed after one year due to lack of leadership. However, some forestry-related groups began to appear, at both the national and state level. A federal Division of Forestry was established in the late 1800s under Bernhard Fernow, though its functions were primarily confined to supplying information rather than advising on management (Ise, 1920). In part, this was a criticism of Fernow who, although possessing a fine background in forestry, was regarded as being too theoretical and lacking the initiative and dynamism actively to promote professional forestry.

In 1905, the forest reserves were transferred from the Department of the Interior to the Department of Agriculture and, with that move, the U.S. Forest Service was established under the leadership of Gifford Pinchot. The new agency held jurisdiction over about 85 million acres. In less than five years, a concerted expansion effort by Pinchot and Theodore Roosevelt brought these reserves to 194 million acres. Although the speed and apparent lack of planning that accompanied this expansion were severely criticized in some quarters (Clar, 1959), Pinchot (1947) argued that establishment of public reserves was the only available means to prevent prime forest lands from passing into private ownership through the numerous avenues Congress had made available.

Although this discussion hardly does justice to the rich history of forestry in the United States, it is important that one has some feeling for the nature of the political and social conditions existing at the time when much of the present day forest policy was conceived. As we have seen, forest lands had been subjected to extensive abuse, both from timber companies and from private settlers. To conservationists and professional foresters alike, the future of the country's forest lands under a continuing laissez-faire policy must have appeared dim. Moreover, there was little evidence to indicate that conditions might change: on the contrary, Congressional actions had accentuated many of the problems. For example, it was not until 1908 that timber lands were required to be appraised and sold at their actual value (Gates, 1968).

It was in this early period, however, that scientific forest land management was born. Despite limited resources and the presence of a generally unsympathetic Congress, foresters were able to make remarkable achievements in their efforts to bring professionalism to forestry. Much of this success must be attributed to Pinchot's political sensitivity and his ability to gain Executive support (at least from Roosevelt). By demonstrating and promoting the application of scientific principles to forest management, professional foresters gained considerable repute. By establishing their qualifications technically to manage and protect the forests, they also acquired the implicit qualifications to determine normative goals; i.e. what it is the public should receive from the forests. It is only recently this latter ability has been challenged, yet it is the very crux of the present controversy.

The Foundations of Current Forest Management Policy

There was early recognition that forests held several important values. In 1897, the Organic Act (the Congressional Act originally establishing the National Forests) specified water and timber as purposes for which the National Forests were to be established. Since that Act, however, a variety of factors—court decisions, administrative and policy directives, Congressional appropriations, and simply the need to handle changing demands—have broadened the original concept (see Alston, 1972, pp. 19–35). Outdoor recreation, range and wildlife have long been recognized as legitimate values of the National Forests in addition to water and timber. The 'multiple-use' concept of forest management became embodied as law in 1960 along with the concept of 'sustained yield'. Sustained yield had been implied even within the Organic Act of 1897 and various administrative regulations and policies have reinforced the application of the concept to the *timber* resource. However, under the important Multiple Use–Sustained Yield Act (1960) the concept was broadened to include *all* the 'various renewable surface resources of the National Forests'. Thus, the Forest Service received wide discretion and authority to allocate and manage the National Forest System.

The dominant position which the Multiple Use–Sustained Yield concept has come to hold in forest management has been criticized on various grounds. One of the major objectives of the sustained yield philosophy is to serve as a regulator of forest harvesting, ensuring that rates of harvest do not exceed rates of replacement, be it timber, water or range. However, critics of the sustained yield regulation model have pointed out substantive conceptual weaknesses (Thompson, 1966; Waggoner, 1969). For example, sustained yield is not sensitive to signals from the market place and thus tends to provide an even flow of goods and services in the face of widely fluctuating demands. This, in turn, results in periods of insufficient inventory to meet demand (and perhaps more importantly, in a lack of flexibility to do anything about it) while surplus supplies are available during periods of slack demand.

The second criticism focuses on the extent to which the Forest Service has received discretion in the management of public forest lands. This issue, as we shall see, is fundamental to understanding the conflicts we experience today.

Obscuration of Goal-setting and Goal-attaining Roles

Today we live in a society of growing complexity and sophistication. The expansion of knowledge and information is awesome. In response to this growth, society has tended to rely ever increasingly upon specialists to assimilate, weigh and respond to conditions with which the lay citizen could not possibly contend. Congress has certainly adopted this strategy. Over the years, it has increasingly delegated legislative powers to administrative agencies charged with various functional responsibilities. However, one consequence of this process has been the gradual passage of decision-making authority from a body subject to public control (through the electoral process) to bodies buffered and insulated from such control (Reich, 1966). Moreover, the highly specialized nation of expertise found within these agencies has often promoted a technological orientation to problem-solving when, in fact, the problems involve changing social values. As Curlin (1972) points out:

> 'We tend to manage by élitist groups and our checks and balances are through peer review: one must possess the proper credentials before assuming a position of governance. But what special knowledge do these credentials impart the holder when the decision is normative and involves, not objective scientific fact, but consensual value judgments?'

Nevertheless, resource management decision-making has become dominated by 'experts' making decisions not only on *how* some objective might be attained (technical decisions) but also on what objectives *should be* attained (normative decisions).

The current involvement of forestry professionals in the formulation of normative decisions has its roots, at least partially, in the historical conditions discussed earlier. It was largely through the efforts of a cadre of professional foresters that conditions such as unrestricted entry to public forest lands and severe overcutting were halted. Had not these individuals seized the initiative in making value judgments about what forest lands *should* provide, it is likely that the forest lands would have continued to be utilized as a 'commons', with possibly irreversible consequences.

Moreover, it can be correctly argued that yesterday's public showed little concern with the normative issues of forest land management. Those conditions no longer prevail. For instance, a nation-wide poll reported that 86% of those surveyed were concerned to some degree with the condition of our natural surroundings (National Wildlife Federation, 1969). However, it is easy to see some of the conflicts this increased public interest has aroused, as resource managers now find decisions being challenged by persons who do not possess the 'proper credentials'. Managers often fail to see these public concerns as legitimate and appropriate within the normative framework of decision-making.

Implications of a Utilitarian Concept of Management

A frequently stated objective in forestry is to 'get the land under management'. Properly speaking, 'managed' refers to land to which scientific, economic and social principles are applied to achieve specified objectives. Thus, according to this definition, wilderness is managed land as would be an area developed for timber production. However, the concept of 'managed land' has become distorted to refer to areas of substantial human development, particularly in terms of access, and generally planned for timber harvest. Certain consequences stem from this notion.

First, it has encouraged the development of marginal timber-producing lands (the extensive margin). Forcing timber production on lands of marginal economic and physical quality might mean that the concept of sustained yield has been violated because such sites are probably incapable of economic regeneration. Harvesting on these sites, it has been suggested, represents 'mining' rather than timber management because of the extensive time and investment requirement to ensure regeneration (University of Montana Select Committee, 1970).

A second consequence of this utilitarian concept has been the rapid decline of *de facto* wilderness lands as well as other primitive recreational opportunities. Because the utilitarian concept of management emphasizes the availability of resources to man, access is a key factor. For instance, trail mileage on National Forests has declined one-third since the end of World War II, largely as a result of accelerated road construction programmes (Lucas, 1971). Undeveloped wild lands are becoming an increasingly scarce resource and are, both technologically and economically, essentially beyond our capability to reproduce (Krutilla, 1967). Moreover, they often provide recreational experiences that are, to a considerable extent, non-substitutable. Thus, resources have been developed which, relative to the value to be derived from them in a developed state, would be excessively costly to reproduce.

The 'Timber Famine' Myth

A recurrent theme in American forest policy has been the Malthusian threat of running out of wood. Expressions of the concern can be traced to the 1860's, but it has surfaced periodically.

The basic confusion stems from the failure to distinguish between *physical* and *economic* supplies. In forestry, supply has been regarded as a naturally given quantity of timber of a certain size and quality while demand has been defined as a quantity consumed, or 'needed', irrespective of price. With demand interpreted as a fixed requirement, supply thus becomes the critical variable affecting price. The concern with depletion was a major factor in Pinchot's efforts to gain public control over American forest lands.

A recent case study of railroad use of timber demonstrates the conceptual weaknesses of the 'timber famine' notion (Olson, 1971). The author analyses how the railroad industry, a major consumer of wood products in the early portion of the century, accommodated changing conditions of supply by improved technology, mechanization and economies of scale. She concludes (p. 81):

> 'It may well be true that the United States had 820,000,000 acres of forests in 1800 and only 495,000,000 in 1933, or 509,000,000 commercial acres in 1963, but these

> figures are wholly irrelevant to the economic facts of supply. Today the nation has a much larger acreage of timberland accessible at the same real cost than it had in 1800 or 1900.'

This issue of a physical definition of resource, which is, in fact, an economic concept, is a vexing problem. The Forest Service definition of commercial timber land, for example, is based on physical capability: land capable of growing 20 cubic feet of wood per acre per year. Realistically, such lands are generally only marginally suited to timber production for commercial purposes. Focusing limited resources of funds and manpower on the extensive margin of timber production spread these scarce resources even more thinly, with less opportunity for significant return. At the same time, timber production could be significantly increased should investments be made at the intensive margin; i.e. on the high-quality timber growing sites (Marty and Newman, 1969). A lack of knowledge about where such sites are has been a constraint. A study of several western forests suggests existing definitions of commercial forest land have significantly overestimated this acreage because of the inclusion of long isolated stringers of trees, areas where logging would create serious resource problems (e.g. steep, unstable soils with high erosion risk) or unacceptable conflicts with other values such as wilderness, watershed, wildlife and scenery (Wikstrom and Hutchinson, 1971).

Closely tied to the problem of a physical definition of the timber resource have been other implicit assumptions regarding the production of timber that make it the weighted favourite compared to other forest land uses. Gould (1962) has outlined these assumptions as: (i) stability; (ii) land scarcity; (iii) certainty and (iv) a closed economy.

Stability refers to the need for continuing stable flows of wood products and we have already noted some of the shortcomings of that assumption in the discussion on sustained yield. There is little evidence to support this assumption: per capita consumption of wood has risen only slightly in this century and what rise has occurred has been primarily in products other than sawtimber, such as pulp and plywood. The assumption of *land scarcity* argues that forest products are so scarce, compared to labour or capital, that land must be managed so as to maximize biological productivity (although nobody expects this for agricultural lands). Here the aspect of substitutability must be considered. Other materials have come to replace wood in some cases. In others, we have discovered the extent to which labour and capital can be used to substitute for land. Advances in silviculture have reduced acreage demands. The issue of *certainty* is perhaps most easily dealt with. Many of our land management activities have been initiated as though we had perfect or near perfect knowledge of future demand, technology and human values. Actually, the uncertainty that surrounds these questions is immense. Finally, the notion of a *closed economy* has led us to ignore outside supplies of forest products and alternative uses for the land, labour and capital at present tied up in the production of these goods. Simply stated, we have often failed to weigh the rather substantial opportunity costs incurred in the production of wood fibre.

Shifting Public Attitudes to U.S. Forests

At the time when many of the foundations of American forestry policy were laid down, this country was a rural, agrarian society, shortly removed from the conquest of the frontier. For example, the Organic Act establishing the National Forest reserves was signed in 1897, only 20 years after Custer's Last Stand. Western politicians wielded considerable political power and were instrumental in influencing policies that promoted and facilitated western settlement (Ise, 1920). Even today, western political figures hold key positions affecting forest policy. An outstanding example was Congressman Wayne Aspinall of Colorado (now retired from office) who was chairman of the prestigious Public Land Law Review Commission (1970) study on the management of all federal lands, most of which lie in western states.

Today we find most of the population in a few urban areas; 80% of the population resides in 200 metropolitan areas, occupying only 2% of the land. At the same time, the Supreme Court's decisions regarding 'one man-one vote' has further shifted political power from rural into urban hands. An increasing proportion of the population living in urban areas, removed from any direct involvement with the land, possesses an increasing ability to influence public policy.

Although populations have become increasingly concentrated in urban areas, the broad geographic relationships between urban centres and the National Forests have been rapidly changing. When many National Forests were created, they were remote from population centres or well-developed access, and in a pre-automobile era. As a consequence, the management practices undertaken on them were frequently never seen by many people. Increasingly, however, the expansion of metropolitan regions has brought the people closer to the forests. For instance, in 1960, 25% of the National Forests were within 100 miles of the CBD of a large metropolitan area, and an additional 25% were located in densely populated states.

Additionally, the vast improvement in communications technology now means that a nationwide audience can witness events that once only a local population saw. National telecasts of scenes of clearcutting and terracing on the Bitterroot National Forest in Montana might have been responsible for much of the concern expressed elsewhere.

The changing spatial and psychological relationships of the public to the forests have been manifested in shifting public demands. The utilitarian value systems of yesterday have gradually given way to value systems that more strongly emphasize recreation, aesthetics and other appreciative values (Wagar, 1968). However, managerial recognition of these shifting demands has been slow.

One of the central themes to emerge from the literature on man's use of natural resources is that professional resource managers hold distinctly different perceptions of resources from those held by clientele groups. Moreover, managerial beliefs of what these clientele groups seek have been demonstrated to suffer from certain systematic biases that result in misjudgments of group motives and interests. For instance, wilderness users are often judged by managers as holding fairly strong opinions opposing use regulation, but surveys of users suggest that a more favourable attitude about behaviour control is held (Hendee and Harris, 1970).

The varying perspective of managers and users is not difficult to understand. The perception of managers is moulded and influenced by technical-educational background emphasizing production, efficiency and a biological perspective. The perception of the resource by users, on the other hand, is influenced by rather different interests, motives and personal experience. There is, of course, no single public view of the forests. For many forest users, however, the forest is a scene for relaxation and recreation; their perception might be summarized as appreciative as opposed to a more utilitarian view held by managers. Moreover, the respective images these groups hold of one another are subject to biases such as selective perception, that tend to perpetuate and reinforce misconceptions. Thus, the relationship between manager and user often becomes one of conflict and debate.

Timber Products, Prices and the Quality of Life

I earlier discussed the issue of non-substitutability and scarcity in forestry. Also discussed was the rather static picture of per capita wood consumption that has existed over several decades. One might conclude, then, that wood and wood products will become relatively less significant in the future, particularly if population growth continues to flatten out. But other factors need to be cited.

First, an historical examination of price trends suggests that modern society has won considerable independence from the natural resource sector (Barnett and Morse, 1963, pp. 7–11). For example, mineral raw materials maintained essentially unchanged price levels between 1877 and 1957 (Potter and Christy, 1962). However, this independence has been won at a cost. Prices have been maintained to a considerable degree by the rapid growth in technological progress, improved methods of resource exploration and the availability of 'common resources' for the disposition of effluent. As a result, although we have to date avoided the dire predictions of Malthus we have merely delayed some of the costs associated with increased affluence. While we see appreciable gains in the material returns of the 'good life', we also have evidence that what we might call the 'quality of life', a clean, beautiful environment, recreational opportunities and so forth, is in decline (Barnett and Morse, 1963, pp. 252–268).

Although past price trends have demonstrated a remarkable degree of stability or even have declined, future demands on the natural resource sector could lead to some rather substantial variations in the pattern of consumption. Increasing demands for certain non-renewable resources have led us to depend heavily on foreign imports, such as nickel.

As the prices of other materials rise in the face of increasing scarcity, we can expect to see a greater emphasis placed on the use of renewable resources, such as wood. Wood prices have increased substantially over the past 100 years and this is a primary reason why we have seen per capita consumption remain relatively constant, as other materials were readily available as substitutes. If prices increase for these alternative materials, other factors will certainly come into play in the choice of raw materials for domestic and industrial consumption. The energy costs associated with conversion of raw material to finished product will be a major factor. Aluminium can be used for siding instead of wood; however, the energy requirements for the conversion of bauxite to alumina will mean more dams or strip-mined coal fields. For every ton of aluminium produced, over 15 tons of raw materials and processing materials are consumed, compared to about 3·4 tons of raw material for every ton of wood (Dane, 1972). The environmental costs are everywhere and we as a society will be faced with tough decisions regarding the trade-offs we are willing to make.

A third variable must be added to this already complex equation. We see implications for the increased utilization of wood and wood products at the same time as we see concern for the declining 'quality of life'. Moreover, many of the values we associate with a quality life are linked to our forests: clean air and water, recreational opportunities, solitude, etc. To a considerable extent, these particular forest products provide goods and services that are non-substitutable, in terms of the particular human needs and motivations they fulfil. This is especially true of the experiences associated with the more extensive forms of recreation, such as wilderness, that are relatively less abundant or susceptible to capital intensive management. Thus, the stage is set for additional conflicts as pressing demands for increased commodity production meet growing pressures for the use of forest goods and services for scarce, irreplaceable amenity values.

The Resolution of Conflicts–What Possibilities?

The polarity and conflict that perhaps reached a peak in the late 1960s left behind a condition of severe social stress. There is, to be sure, some benefit gained from any confrontation. The overall programme of forest management in the United States has seen some dramatic changes. One recent Forest Service publication, *Timber Management for a Quality Environment* (U.S. Forest Service, 1971) calls for fundamental changes in the entire organization's planning efforts, with particular attention given to the creation of multidisciplinary teams.

Nevertheless, there are still fundamental issues to be decided. At present, the judiciary branch of government has found itself the principal arbiter in resolving many of the conflicts, but the judiciary was never intended to fulfil such a position under the U.S. Constitution. Moreover, relying upon the courts to decide these issues is time-consuming, costly and an inefficient way to manage the public lands. That the courts are used so extensively, however, attests to the extent to which the traditional avenues of decision-making are viewed as unresponsive.

Strong support exists today for increased legislative control over the resource management bureaucracies. Proposals of a 'Blue Ribbon Commission on Timber Management in the National Forests' have been made. Recently, one student of public administration (Kaufman, 1969) has suggested that unless Federal agencies (e.g. the Forest Service) are able to undertake programmes that more effectively reflect public needs and desires, major re-organization might be called for.

Perhaps the most significant and controversial notions for changes in land management programmes stem from the report of the Public Land Law Review Commission (1970). This Commission was established to make a comprehensive review of the public land laws and 'to determine whether and to what extent revisions thereof are necessary'. The report set forth 137 recommendations, of which one in particular has aroused much interest and concern. Recommendation 4 reads: 'Management of public lands should recognize the highest and best use of particular areas of land as *dominant* over other authorized uses' (emphasis added). Furthermore, 'As to land set aside for primary uses, Congress should direct the agencies to manage them for secondary uses that are compatible with the primary purpose'.

The concept of 'dominant use' is, at present, only a recommendation, but it has generated much discussion. By assigning a top priority use to a tract of land, there is, at least conceptually, the benefit of greatly reducing the potential for conflict. In reality, there are a number of flaws. Perhaps one of the most significant is that which currently vexes land managers; how does one arrive at the normative decision regarding what is 'best' or 'highest'? No criteria are provided to assign priorities and it is reasonable to expect that, in the absence of such guidelines, traditional values will continue to predominate (Pyles, 1970). Second, the dominant-use philosophy assumes an unrealistic level of knowledge. Third, there is static notion to the concept; for example, key winter elk range might also be key summer sheep range. It would be rigid and unresponsive to changing values over time. Finally, the application of dominant-use zoning in only those areas where no reduction in the dominant use would be permitted realistically, means there would be few locations where it could in fact be practised.

It is entirely consistent within the current interpretation of multiple use that certain uses 'dominate' in some locations. Much of the controversy over dominant use versus multiple use is one of semantics more than philosophy (Hagenstein, 1972a, 1972b). However, neither concept solves the basic dilemma of establishing a normative framework for the decision as to what shall be done on the land. Although both are expressions of efforts to optimize the flow of benefits from forest lands, the substantive issues of defining the mix of benefits desired and what costs society is willing to incur to obtain those benefits still remain. Resolving these issues will require basic changes in how management programmes are formulated, staffed and funded.

Involving the Citizen in Project Management

Forest policy and management programmes have traditionally evolved within the ranks of forestry professionals; public participation in the development of these has generally been limited, selective and, typically, *post facto*. The debate over forest management gives clear evidence that this situation must be altered.

One major shift toward this end is reflected in the growing efforts by the U.S. Forest Service to solicit and incorporate citizen participation in resource decision-making. Although a variety of reasons is attributed to this effort, one principal value stands out. As a recent administrative study of the Forest Service's public involvement effort noted:

> 'public input serves an especially important role for decision-makers for it is the principal source of information about what values the public holds regarding the National Forests. The "best" use of forest resources is never evident from the resources themselves . . .' (U.S. Forest Service, 1973)

It is through adequate citizen participation that the public can assert its rightful role in formulating normative goals. There are obviously constraints on such goals; resource capability, legal and administrative considerations, budgets and so forth. The role of the resource manager in the relationship between citizen and bureaucrat is to provide basic inventory-level information, define the constraints (legal, budgetary, etc.) within which he must operate and define the probable consequences of alternative courses of action.

With meaningful participation by the public in the decision-making process, we would expect to see the evolution of programmes that reflect changing public values. However, also needed are personnel capable of putting into operation the goals and objectives set forth in such programmes. As I have suggested, there is a strong thrust of public sentiment that views forests as the source of non-utilitarian goods and services. Certainly there is strong pressure to redress the imbalance of management programmes where timber has been the predominant value. For instance, the National Environmental Policy Act (NEPA) calls on Federal agencies to 'utilize a systematic, interdisciplinary approach which will insure the integrated use of the natural and social sciences and the environment design arts in planning and decision-making, which may have an impact on man's environment'. Although multi-disciplinary teams are not being utilized in Forest Service planning efforts, there are large gaps in the skills necessary to meet adequately the obligations imposed by NEPA. This is particularly true with regard to the social sciences. Less than 1% of Forest Service personnel hold degrees in the social sciences. Moreover, most of these persons are within the research branch of the agency and their input to land-use planning is on a limited, *ad hoc* consulting basis. The perspective of the social and behavioural sciences needs to be formally incorporated into the planning process and this will require some marked changes in both traditional forestry education programmes and in agency hiring practices.

Funding is another area where the subordination of values other than timber stands out. Revenues generated from the National Forests are returned to the U.S. Treasury rather than to the agency. Receipts from timber sales predominate; in fiscal year 1972 they constituted 92% of the total receipts. Recreation, on the other hand, provided only 1%. The operating funds of the Forest Service are derived from appropriations from Congress, which in turn are based on budget requests from the Forest Service that have been reviewed (and perhaps revised) by both the Department of Agriculture and the Office of Management and Budget. Congressional priorities in funding are clear from a review of appropriations between 1955 and 1971 (Alston, 1972), when 97% of the funds requested for timber sales and administration during this period were granted. However, only 72% of the funds for recreation were made, even though these two functional areas comprise about the same proportion of the total Forest Service budget request (about 20% each). Not only has funding for other values, such as wildlife habitat and soil and water management, been funded significantly lower than Forest Service requests, but these functions have constituted only minor proportions of the total budget request.

Redressing the imbalance of resource management programmes in the face of severe budget inequities is virtually impossible. Pressure has been placed on the U.S. Congress to recognize other resource values of the forests more fully and to modify appropriations accordingly. Various pieces of legislation have been introduced to provide statutory guidelines to resource managers so that more equitable management programmes and policies can be developed. Manpower cutbacks at a time when public attention is focused on the forests add an additional burden on administrations striving to do a better job. However, it is important to recognize that more money and more people are not the solution to the forest management crisis. The major challenge lies in sensitizing the bureaucratic structure to public values and this will involve the development of formal avenues for citizen participation in decision-making, coupled with a willingness on the part of agency personnel to be receptive and responsive to changing public images of the forests.

References

Alston, R. M., 1972. *Forest—Goals and Decisionmaking in the Forest Service*. Ogden, Utah: U.S.D.A. Forest Service, Research Paper INT-128.

Barnett, H. C. and Morse, C., 1963. *Scarcity and Growth: The Economics of Natural Resource Availability*, Baltimore: Johns Hopkins Press.

Burk, D. A., 1970. *The Clearcut Crisis: Controversy in the Bitterroot*, Great Falls, Montana: Jursnick Printing Co.

Clar, C. R., 1959. *California Government and Forestry*. Sacramento: Division of Forestry, Department of Natural Resources.

Curlin, J. W., 1972. Trans-forestry: another dimension to resource management. Paper presented to the Annual meeting of the Society of American Foresters, Hot Springs, Arkansas.

Dane, C. W., 1972. The hidden environmental costs of alternative materials available for construction. *Journal of Forestry*, **70,** 734–736.

Gates, P. W., 1968. *History of the Public Land Law Development*, Washington, D.C.: Government Printing Office.

Gould, E. M., Jr., 1962. Forestry and recreation. Petersham, Mass.: *Harvard Forest Papers.*

Hagenstein, P. R., 1972a. 'One Third of a Nation's Land'—evaluation of a policy recommendation. *National Resources Journal*, **12,** 56–75.

Hagenstein, P. R., 1972b. The Public Land Law Review Commission and its approach to land use conflicts. *Journal of Forestry*, **70,** 610–611.

Hendee, J. C. and Harris, R. W., 1970. Foresters' perception of wilderness user attitudes and preferences. *Journal of Forestry*, **68,** 759–762.

Ise, J., 1920. *The United States Forest Policy*, New Haven, Conn.: Yale University Press.

Kaufman, H., 1969. Administrative decentralisation and political power. *Public Administration Review*, **29,** 3–15.

Krutilla, J. V., 1967. Conservation reconsidered. *American Economic Review*, **67,** 777–786.

Lucas, R. C., 1971. Hikers and other trail users. In *Outdoor Recreation Symposium: Proceedings* (Ed. W. Doolittle), Upper Darby, Pa.: U.S.D.A. Forest Service, Northeast Experimental Station, pp. 113–122.

Marty, R. and Newman, W., 1969. Opportunities for timber management intensification on the National Forests. *Journal of Forestry*, **67,** 482–485.

National Wildlife Federation, 1969. The U.S. public considers its environments. Princeton: National Institute of Public Opinion.

Olson, S. H., 1971. *The Depletion Myth. A History of Railroad Use of Timber*, Cambridge, Mass.: Harvard University Press.

Packer, P. E., 1967. Criteria for designing and locating logging roads to control sediment. *Forest Science*, **13,** 1–18.

Pinchot, G., 1947. *Breaking New Ground*, New York: Harcourt Brace and World.

Potter, N. and Christy, F. T., 1962. *Trends in Natural Resource Commodities*, Baltimore: Johns Hopkins Press.

President's Timber Supply Advisory Committee, 1973. *Report on Timber and the Environment.* Washington, D.C.: Government Printing Office.

Public Land Law Review Commission, 1970. *One Third of the Nation's Land.* Washington, D.C.: Government Printing Office.

Pyles, H. K., 1970. *What's Ahead for our Public Lands?* Washington, D.C.: Natural Resources Council of America.

Reich, C. A., 1966. *Bureaucracy and the Forests*, Berkeley: Centre for the Study of Democratic Institutions.

Subcommittee on Public Lands, 1971. Management Practices on the Public Lands. Washington, D.C.: U.S. Senate Committee on Interior and Insular Affairs, 5–7 April, 7 May, 29 June, 1971.

Thompson, E. F., 1966. Traditional forest regulation model: an economic critique. *Journal of Forestry*, **64,** 750–752.

University of Montana Select Committee, 1970. *A University View of the Forest Service.* Washington, D.C.: Government Printing Office.

U.S. Forest Service, 1970a. *Management Practices on the Bitterroot National Forest.* Missoula, Mont.: U.S.D.A. Forest Service.

U.S. Forest Service, 1970b. *Even-Aged Management on the Monongahela National Forest.* Washington, D.C.: U.S.D.A. Forest Service.

U.S. Forest Service, 1971. *Timber Management for a Quality Environment.* Washington, D.C.: U.S.D.A. Forest Service, C.I. Report No. 6.

U.S. Forest Service, 1973. *Public Involvement and the Forest Service: Experience Effectiveness and Suggested Direction.* Washington, D.C.: U.S.D.A. Forest Service.

Wagar, J. A., 1968. Non-consumptive uses of the coniferous forest, with special relation to coniferous forest. In *Coniferous Forests of the Northern Rocky Mountains* (Ed.

R. D. Taber), Missoula, Mont.: University of Montana, Centre for Natural Resources, pp. 255–265.

Waggoner, T. R., 1969. Some economic implications of sustained yield as a forest regulation model. Seattle: University of Washington, Department of Forestry Contemporary Forest Paper No. 6.

Wikstrom, J. H. and Hutchinson, S. B., 1971. *Stratification of Forest Land for Timber Management Planning on the western National Forests*. Ogden, Utah: U.S.D.A. Forest Service Research Paper, INT-108.

Wood, N. C., 1971. *Clearcut: The Deforestation of America*, San Francisco: Sierra Club Battle Books.

Wyoming Forest Study Team, 1971. *Forest Management in Wyoming*, Ogden, Utah: U.S.D.A. Forest Service.

18

Outdoor Recreation and the American Environment

TIMOTHY O'RIORDAN AND JOHN DAVIS

During the third quarter of the 20th century, America has become not only a leisure-orientated society but also a leisure-spending one. In 1900 the average work week was 60 hours of arduous manual labour undertaken in exacting environmental conditions. Paid vacations were almost unknown and few but the wealthy had the energy, let alone the time, to indulge in outdoor recreation (Zeisel, 1958). Today the average American works 36–38 hours per week (probably about 20% choose a three-day working week), enjoys two or three weeks of paid vacation and can afford to spend as much as 25% of his income on leisure time activities—about one-third of that on outdoor recreation alone.

Work is no longer cherished for its discipline and moral value as it was less than a lifetime ago. Leisure pursuits provide the focus for the modern American life style: for many, work is regarded simply as a necessary and generally unpleasant means of making leisure activities financially possible. Labour unions are blocking the concept of compulsory overtime (a major point of contention in the 1973 contract negotiations by the United Auto Workers) and are pressing for early retirement. Already for many the retirement age has dropped to 62 and there is growing pressure to permit voluntary retirement after 30 years at the job, i.e. around the age of 55 (Townsend, 1974, p. 7).

Higher disposable incomes, greatly increased leisure time, widespread mobility and a whole life-style centred on leisure have left their impact not only on the American landscape but also upon the American economy. The leisure industry has grown from around $25 billion in 1960 to as much as $83 billion in 1970 (Cunniff, 1971). Consumer society has hit the recreation business with a flourish. In 1970 Americans spent over $20 billion on recreational equipment and services, a figure doubled in 1975. In 1971 sales of sports equipment topped $17 million, motor boat sales exceeded $800 million and ski equipment sales ran as high as $70 million. Innovations in technology have made recreational equipment more durable and less expensive (for example, the use of fibreglass for boats, yachts and skis) and have opened up whole new markets: for example, the snowmobile, the trail bike, the lightweight tents and backpack and the modern, comfortable recreational motor home. Recreational expenditures per person have risen from around $30 to $150 in constant dollars between 1950 and 1970. As might be expected, such expenditures vary considerably with family income. The typical American family earning $10,000 per year or more, spends $500 on recreation (excluding travel costs) including $90 per year on participant sports. But for families earning less than $5000 annually, recreational expenditures amount to only $125 with about $14 going to participant sports (Cunniff, 1971).

The Rise in Outdoor Recreation Demand

In 1962 the Outdoor Recreation Resources Review Commission (ORRRC) published a most comprehensive report on the recreational activities and needs of the American people. Supported by 27 detailed Study Reports, this document remains today as a classic in the analysis of outdoor recreation supply and demand. The ORRRC listed 14 activities in order of popularity with Americans (Table 18.1) to which it appended this commentary:

> 'At present, it is the simple pleasures Americans seek most. By far the most popular are pleasure driving and walking: together they account for 42 per cent of the total annual activity . . . The Sunday drive through the countryside is one of the great experiences that families share, and for those who live in the city it is anything but passive; they will often put up with an extraordinary amount of intervening traffic to break their way out'.

Table 18.1. Number of days in which persons over 12 years old took part in specified types of outdoor recreation, 1960–70

Recreational activity	Total recreational days (in thousands)		Number of recreational Days per year per person	
	1960	1970	1960	1970
Driving for pleasure	872	—	20·7	—
Walking for pleasure	566	1759	17·9	11
Playing outdoor games and sports	474	1929	12·7	16
Swimming	672	1423	6·5	10
Cycling	228	1035	5·2	10
Fishing	260	514	4·2	3
Attending sports events	172	563	3·8	4
Picnicking	279	479	3·5	3
Nature walks	98	337	2·7	2
Boating	159	388	2·2	3
Hunting	95	210	1·9	1
Horse riding	55	177	1·2	1
Camping	66	354	0·9	3

Source: ORRRC (1962); Bureau of Outdoor Recreation (1972a)

Table 18.2. Past and projected outdoor recreation trends in the United States in terms of numbers of occasions

Activity	% change 1960–65	Projected % change 1960–2000
Picnicking	62	266
Pleasure Driving	8	146
Swimming	44	344
Walking	82	356
Fishing	24	121
Boating	38	336
Studying nature	19	180
Hunting	− 1	81
Camping	62	447
Horseback riding	40	225
Hiking	47	368
Water skiing	44	564
Snow skiing	115	—
Wilderness use	68	859
Total, all activities	51	293

Source: Bureau of Outdoor Recreation, 1967

The ORRRC predicted a fourfold increase in outdoor recreation demand by the end of the century (Table 18.2). Some indication of the accuracy of their predictions is already evident. For comparative purposes Table 18.1 shows increases in actual participation in selected activities between 1960 and 1970. In many cases there has already been more than a 300% increase in activity days within these ten years with more than a fivefold increase in camping and walking–hiking. These remarkable statistics reflect not only an increasing number of people participating in outdoor recreation, but also greater frequency of participation by many individuals. It is worthy of note that the most spectacular increases are in those activities which either require a lot of open space and/or expensive equipment which heretofore was beyond the wallets of many Americans. Camping, skiing, sailing, snowmobiling and wilderness hiking are typical of the more rapidly growing pursuits: all of which require large areas of land or water to be most fully enjoyed.

What factors have led to this rising demand? Obviously increased incomes and more leisure time are highly important, but probably of more significance are the love of mobility and a basic desire to 'get away from it all' by visiting the wilder areas of the nation. Americans have always expressed a love for untamed land. 'The most impressive characteristic of the American scenery is its wildness', wrote Cole in 1836. Emerson's essay on *Nature* published during the same year profoundly influenced American intellectual thought throughout the remainder of the decade. To Emerson nature represented 'the Spirit, a guiding Unity which permits man to enjoy an original relation to the universe'. The forest holds particular appeal. 'Within these plantations of God', he noted, 'a decorum and sanctity remain, a perennial festival is dressed, and the guest sees not how he should tire of them in a thousand years. In the woods we return to reason and faith'.

Outdoor recreation undertaken in a natural setting is regarded by many Americans as a spiritual necessity, by some as of fundamental psychological importance (Searles, 1960) and by others as a desirable mechanism of diffusing pent-up frustrations and of building moral character. In 1942 the first National Conference on Outdoor Recreation resolved that 'outdoor recreation has a direct beneficial influence on the formation of sturdy character by developing those qualities of self control, endurance under hardship, reliance on self, and cooperation with others on team work which are necessary to good citizenship' (quoted in Revelle, 1967, pp. 1173–1174). Philosophies aside, to the average American, trapped in the noisy crime-filled city and forced to live a stressful life dominated by the clock, the telephone and a variety of social and business obligations, the call of the wild is strong, and its effects are staggering on the modern American landscape.

Mobility and the Landscape

The car has provided the means whereby this longing for the outdoors can be satisfied. The essence of recreation is delectable spontaneity: freedom of choice and off-the-cuff decision-making. The car provides this versatility and therefore it is hardly surprising that over 90% of all outdoor recreation trips make use of the car and that driving for pleasure is the single most popular recreational activity in the United States (Table 18.1).

Not satisfied even with the mobility of the car, Americans are turning to other forms of motorized recreation. The natural blend of a desire to camp and a dependence on the automobile is the recreational motor home: a veritable house on wheels complete with colour-coordinated fittings, full bathroom facilities, water, electrical and sewerage hook-up arrangements and air-conditioning. The first motorized home was built by Ray Frank of Brown City, Michigan in 1959 (Friedlander, 1973). By 1966 over 285,000 recreational vehicles (including camper-trucks and caravans as well as motor homes) were made, having a value of $567 million. In 1973 some 725,000 were constructed at a value of $2·7 billion and manufacturers are confident of a $5 billion industry by 1975 with over 8 million vehicles on the road. Roughly 13 million American families have joined a motorized camping organization of one kind or another and the sight of recreation vehicle 'caravans' of up to 100 vehicles is not uncommon on American roads during summer months. (It is interesting to record that during the fuel shortages of 1974 sales of recreation vehicles dropped by 10% though manufacturers remained confident of the buoyancy of the industry (Hohnstrom, 1974).)

While we shall see that the growth of the recreation vehicle industry has created serious problems for outdoor recreation managers in America, the rise of the all-terrain vehicle (ATV) has produced far more serious problems. The Bureau of Outdoor Recreation (1972b) reported that there were over

5 million ATVs in the U.S. in 1972—motor cycles, trail bikes, snowmobiles, dune and swamp buggies and airboats—but these figures give no indication of the recency and rapidity of growth in the numbers of these vehicles. In 1961 there were only 10,000 snowmobiles in the U.S.; in 1973 there were over 1·5 million, mostly confined to the snowy northern states. By 1971, 730,000 motor cycles and 700,000 mini-bikes were bought annually by Americans and these figures are expected to rise to 850,000 for each by 1980 (Stupay, 1971). In 1971, projected sales of motor cycles were estimated at $900 million; snowmobile sales reached $376 million (Stupay, 1971, p. 17). In addition to stimulating a new industry, other economic benefits of the ATV boom are also evident: the introduction of the snowmobile has boosted the depressed economies of northern rural states by encouraging summer resorts to remain open all year and hire full-time staff. The Upper Great Lakes Regional Commission reported that snowmobiling was worth $212·2 million to the states of Wisconsin, Minnesota and Michigan in 1970–71.

Despite the economic benefits, however, the ecological damage caused by these vehicles is devastating (Baldwin, 1970). Roaming at will, ATVs have destroyed ecosystems previously inaccessible to man. The effect on wildlife, vegetation and soils has been so noticeable that, in 1972, President Nixon issued an historic Executive Order to urge various agencies responsible for public open space to designate zones from which such uses would be prohibited. As can well be imagined, in a nation that deeply cherishes a frontier tradition of freedom and mobility, reactions to this order were mixed and intense. The problem faced is that of providing balanced policies and procedures that will best protect the resources, promote the safety of both users and non-users and minimize conflicts between various legitimate users of public lands.

An example of the problem is that of trail bike use in the southern California desert (Carter, 1974). Though the desert is vast (25,000 square miles), it is regarded by over 11 million Californians as 'their own back yard'. In 1972 there were over 150 trail bike rallies involving some 67,000 vehicles and 50,000 spectators. On one occasion the bikes obliterated all vegetation over an area 3 miles long and 3–6 feet wide: the creosote bush destroyed in this one race will not recover this century, yet the area is part of a tortoise preserve. The management aim is to restrict use to back-country roads and to design special areas for circuit racing over natural terrain. The California legislature has proposed a $15 licence fee covering two years for all off-the-road vehicles: $6 of this is to be used to provide suitable areas for ATV use. The question remains whether such a policy will satisfy the cyclists' craving to roam freely from horizon to horizon.

The Supply of Recreational Land

Increases in population, affluence, leisure, vacations and car ownership, a good road system and a general awareness of the availability of and pleasure to be gained from recreation have produced these great increases in demand.

Table 18.3. Acreage of and attendance at outdoor recreation areas, by level of government and private sector

	Acres 1965		Attendance 1970	
	Million visits	%	Million visits	%
National park service	26·7	3	172·0	4
Other Federal agencies	419·9	43	665·3	14
State agencies	39·7	4	482·5	10
County and local agencies	5·0	[a]	1500·0	31
Private lands	491·0	50·0	1950·0	41
Total	982·3	100		

[a] Less than 1%.
Source: Council on Environmental Quality (1972, p. 322)

What of supply? The ORRRC noted that the supply of recreational open space was neither being enlarged nor was it in the right place. In 1971 the Council of Environmental Quality commented that less than 3% of public open space is situated within one hour's drive of the large metropolitan centres where nearly 80% of Americans live. In fact, between 1964 and 1970 over 22,000 acres of urban parkland were lost to housing and other uses, and many other areas of open space were effectively cut off from local residents due to the location of freeways, railways and industrial parks. Yet from Table 18.3 it can be seen that 31% of all outdoor activities take place either within or near the city, or less than 1% of all public outdoor space in America.

Nor has the extent of large-scale public open space increased significantly over the past 25 years. Part of the problem here lies with policies which have not stressed the need for acquiring more space for public recreational use, but probably more serious are the costs of acquiring such lands. It is calculated that it would cost over $250 million to purchase the remaining private lands in existing National Parks (a price that is probably doubling every ten years of delay): to protect the critical water source supplying the Everglades National Park would require the purchase of land currently estimated at $170 million.

The increasing sprawl of suburbia has raised the price of an acre of suburban land to over $22,000. In 1969 the Bureau of Outdoor Recreation reported that 'it would require in excess of $25 billion above existing (1969) expenditure levels to give urban dwellers the same amount of nearby recreation opportunity by 1975 that was available on the average nationwide in 1965' (Conservation Foundation, 1972, p. 6). A Task Force of the Conservation Foundation urged a $100 billion 'Buy Back America' programme to acquire 52 million acres of land including 8608 acres of 'vest pocket' parks with a residue of $32 billion for development and improvement of such areas (Conservation Foundation, 1972, p. 74).

The ORRRC anticipated these problems when it recommended the establishment of a fund to be used for the preservation of open space, especially near large metropolitan centres. In 1964 Congress authorized such a fund, the Land and Water Conservation Fund, which was to be financed through gasoline taxes on motor boat fuel and from the sale of surplus federal lands. Between 1965 and 1969 some $500 million were made available for matching grants at both the state and local levels. However, contrary to hopes for an increase in appropriations, President Nixon reduced the allocation from $500 million to $300 million for fiscal year 1973–74. Yet land values in peripheral metropolitan areas continue to rise at between 10 and 20% annually, and state and local governments are more cautious about allocating the necessary matching funds (Council on Environmental Quality, 1973, p. 323).

A considerable amount of outdoor recreation in the U.S. (about 41% in fact) takes place on private lands (Table 18.3). Much of this is leased to clubs and other organizations for a fee. The result is additional income to the landowner and guaranteed rights of access to wealthy recreational groups. In 1965 the Bureau of Outdoor Recreation estimated that there were 2·6 million private enterprises involved in outdoor recreation, using up about 491 million acres of land. About one million farms are open to fishing and hunting. Not all of this is beneficial: sometimes farmers have ceased to permit the recreational use of their lands due to the increasing incidence of vandalism and property damage.

Another aspect of the supply of recreational open space has been the opening up of lands and reservoirs by public agencies which originally provided such facilities for other purposes. For example, neither the Tennessee Valley Authority nor the other two federal dam-building agencies, the Corps of Engineers and the Bureau of Reclamation, had recreation in mind when they flooded vast tracts of land for flood control, hydro-power and irrigation schemes. Yet visits to TVA reservoirs for recreational uses rose from 7 million in 1947 to 48 million in 1964 with boat use increasing from 9600 to 52,700 during the same period (Clawson and Knetsch, 1966, p. 190). Visitors to Corps reservoirs rose from 5 million in 1946 to 275 million in 1970, while recreational use of Bureau of Reclamation lakes increased from 6·5 million in 1950 to over 70 million in 1970 (Clawson and Knetsch, 1966; Council on Environmental Quality, 1972, p. 325).

In past years, both the Corps and the Bureau of Reclamation have been severely criticized by conservationists and economists for the callous manner in which they have flooded scenic valleys and fertile farmlands. (For example, see Maass, 1951; Freeman, 1967; McPhee, 1972.) Yet in many states their reservoirs provide the major supply of water-based outdoor recreation and consequently are very popular. For example, the Dennison Reservoir on the Red River in Oklahoma and Texas recorded nearly 7 million visitor days in 1962 (Corps of Engineers, 1964): ten years later, the demand was probably double that figure. As a result the use of such reservoirs is enormous, and agency officials have wasted little time in publicizing the public benefits generated by their reservoirs. Indeed, many controversial proposals (both large and small) have been advanced by a variety of state and federal agencies because of the potential recreational benefits that are claimed.

Such proposals have included attempts to dam the Colorado River to the point where back-up water might have penetrated the famous Grand Canyon, resulting in irreparable damage to the Canyon with little in the way of local benefits (Freeman, 1967). The proposal was blocked by a Congress spurred by the protests of many environmental groups, but the attitude of the Bureau of Reclamation about such matters can be seen in a comment made by a former Commissioner to David Bower, former President of the Sierra Club, regarding the flooding of a beautiful canyon by another dam in the same area, Glen Canyon Dam. 'I'm a fair man', he confided. 'When we destroyed Glen Canyon we destroyed something really beautiful. But we brought in something else, equally beautiful, more accessible . . . you conservationists say we are destroying Rainbow Bridge simply because we are making it available to people (McPhee, 1972, pp. 180–182).

U.S. National Park Policies

As Runte has noted elsewhere in this volume (Chapter 4), a major spur to the formation of National Parks in America was the desire to preserve areas of magnificent scenic splendour for all Americans. This concept was extended to the protection of particularly important sites of archaeological and geological interest known as National Monuments, and to the preservation and presentation of historic buildings and other features of American heritage in the form of Historic Sites. In 1973 the National Parks Service administered 38 National Parks, 36 National Monuments, 172 Historic Sites and 36 recreational areas.

The legislation establishing the National Parks Service (the 1916 National Parks Act) was designed 'to conserve the scenery and the natural and historic objects preserved therein in such a manner and by such means as will leave them unimpaired for the enjoyment of future generations'. Preservation of intrinsic landscape qualities alongside enjoyment by present and future populations of Americans is a difficult and contradictory goal to pursue. In 1972 there were 214 million visits to the whole National Park system compared with only 121 million visits seven years before (Council on Environmental Quality, 1972, p. 320). Visits to U.S. National Parks accounted for about 20% of the 837 million visits to all federal recreation areas in 1971.

The problem is not so much people *per se* but the urban culture, especially the automobile, that they bring with them. One car uses almost 100 square feet of surface that could comfortably accommodate a dozen people on foot. Bumper-to-bumper traffic, congested campgrounds and over-used trails have created enormous headaches for park officials who must balance a mandate to encourage park visitation while protecting park values. The resulting situation is rather chaotic as Rosak's (1972, p. 40) description of a trip to Yellowstone indicates:

> 'The traffic for forty miles leading into the park and all the way through the park was packed bumper to bumper. There was not an inch of solitude or even minimal privacy to be found anywhere during the two days we stayed before giving up and leaving. Never once were we out of earshot of chattering throngs and transistor radios or beyond the odour of automobile exhausts. We spent a major part of each day waiting in line to buy food or to get water. And at night the coming and going of traffic made sleep nearly impossible. Ah, wilderness . . . !'

The obvious solution is either to limit or ban the use of the car. Take the case of Yosemite Valley in Yosemite National Park near San Francisco. Over 100 years ago Frederic Law Olmstead, the creator of the road system which permits public access to this simply magnificent valley, prophesied:

> 'It is but 16 years since the Yosemite was first seen by a white man, several visitors have since made a journey of several thousand miles at large cost to see it, and not withstanding the difficulties which now interpose, hundreds resort to it annually. Before many years if proper facilities are offered these hundreds will become thousands; in a century the whole number of visitors will be counted by the millions. An injury to the scenery that may be unheeded by any visitor now, will be one of deplorable magnitude when its effect upon each visitor's enjoyment is multiplied by these millions. But again, the slight harm which the few hundred visitors of this year might do, if no care were taken to prevent it, would not be slight should it be repeated by millions. (Quoted in Council on Environmental Ouality, 1972, p. 313).

In 1970 this valley, only some five miles long, took the brunt of as many as a million cars on a holiday weekend. There are 750 miles of roads, 2100 permanent buildings, 7 amphitheatres, 24 water systems, 30 sewer systems, 10 electric generators, 93 miles of transmission lines, 54 picnic areas, 3143 campsites, and hotel and cabin accommodation for 8586 people daily (Council on Environmental Quality, 1972, p. 326).

In 1970 the National Parks Service inaugurated an experimental policy of banning the private car and instituting a one-way public transport system. Visitors entering the Valley must now park their cars at the perimeter and board one of many shuttle buses that drive through the area. Now there is room for the equestrian, the pedestrian and the cyclist in an area that was once essentially barred from them. Moreover, with many of the campgrounds relocated at the edge of the valley the campfire smoke and automobile smog that obliterated many of the scenic vistas have all but disappeared. Armed with this success in Yosemite, the Service inaugurated a similar system of public shuttle buses for Everglades National Park (Florida) in 1972 and is proposing further limitations on private automobile use in the village area of the Grand Canyon National Monument (Arizona) and Mount McKinley National Park in Alaska. (In the last case the problem is somewhat different: traffic densities are not substantial in Mount McKinley but officials are concerned that an increase in cars will frighten the wildlife that many travel to the park to observe.)

Much of the increase in National Park visitation has come from a new, urban-orientated camping culture that has placed a great strain on camping facilities and management policies. In most parks, campgrounds are full during July and August and long waiting lines of frustrated motorists await campsites. The majority of campers come in a recreation vehicle and regard the large busy campground as a 'natural setting' in which to enjoy 'back-country camping'. Clark and his colleagues (1971) demonstrated that many such campers are quite tolerant of noise, littering and vandalism. Indeed, both theft and senseless vandalism are serious problems in U.S. public parks generally, with the result that scarce funds must be diverted to security, maintenance and more expensive facilities. In 1972 the National Park Service reported that serious crime had risen by 150% between 1966 and 1970 (from 2336 to 5904 cases) compared with a national increase of 71% (Council on Environmental Quality, 1972, pp. 326–332). In one year 361 people (probably only a fraction of all those culpable) were caught stealing objects from the Petrified National Forest. Repairing damaged facilities now costs $1 million annually and defacement of Indian carvings in some areas is irreparable. Litter clean-up costs in American public spaces exceed $500 million each year. Much of this damage could be avoided if people were to accept responsibility for protecting the quality of their natural surroundings: however, such an ethic does not appear to exist among many modern recreationists.

In an attempt to grapple with the problem of overcrowding, the National Park Service established an experimental campsite registration system in 1973, covering six of its most popular parks: Grand Canyon, Grand Teton, Mt. Rainier, Everglades, Yellowstone and Acadia. The system guarantees a campsite at a surcharge of $1·50 per night (unreserved sites are $3·00 nightly). In addition, the Service is encouraging private camp-ground operators to locate facilities near parks to take up much of the park overflow. Furthermore, the Service has constructed a large number of more primitive 'walk-in' campgrounds in less accessible areas at nominal nightly cost (25 cents) to cater to the demands of an ever-growing and an increasingly vociferous group of young park visitors who wish to enjoy a natural experience on the terms offered by the park itself, not the automobile.

The National Park Service is conscious of its educational role and the need to preserve natural areas not only as a showcase for urban Americans but as a living example of how man can act as steward of the natural landscape. Accordingly it is pressing for expanded educational programmes, ranger interpretation services and improved self-guided nature trails. In 1971 it inaugurated three interrelated educational programmes: National Environmental Educational Development (NEED), an integrated curriculum package available to every school in the U.S.; a system of National Environmental Study Areas (NESAs) which provide interpretive laboratories for intensive field study; and National Environmental Educational Landmarks (NEEL), which have a similar educational function to NESAs but which lie outside the National Park System. The aim of these programmes is to link young Americans to their natural landscape so that modern U.S. outdoor recreation policy meets one of its basic objectives: to improve the understanding of man's relationship with his surroundings.

Wilderness Recreation

Although National Forests were officially designated in 1905, as early as 1843 camping expeditions by the wealthy, accompanied by servants, were reported in many forest areas. Hunting and fishing trips became so common that luxurious hotels sprang up in remote districts. The woodland had become popular. Nevertheless the 1905 Forest Reserve Act was designed not to cater to the demands of outdoor recreation but to protect watershed areas and preserve the nation's timber for continuous productive use. In 1971 an extensive study of recreation in the National forests pointed out that while some areas were quite popular, the main need was to acquaint people with the great recreational opportunities available in the forests generally.

Today the problem is almost the reverse. Visits to the 181 million acres of National Forests rose from 250,000 in 1950 to over 180 million in 1970. The upsurge in demand has been so great that the Forest Service is faced with a serious backlog in its provision of recreational facilities. By 1970 only 4480 of a projected 32,000 specific projects (campsites, launching ramps, picnic places, etc.) had been completed (Hill, 1971, p. 1).

The Forest Service is primarily responsible for administering the nation's wilderness areas, following the passage of the 1964 Wilderness Act (some wilderness areas lie in National Parks and other federally-owned lands). Although fewer than 5% of Americans visit such areas, wilderness use is growing at about 16% per year as more and more people seek the solitude, the physical exercise and the uninhabited natural surroundings that the wilderness system is designed to provide. To date, all this growth has taken place on a relatively fixed acreage: between 1940 and 1969 wilderness areas were enlarged by only 5% to about 10 million acres (Stankey, 1972). Consequently, signs of damage to the fragile wilderness ecosystems are already apparent.

Thus the problem facing wilderness managers is to define the ecological carrying capacity of such areas. This will depend upon the robustness of the flora and fauna, the opportunity to divert users to other areas and upon public education campaigns. But there is another form of carrying capacity: the tolerance which wilderness users have to meeting each other. The wilderness experience depends upon solitude: excessive contact with other recreationists or other kinds of uses can easily destroy the value of this experience. The problem therefore is to define this 'psychological' carrying capacity (as well as the ecological carrying capacity) and to limit use accordingly. While managers agree that some kind of limitation on wilderness use is necessary for certain areas, they also recognize both the practical and political difficulties of executing such a policy, because wilderness areas have large unsupervised boundaries and freedom of access is a right deeply cherished by many users. However, in 1972 the Forest Service began to make the registration of all trail use mandatory as a prelude to limiting use through some form of permit system (Hendee and Lucas, 1973). In 1973 wilderness use was actually restricted in four areas: Rocky Mountain and Great Smokey Mountain National Parks and Sequoia and Kings Canyon National Forests. Such drastic moves are necessary to protect the very landscape that users seek to enjoy, but which in aggregate they unwittingly destroy.

Under the terms of the 1964 Wilderness Act various federal agencies were authorized to designate potentially suitable areas for inclusion into the National Wilderness Preservation System by 1974. About 47 areas are being actively considered (some of these in the wilderness-deprived eastern states) and the battle for their inclusion should be a warm one as traditional lobbies interested in developing many of these areas in view of increasing natural resource scarcities remain firmly entrenched in the American political scene. A similar struggle is expected over the completion of a National Wild and Scenic Rivers System (launched in 1968) where only three of eight specially designated rivers have so far been granted full protection. Nevertheless, a five-year moratorium on construction projects remains in effect for all eight river basins (Council on Environmental Quality, 1973, pp. 223–224).

Urban Recreation

It has already been noted that there is an imbalance of supply and demand in outdoor recreation in America, but nowhere is this more evident than in the city. Despite the enormous increase in the use of non-metropolitan open space, the fact is that probably 50% of the population remain in the city in a typical summer weekend, and that only 16% take more than two weekend trips away from the city (National and Provincial Parks Association of Canada, 1971, p. 11). But a paradox exists here, for most urban neighbourhood parks are hardly ever used. Gold (1972) attributes this to poor location vis-à-vis the needy population, poor maintenance, hooliganism and crime and the fact that many urban recreation planners are not very familiar with the particular recreational needs of inner city residents.

Many children desert the park and the playground for the readily accessible neighbourhood street where the opportunity for spontaneous play is made more interesting by contact with new acquaintances. 'Perhaps the most fundamental reason for the non-use [of urban open space],' observed Gans (1957, p. 312), 'is the fact that much of the leisure offered by public recreation is self-oriented—encourages self-improvement, self-expression or interaction. Individual leisure preferences tend toward involvement with the person in vicarious role-playing and various sorts of fantasy, that result in a different orientation towards the self.'

It would appear that urban recreation is neither sufficient for nor appropriate to the needs of the local residents. This is the conclusion of Cicchetti (1971) who notes that inner city people are actually participating in less recreation than they used to. The National Commission on Civil Disorders (Kerner Commission, 1968, p. 1) reported that in 75% of cities where disturbances occurred in 1967 lack of recreational facilities was a major grievance, and in 67% of these cites a lack of organized recreational programmes was also cited as a cause of unrest.

The wish of many community leaders is to be given the opportunity to work with city officials in the provision of suitable recreational facilities to meet the particular needs of their communities. Leisure and recreation activities can offer a focus for community involvement, and city politicians throughout the nation are considering proposals to encourage greater community participation in local area planning. In addition, proposals are now being advanced to extend public transport to local public parks in an effort to encourage the otherwise immobile to benefit from the pleasures of open space and exercise for which many yearn. Denied the reality of such activities, many simply dream about what they would like to do. For example, Craig (1972) found that Negroes in a southern community aspired to the same recreational pursuits as whites but that they were barred either socially or economically from participating in such activities. In 1972 two major metropolitan parks were designated after a number of years of debate: both of these, the 23,000 acre Gateway National Recreation Area in nearby New York City and the 24,000 acre Golden Gate National Recreation Area beside San Francisco, will be serviced by public transport. The designation of these parks represents an important breakthrough in the drive to ensure more equality of outdoor recreational opportunities for all Americans.

Recreational Housing and Land Use Control

The desire for a country residence as a retreat from the stress, noise, fear and general unhealthiness of modern urban America is strong. In 1970 over 10% of all houses built in the U.S. were second homes (1·5 million in all). The most popular areas are the scenic fringes of the great metropolitan centres—New England, the Smokey Mountains, the Rockies and the Sierras—followed by peripheral regions favoured by sunshine, such as Florida, Texas, Arizona and the Californian coastline. Where possible, waterfront land is most desired. One outcome of this demand has been a noticeable increase in rural land values. In 1967 a report by the Bureau of Outdoor Recreation (1967) estimated that land values throughout the nation were rising by between 5 and 10% annually (or doubling every seven to fourteen years). Farm land in California increased by 71% between 1960 and 1966, Texas 65% and Wyoming 45%. Waterfront land was worth $4400 per acre undeveloped in 1965: today it is worth $11,000.

Development companies entice prospective buyers to free dinners and slide shows, then offer to fly interested purchasers to potential homesites. Sometimes the buyers do not know what they are getting for their money and lots worth up to $5000 are sold to unsuspecting couples who subsequently find that all they own is a piece of desert scrub or swamp. For example, the Gulf American Land Corporation began development of 6000 homes at Cape Coral, near Fort Meyers, Florida. The raw land was worth perhaps $500 per acre. Today the Corporation is selling one-third acre lots for $3–5000 each without any water, sewers or even road access (Cahn, 1973a). The Housing and Urban Development Agency estimates that 'upwards of 10,000' development companies are in existence, operating over 4400 sub-divisions. (These are conservative figures as many developers simply do not bother to register.) Florida has over 500 developers selling about 200,000 lots per year, Arizona has 461 recreational sub-divisions, and 561 such sub-divisions are reported in California where one company (Boise Cascade) has sold more than $300 million worth of lots at 18 recreational sites over the past five years.

With the price of second homes outstripping the financial resources of many middle-income Americans who still yearn for a piece of waterfront land, the condominium campground has recently attained much popularity. Here, for the price of a few thousand dollars, an individual can buy a campsite fully fitted for water, electricity and sewage, on to which he can drive his recreational vehicle at any time he pleases. Resorts of America, Inc. now own condominium campgrounds in Florida, Tennessee and Georgia, and plan to develop a chain of 50 more around the country, each with an adjacent franchised Sheraton Inn. Individual plots sell at $7500–13,000 with a nightly rental charge of $10. Plot owners rent their sites to less fortunate campers during the period they are not in occupancy, thereby earning enough money to pay off their investment.

The effect of recreational land speculation is staggering, not simply on land prices (whether urban or rural) but upon the quality of the landscape. Prime scenic areas are increasingly alienated from public use as lakeshores, coastlines and pleasant hillsides and vales are being gobbled up by private interests. It is therefore inevitable that some sort of control on land development should be forthcoming. This is taking two forms: the enforcement of appropriate land zoning codes and the likelihood of legal restrictions on the private use of privately owned property in order to protect the public interest.

During 1973 Congress failed to agree over the merits of a National Land Use Policy Bill which would assist states in making inventories of remaining open space and encourage them to protect environmentally critical areas. Even the report of a special task force which thoroughly reviewed the problem of land use legislation and recommended much stiffer land use controls (Reilly, 1973) did not spur Congress into action. The debate over the Bill illustrates the conflict between those advocating some sort of public control of all land and those who believe that private property should be enjoyed with minimum obligation to the public interest. Passions are strong and the Bill is long delayed. Meanwhile a number of states have passed strict land use development control codes. Probably the most notable is Vermont's Development Control Act which states that a Land Capability and Development Plan must be prepared for the whole state and that developers must furnish environmental impact statements for all new housing schemes. (It should be emphasized that this legislation is particularly aimed at recreational housing schemes. The states of Oregon, Wisconsin, Hawaii and Florida have recently passed similar legislation for similar purposes (Council on Environmental Quality, 1973, pp. 215–222).)

In addition to statewide zoning codes, a number of communities are at present endeavouring to stop the development of new housing (again, particularly recreational housing) in their areas. For example, faced with the prospect of a population of 200,000, the town of Boca Raton, Florida (present population 35,000) passed an ordinance limiting the total number of dwelling units to 40,000, or a population of about 105,000 (Council on Environmental Quality, 1973, p. 220). A number of other communities in Florida have either adopted street limits on residential densities or, like Miami, have simply refused to permit new building in certain districts under their jurisdiction. In 1972 the citizens of Colorado voted down a bond issue which was necessary to fund the proposed Winter Olympic Games of 1976 since they did not wish the inevitable recreational boom that would follow such a development. Governor McCall of Oregon has publicly commented that tourists are welcome as long as they do not plan to stay in the state already burdened with second-home development in prime scenic areas.

Linked to this no-growth motive is the issue of protecting recreational open space. In 1972 the residents of California voted in a $100-million bond issue to protect the remaining areas of the Pacific coastline for public access. New York State residents voted in favour of a $1·14-billion bond issue to protect wilderness, wetlands and other wild areas for public open space. Environmentalist lobbies forced Governor Rockefeller to control all further private development in the Adirondack State Forest Preserve: a highly controversial decision since much of the Preserve is owned privately (Cahn, 1973b). The Adirondack initiative means that, on over half the private lands in the area, development is limited to a density of not more than one building per eight acres, that second-home development is to be severely curtailed and that spot-uniform zoning is to be discouraged in favour of cluster development in specially designated 'nodes'.

Limitations on development and the preservation of critical areas of amenity imply increasing public regulation of private property. This is indeed the trend in U.S. land use policy and it has raised very important legal questions regarding the authority of states to restrict the use of private property without due compensation. The basic issue here is whether land is to be regarded as a commodity for private consumption or as a resource which entails certain limitations on private use in the interests of the public welfare (Bosselman and Callies, 1971). For example, if a developer has purchased rights to an area of coastal marshland in the expectation of filling in the area for an industrial park, only to find that the state prohibits the use of fill-in in critical wetland areas, should he be compensated for his potential loss? Similarly, if a developer plans a middle-income, second-home scheme on a scenic hillside only to be faced with a state moratorium on second-home developments should he likewise be compensated? Such questions are very difficult to answer since there are no clear legal guidelines here but many states are facing precisely such claims

A number of test cases in U.S. courts have not resolved this issue, since there are important constitutional questions at stake, but it does appear that, politically at least, the American public is slowly recognizing the fact that public interest considerations must be formally integrated into the land development process and the private property owners must conform to certain minimum codes of social obligation. Nevertheless, the issue remains largely unsettled and illustrates the problem of safeguarding recreational amenity against the many pressures to expropriate scenic areas for private recreational enjoyment (Council on Environmental Quality, 1973, pp. 121–150, 214–222).

Conclusion

The outdoor recreation demands of modern Americans have had a significant effect on the American landscape and pose complex problems for recreation managers and policy makers. On the one hand, the growing pressures for low-density recreational use of a declining supply of land make it all but impossible to enlarge substantially the existing supply of open space. Consequently, any attempt to limit the supply of such areas is greatly contested by recreational and environmental groups. Organizations such as the Conservation Foundation, the Sierra Club, Friends of the Earth and the Environmental Defence Fund are devoting much of their attention to the protection of America's wild spaces. Yet the demand is already so great that personal freedoms as to the use of such areas are increasingly restricted.

On the other hand, for another minority of Americans, namely those who cannot find either the time or the mobility to leave the city, there is also an imbalance of supply, particularly a quality of supply, in relation to demand. The problem here is one of conflicting views as to genuine needs and to the appropriate mechanisms through which inner city recreational needs can best be provided. City and local government officials are desperately trying to eke out their precious budgets to purchase more open space on the metropolitan peripheries yet they find that many of their existing parks are either underused or misused.

At the intermediate level the almost frantic desire to 'get away' has resulted in a massive increase in recreational pleasure driving, in the widespread popularity of the recreational vehicle and in increasing use of the all-terrain vehicle. A hyper-productive society demands hyper-productive leisure pursuits and is not easily persuaded to change its ways. The problems of the urban environment—congestion, smog, crime, anger, frustration—are all too readily transposed to the national settings where many seek to escape such displeasures.

In so many ways an analysis of outdoor recreation in modern America reveals many of the more brutal traits of the American culture: its latent selfishness, its ruthless demand for personal satisfaction, its unwitting dependence first on materialism and subsequently upon private expropriation of environmental quality, its lust for mobility and its basic yearning for 'natural' landscapes. Something has to give, and in this case it seems to be a little of everything.

References

Baldwin, M. F., 1970. *The Offroad Vehicle and Environmental Quality*, Washington, D.C.: Conservation Foundation.

Bosselman, F. and Callies, D., 1971. *The Quiet Revolution in Land Use Control*, Washington, D.C.: Government Printing Office.

Bureau of Outdoor Recreation, 1967. *Recreation Land Price Escalation*, Washington, D.C.: Department of the Interior.

Bureau of Outdoor Recreation, 1972a. *The 1970 Survey of Outdoor Recreation Activities: Preliminary Report*, Washington, D.C.: Department of the Interior.

Bureau of Outdoor Recreation, 1972b. Draft environmental impact statement pertaining to the use of off-road vehicles on the public lands. Washington, D.C.: Department of the Interior.

Cahn, R., 1973a. Land in Jeopardy, Boston: *The Christian Science Monitor*.

Cahn, R., 1973b. Where do we grow from here? Boston: *The Christian Science Monitor*.

Carter, L. J., 1974. Off-road vehicles: a compromise plan for the California desert. *Science*, **183,** 396–400.

Cicchetti, C. J., 1971. Some economic issues in planning urban recreational facilities. *Land Economics*, **47,** 15–23.

Clark, R. N., Hendee, J. C. and Campbell, F. L., 1971. Values, behaviour and conflict in modern camping culture. *Journal of Leisure Research*, **3,** 143–159.

Clawson, M. and Knetsch, J. L., 1966. *Economics of Outdoor Recreation*, Baltimore: Johns Hopkins Press.

Conservation Foundation, 1972. *National Parks for the Future*, Washington, D.C.: Conservation Foundation.

Corps of Engineers, 1964. *Recreation*, Washington, D.C.: Department of Defence.

Council on Environmental Quality, 1971. *Environmental Quality: Second Annual Report of the President's Council on Environmental Quality*, Washington, D.C.: Government Printing Office.

Council on Environmental Quality, 1972. *Environmental Quality: Third Annual Report of the President's Council on Environmental Quality*, Washington, D.C.: Government Printing Office.

Council on Environmental Quality, 1973. *Environmental Quality: Fourth Annual Report of the President's Council on Environmental Quality*, Washington, D.C.: Government Printing Office.

Craig, W., 1972. Recreational activity patterns in a small Negro urban community: the role of the culture base. *Economic Geography*, **48,** 107–115.

Cunniff, J., 1971. Leisure time spending climbs to new heights. *The Vancouver Sun*, 3 October, p. 17.

Freeman, A. M. III, 1967. Six federal reclamation projects and the distribution of income. *Water Resources Research*, **4,** 877–889.

Friedlander, P. C., 1973. Recreational vehicles: a roundup. *The New York Times*, 11 March, pp. 10, 47.

Gans, H. J., 1957. *Recreation Planning for Leisure Behaviour: A Goal-Oriented Approach*. Philadelphia: University of Pennsylvania, Unpublished Dissertation.

Gold, S. A., 1972. Non use of urban parks. *Journal of the American Institute of Planners*, **38,** 369–378.

Hendee, J. C. and Lucas, R. C., 1973. Mandatory wilderness permits: a necessary management tool. *Journal of Forestry*, **71,** 206–207.

Hill, G., 1971. National Forests: timbermen vs conservationists. *The New York Times*, 15 November, pp. 1, 48.

Hohnstrom, D., 1974. Recreational vehicle forms challenge energy crisis. *The Christian Science Monitor*, 4 February, p. 7.

Kerner Commission, 1968. *Report of the National Advisory Commission on Civil Disorders*, New York: Bantam Books.

Maass, A., 1951. *Muddy Waters: The Army Engineers and the Nation's Rivers*, Cambridge, Mass.: Harvard University Press.

McPhee, J., 1972. *Conversation with the Archdruid*, New York: Ballantine.

National and Provincial Parks Association of Canada, 1971. Brief to the public hearings on Banff, Jasper, Yoho and Kootenay National Parks. Toronto: National and Provincial Parks Association of Canada.

Outdoor Recreation Resources Review Commission, 1962. *Outdoor Recreation for America*, Washington, D.C.: U.S. Government Printing Office.

Reilly, W. K. (Ed.), 1973. *The Use of Land: A Citizen's Policy Guide to Urban Growth*, New York: Crowell.

Revelle, R., 1967. Outdoor recreation in a hyper-productive society. *Daedalus*, **96,** 1172–1191.

Rosak, T., 1972. *Where the Wasteland Ends: Politics and Transcendence in Post-industrial Society*, Garden City: Doubleday.

Searles, H. F., 1960. *The Non Human Environment*, New York: International Universities Press.

Stankey, G. H., 1972. Myths in wilderness decision making. *Journal of Soil and Water Conservation*, **26,** 184–188.

Stupay, A. M., 1971. Growth of powered recreational vehicles in the 1970s. In *Proceedings of the 1971 Snowmobile and Off the Road Vehicles Research Symposium*. East Lansing, Mich.: Michigan State University, Department of Parks and Recreation Resources, pp. 14–18.

Townsend, E., 1974. Unions push earlier retirement. *The Christian Science Monitor*, 1 February, p. 7.

Zeisel, J. S., 1958. The workweek in American industry. In *Mass Leisure* (Eds. E. Larrabee and R. Meyersohn), New York: Free Press, pp. 145–153.

19
The Role of Environmental Issues in Canadian–American Policy Making and Administration

TIMOTHY O'RIORDAN

On 24 October 1969, the Canadian Prime Minister, Mr. Trudeau, rose in the House of Commons to state that, in developing its policies for the Canadian north, Canada had four primary interests: the security of Canada, the economic development of the north, the preservation of ecological balance and the continued high stature of Canada in the international community. Less than a month later the Canadian Parliament acted upon these principles when the Standing Committee on Indian Affairs and Northern Development recommended 'that the government of Canada indicate to the world without delay that vessels, surface or submarine, passing through Canada's Arctic Archipelego are and shall be subject to the sovereign control and regulation of Canada (*Hansard*, 1969, p. 5943). By April 1970, the government had approved the second reading of the Arctic Waters Pollution Prevention Act which established Canadian sovereignty over all waters north of the 60° parallel lying within 100 miles of the Canadian shoreline. The implications of the legislation were that all ships navigating these waters must meet stiff Canadian regulations, including adequate mechanical devices to avoid inadvertent waste discharge, the presence at all times of a Canadian pilot and the deposit of a bond to cover liability in the case of accident, to safeguard against pollution of Arctic waters (*Canada Gazette*, 1972a).

The immediate Canadian concern was pollution by oil, since the discovery of large oil reserves along the northern Alaskan shoreline had prompted a number of U.S. oil companies to consider the shipping of oil by tanker through the North-west Passage to eastern ports. The most spectacular indication of this intent was the voyage through the north-west Passage in July 1969 of the empty tanker 'Manhattan' sponsored by the Humble and Atlantic Richfield oil companies. The 'Manhattan' was accompanied throughout its journey by a Canadian icebreaker. The Canadian government was determined to protect the vulnerable yet still undisturbed Arctic ecology from depredation by man, and felt responsible to all mankind when undertaking this task. Mr. Trudeau made this quite clear in the House. 'We do not doubt for a moment that the rest of the world would find us at fault, and hold us liable,' he said, 'should we fail to ensure adequate protection of that environment from pollution or artificial deterioration. Canada will not permit this to happen' (*Hansard*, 1969, p. 39).

Nevertheless, extension of Canadian territorial sovereignty to 100 miles offshore was an act of great daring on the part of the Canadian government. International conventions have restricted the formal extension of national sovereignty to 12 miles (see Goldie, 1970), but Trudeau was insistent on his nation's rights in the matter. 'Canada will not submit that legislation to the Hague court,' he commented, 'as long as international law has not caught up with technological developments' (*Hansard*, 1972, p. 4324). The United States was most disturbed by this act of daring and in a diplomatic note to the Canadian Minister of External Affairs, Mr. Sharp, it made its position quite plain. 'The enactment and implementation of these measures would affect the exercise by the U.S.A. and other countries of the right to freedom of the seas in large areas of the high seas and would adversely affect efforts to reach (international) agreement on the use of the seas . . . (International) law provides no basis for this proposed unilateral extension of jurisdiction on the high seas, and the U.S.A. can neither accept nor acquiesce in the assertion of such jurisdiction (*Hansard*, 1972, p. 5923).

Mr. Sharp was unperturbed by this rather fierce diplomatic language. 'The problem of environmental preservation,' he replied, 'transcends traditional concepts of sovereignty and requires an imaginative new approach oriented towards future generations of men and the plant and animal life on which their existence and the quality of that existence will depend. The problem of environmental preservation moreover must be resolved on the basis of the objective considerations of today, rather than the historical accidents and territorial imperatives of yesterday' (*Hansard*, 1972, p. 5948). His Parliamentary Secretary, Mr. Paul St. Pierre, clarified the Canadian position on this matter even more fully: 'International law in relation to the sea, made largely by shipping nations for shippers, has ignored and overlooked the rights of third parties, the coastal states, and until international law has developed to the desired level, this country is obliged to act unilaterally. The prime reason is the new and unexpected threat of massive pollution' (*Hansard*, 1972, p. 5963). The Canadian government promulgated this legislation on July 1972, an action which heralded a new era of international relations between Canada and the United States.

The Canadian–American controversy over the Arctic Waters Pollution Act is indicative of the extent to which public concern over environmental protection has influenced the relations between the two countries. The purpose of this chapter is to review and assess how the whole question of 'the environment' has created new demands upon both legislative and administrative performance in the two countries. Of primary interest is the manner in which each country has responded to actual or potential environmental damage to their lands and peoples caused by unilateral actions undertaken by the public or private sector of the other nation. However, before pursuing the nature of the conflicts that have arisen between the two countries, it is useful to look briefly at the nature of modern environmental issues that have strained international relations, and have forced all developed nations of the world to reassess the legal and administrative arrangements through which they handle environmental questions.

Peculiar Features of Modern Environmental Issues

Environmental issues have a number of peculiarly intrinsic features which have shown up the inadequacies of traditional governmental administration. In summary, these are: (a) the interconnectedness of man–environment relations and the increasingly pervasive incidence of unintended side-effects; (b) the difficulty of making sound social choices; (c) the rise of the administrative state in monitoring and regulating environmental impact and the compensating need for changing the legal status of the citizen to defend his environmental rights; (d) the effect of transnational environmental damage on traditional concepts of national sovereignty and (e) the effect of public concern over the environmental impact of resource exploitation, particularly by alien interests.

(a) The interconnectedness of Man–Environmental Relations

Widening public concern over environmental damage (defined both in biological and in social terms) has brought to the attention of many policy makers the enormous interrelatedness and complexity of linkages that make up the human and biophysical environment. No matter how careful the analysis, any major decision involving the use of natural resources will produce a series of unintended consequences which may be so substantial as to create serious political repercussions. An example of this, which will be discussed more fully later, is the proposal by the Canadian government to help private industry construct a road–pipeline complex to bring both American and Canadian natural gas from the high Arctic to the great metropolitan centres of Canada and the United States. While the biophysical impact of such a pipeline on the northern ecology is not really well-known but could be substantial, the social and cultural repercussions on the native Indian and Eskimo populations will probably be far more catastrophic. No one really knows, however, just what the nature and extent of these spillover effects will be because there are no appropriate accounting mechanisms with which to measure them. In any case, there is only sparse knowledge of all the delicate processes that create northern ecosystems, and few southern Canadians really understand the values and particular life styles of the Indian and Eskimo populations. Changing priorities amongst the Canadian people, combined with the increasing political potency of the National Indian Brotherhood, have resulted in a widespread controversy about the total effects of this proposal. Consequently, the government has been forced into making many more studies, into devising legislation which makes environmental impact analysis mandatory and into considering more carefully whether the question of pipeline development should even be considered at all at this time. This searching hesitation contrasts sharply with the former Canadian Prime Minister John Diefenbaker's "northern dream' of the late '50s, when northern development was regarded as the key to Canadian economic growth.

(b) The Difficulty of Making Sound Social Choices

The example above serves to point out two features of the modern environmental movement. One is the rapidity with which national goals can change, and the other is the enormous difficulty of gathering all the necessary evidence. This latter problem is compounded by the fact that 'base line' data from which to compare various impacts are not generally available, so that most of the estimated outcomes of different courses of action can only be guessed. Traditional methods for making social choices and institutional arrangements for carrying these out are now under critical investigation (Haefele, 1971; MacNeill, 1971; Senate Committee on Interior and Insular Affairs, 1973). Both the Canadian and American governments have responded to these changing needs in a variety of interesting and quite different ways, and the comparison makes interesting reading, for it illustrates how two countries with different political and cultural heritages tackle similar problems.

(c) The Rise of the Administrative Regulatory Agency

The interrelatedness of environmental phenomena has inevitably led to increasing public intervention in all areas of resources decisions. The traditional division of the public and private sectors has all but disappeared as the activities of almost all major private corporations are regulated, monitored and generally scrutinized by government 'watchdog' agencies. Increasing public regulation of the private sector, while necessary, brings with it a number of drawbacks which are being carefully analysed by environmentalists (Sax, 1970a; Lucas, 1971; Crampton and Boyer, 1972). The rise of regulatory agencies, such as pollution control boards and various utility commissions, has emphasized the administrative state at the expense of the political state: politicians represent the various public interests and are accountable, administrators represent no interests, except possibly the clients they regulate, and are not accountable. Students of public administration are justly concerned that the 'administrative middleman' is becoming a broker for the public interest and that he is granted too much discretionary power with which to act. Consequently, environmentalists in particular are pressing for improved citizens' rights of legal redress to counterbalance the dominance of the administrator in the implementation of environmental decisions. The legal status of the citizen vis-à-vis environmental issues both in the United States and Canada will be compared.

(d) Transnational Environmental Damage and National Sovereignty

Environmental issues have a particularly interesting effect on international relations. This has already been exemplified in the debate over the Arctic Waters Pollution Prevention Act, but it is of more general significance when it is realized that the most vulnerable areas to environmental misuse are those which lie between national jurisdictions. The oceans, outer space, international fisheries, etc. are all subject to misuse even when complicated international treaties have been signed. One problem here is the question of sovereignty. Many *causes* of international environmental damage occur within national bounds, but their *effects* are international in the sense that they impinge upon other parties. International agreements to clean up after-effects never get at the root causes, but to eliminate the root causes would require international supervision over the internal affairs of a nation. The result is that symptoms are abated by cumbersome and costly regulation while the real culprits may remain uncensored. It was in this sense that in passing the Arctic Waters Pollution Prevention Act, the Canadian government was essentially extending its national sovereignty over international waters to assume total control and responsibility in the event of ecological damage.

(e) Environmental Impact and Resource Exploitation

The whole question of environmental misuse has recently become linked to the much larger questions of economic growth, international resource development and economic imperialism (see, for example, Weisberg, 1971; Hall, 1972; Daly, 1973). At the international level this was clearly demonstrated at the U.N. Conference on the Human Environment, held in Stockholm in June 1972, where many delegations of lesser developed nations expressed bitterness over the self-righteous concern of the 'have' nations regarding the protection of the global ecosystem when they felt that they had been exploited and contained in underdevelopment for over 100 years. This bitterness was clearly expressed in a preparatory conference to the main Stockholm meeting, where delegates from the developing nations were very antagonistic to the idea of global environmental standards, which, they felt, were merely a means of perpetuating underdevelopment by inhibiting industrialization and economic growth in the Third World (Hawkes, 1972).

Similar sentiments of economic subservience have been expressed by Canadians who feel that their resources are being exploited by the United States. The U.S., they feel, is exporting its polluting practices to Canada, importing unprocessed resources, then re-exporting finished goods, rich in added value and labour input, back to Canada. The net result: Canadians suffer environmental damage while the U.S. gains employment. Anti-American sentiments linked to Canadian resource exploitation by American-owned corporations rose to a crescendo following the announcement by President Nixon in August 1971 of a 10% surcharge on all imports to the U.S. Canada is America's biggest trading partner (Canadian exports to the U.S. in 1971 amounted to $10·71 billion or 13% of Canadian GNP), so the refusal by the U.S. to grant her special dispensation brought the angry remark from Prime Minister Trudeau that Canada cannot merely be regarded as a 'hewer of wood and drawer of water' to subsidize U.S. industrial growth (Head, 1972). All this had a remarkable political impact: in the Canadian general election of October 1972, both the New Democratic Party and the Progressive Conservative Party developed this theme repeatedly, and both parties made gains over the ruling Liberal Party. As as a result, the minority Liberal government was forced to strengthen its policy regarding takeovers of Canadian firms by non-Canadian interests, particularly in the resource and energy sectors (for examples, see Waterton, 1970; Laxer, 1971).

Contrary to certain expectations, the environmental movement in North America has not proved to be a short-lived fad. Indeed, the issue of ecological protection has cut deeply through the economic and political fabric of both the United States and Canada with quite profound repercussions on their internal policies and upon their international relations. Before analysing the international aspect of shifting public concern over the environment, it is desirable to compare how each country in turn has responded both legislatively and administratively to the particular kinds of pressures brought to bear by environmental concerns.

Recent Administrative and Legislative Responses in the United States

Although both Canada and the United States enjoy federalist political systems, there are important constitutional differences between the two countries regarding the respective roles of the federal government and the provinces or states concerning environmental issues. In the United States, the federal government has established considerable control over the states in matters such as air and water pollution, the dumping of solid wastes and improving the rights of citizens to seek adversary proceedings in the courts, largely because it has successfully pursued its constitutional powers over all matters pertaining to interstate affairs. Since most air, at one time or other, crosses state boundaries and most major rivers eventually join interstate water bodies, and generally many major federally financed and/or regulated developments (airports, power schemes, superports and the like) involve the national rather than a local public interest, the U.S. federal government has also been remarkably astute in pressing its control over many intrastate matters of significant environmental concern.

Administrative Innovations

At the administrative level, in December 1970 the U.S. coalesced many of its agencies responsible for regulating air, water and land pollution into one 'super agency' known as the Environmental Protection Agency (EPA). The EPA is an independent executive organization not responsible to any intermediate Departmental Secretary or Interagency Commission. Though this in itself is a remarkable administrative precedent in the matter of environmental management, probably of more significance is the fact that the EPA has not duplicated the activities of existing federal agencies: it simply plucked these agencies out of their respective Departments to forge a new administrative complex (Gillette, 1971). For example, the Federal Water Quality Administration, formerly in the Department of the Interior, and the National Air Pollution Control Administration, the Bureau of Water Hygiene and the solid waste management programme of the Department of Health, Education and Welfare were transferred to EPA. EPA also amalgamated most federal agencies responsible for establishing environmental protection standards for radiation, food additives and pesticides. It is also responsible for the bulk of research and monitoring operations for interstate air and water bodies and federal lands.

The recent concern over energy availability in the U.S. has given a new lease on life to an old refrain, the need for an even larger agency covering all resource matters. In 1971 President Nixon recognized this need and in his State of the Union Message he called for a 'sweeping reorganization' of the federal bureaucracy into four 'super departments', namely Natural Resources, Economic Affairs, Community Development and Human Resources (Senate Committee on Interior and Insular Affairs, 1973, pp. 823–832). His proposals were not enacted by Congress, but in 1973 he established the positions of four 'Councillors to the President' in these areas. President Nixon also resurrected the idea of a Department of Energy and Natural Resources but, in the meantime, spurred by the late 1973 'energy crisis', he established a National Energy office specifically designed as a co-ordinating agency, responsible for all aspects of energy supply, demand, research and policy-making.

While the EPA is essentially a comprehensive pollution-control monitoring and enforcement agency, broader questions of environmental policy such as phased resource development, co-ordination of federal activities, population size and distribution are handled by the President's Council on Environmental Quality (CEQ), established in January, 1970. The Council is an independent, cabinet-level body consisting of three members, responsible for gathering information, for reviewing and appraising federal policies and programmes for their potential environmental impacts, for documenting changes in the environment by means of an annual report to Congress on the state of the environment, and for developing new national policies, where appropriate, to safeguard existing environments of high quality and to improve environments of lower quality. Under the Environmental Quality Improvement Act of 1970 an Office of Environmental Quality was created serving as an administrative arm of the Council.

Prima facie, the Council is armed with considerable investigatory powers. For example, in its first year of operation it set up three advisory committees dealing with the impact of various kinds of taxes on the environment, the effect of law on environmental questions, and the production of a virtually non-polluting automobile. It has also investigated the environmental impact of urban and other land use policies at both the federal and the state levels, and is currently engaged in studies of the environmental impacts of energy activities, deep-water ports for supertankers, stream channelization and recycling (Council on Environmental Quality, 1972, pp. 151–153).

U.S. Environmental Legislation and Environmental Law

The CEQ was established through part of one of the most comprehensive and controversial pieces of environmental legislation that has ever passed through Congress. This is the National Environmental Policy Act (1969) (NEPA) which in many respects placed environmental quality as a national goal on a par with national economic growth and the redistribution of economic and social equality.

The legislative history of the Act makes clear that it was the primary purpose of Congress to establish a policy which enforced a substantive review of the effects on the *total* human environment of all federal actions. Section 102 (1) authorizes and directs that 'to the fullest extent possible' the policies, regulations and public laws of the United States shall be interpreted and administered in accordance with the Act, and Section 102 (2) (B) directs agencies to give 'appropriate consideration' to environmental values in all decisions. The result is that federal agencies, whose statutory mandates previously did not require any appraisal of possible environmental impacts, must now specifically take such effects into account before embarking on any course of action.

The Act is most famous for establishing the principle (in section 102) that all agencies of the federal government shall include in every recommendation or report on proposals for legislation and other major federal actions 'significantly affecting the quality of the human environment' a detailed statement (known as an environmental impact statement) documenting the following:

(i) the environmental impact of the proposed action;
(ii) any adverse environmental effects which cannot be avoided should the proposal be implemented;
(iii) alternatives to the proposed action;
(iv) the relationship between local short-term uses of man's environment and the maintenance and enhancement of long-term productivity;
(v) any irreversible and irretrievable commitments of resources which would be involved in the proposed action should it be implemented.

Taken literally, and a number of environmental lawyers are devoting much of their time to ensure that this is done, these requirements of NEPA represent an enormous undertaking (Green, 1972; Anderson, 1973). The problem, of course, is to ascertain what exactly constitutes an 'acceptable' environmental impact statement, and to ensure that the provisions of such statements are adhered to in implementing policy. The impact statements do provide a critical 'action forcing' provision to an important statement of principle introducing the Act, that it is the responsibility of the federal government to use all practicable means, consistent with other essential considerations of national policy, to assure for all Americans 'safe, healthful, productive and esthetically and culturally pleasing surroundings'. Furthermore, the Act specified that 'Congress recognizes that each person should enjoy a healthful environment and that each person has a responsibility to contribute to the preservation and enhancement of the environment'.

This is the nearest the U.S. Congress has come to recognizing the environmental rights of the citizen, to providing the citizen, at least in principle, with the right to fight for a clean and healthy environment where he feels that right is threatened, and with public safeguards embodied in publicly documented environmental impact statements as to possible detrimental but heretofore unmeasurable and largely unconsidered consequences of large-scale federal proposals.

NEPA really opened up a Pandora's box. By providing the citizen with legal rights to safeguard his environment, the Act has exposed to public scrutiny a veritable maze of federal agency operations. This in turn has given new impetus to the regular lobbying activities of well-established conservation groups such as the Wildlife Federation, the Conservation Foundation and the Audobon Society. But it has also led to the spawning of a number of eco-activist lobbies, both at the federal and the state level, whose prime motivation is to use the legal adversary process to the advantage of the environmental movement. While traditional conservation lobbies enjoyed a tax-deductible status for public contributions, these lobbies, being overtly political in their operations, do not benefit from the tax-free clause. Such groups include the Sierra Club, the Environmental Defence Fund, Zero Population Growth, Friends of the Earth, Natural Resources Defence Council, consumer advocate groups such as 'Nader's Raiders' (a group of lawyers working under Ralph Nader, a very active and committed consumer lawyer), plus various public interest lobbies within professional groups including lawyers, scientists and the medical profession.

The result of this legal activity has been a notable shift in the interpretation of the common law regarding the status of the citizen in environmental litigation. Under traditional tenets of common law the overriding principle of nuisance is 'sic utere tuo et alienum non laudes', which, loosely translated, means use your own property in such a manner as not to injure that of another. In other words, as long as one property owner, A, is not creating unreasonable damage on the legitimate rights of an adjacent property owner, B, and as long as B does not demand such safeguards as to limit A's reasonable use of his property, the amount of nuisance created may be tolerable. After all, the common law is essentially a reflection of society's values: property ownership is regarded by North American society as a desirable and legitimate ideal, therefore society does grant certain rights to individuals by virtue of owning property. These rights are not indiscriminate, however, and the law is very prepared to control the unreasonable use of property, but in so doing it seeks to ensure that there is a balance between the rights of individuals and the public interest at large.

With regard to modern environmental matters, there was some doubt as to whether an individual, or class of individuals, who could not prove that the injury he or they suffered was unusual or extraordinary to himself or themselves vis-à-vis the rest of society, had access to the courts. For example, in a situation where a municipality discharges effluent into a lake, thereby creating a hazard to swimming, it was not very clear whether any particular individual owning property around the lake could sue the municipality on grounds of nuisance. The traditional recourse in such cases was for a representative of the people, usually the Attorney General, to take action on behalf of all the people, though the likelihood of a public official seeking action against his administrative colleagues was understandably slim. This issue centres around the question of legal standing to sue. Since NEPA this issue has been clarified considerably in the United States, largely as a result of a number of test cases and by interpretations by judges and legal authorities.

One of the most outstanding authorities in the field of environmental legislation is Joseph Sax, a Michigan lawyer who has fought hard for the principle that the public right to a decent environment is an enforceable legal right, and that the legal adversary process is the best means for ensuring that the 'administrative middleman', as he calls the regulatory civil service, is kept honest (Sax, 1970a, pp. 158–174). The people, he feels, should not be supplicants to authority but claimants in their own right. Sax personally carried this philosophy through the Michigan legislature where, under the 1970 Michigan Environmental Protection Act, any citizen has the right to go to court when he feels his environmental rights are harmed, irrespective of the nature of the impact upon others (Sax and Connor, 1972). In effect, Sax is developing a common law of environmental quality: definitions such as 'pollution' and 'environmental quality' are purposely avoided in the Act so that the courts are given great flexibility to adjudicate on a broad range of problems.

A similar piece of legislation was brought before the U.S. Congress and a number of State legislatures including New York, Massachusetts, Colorado, Pennsylvania and Tennessee (see Crampton and Boyer, 1972). It is likely that a number of states will adopt some version of this legislative principle in the near future, for the legal status of the citizen to a healthy environment is at present very unclear and therefore courts are unable to prosecute in fairness to all interests involved.

Nevertheless, there is always the danger of going too far in the other direction. The Michigan legislation, for example, permits the court to penalize a polluter who is acting correctly under existing administrative guidelines, if it can prove that the pollution thus created is still exerting an undue burden on the environment. In effect, this is to deny the administrative agencies a primary role even in the areas of their special technical competence. The courts might then become the administrators, a situation which is not only beyond the bounds of constitutional administrative law, but which would place the courts in the position of arbitrating on issues well beyond their special area of expertise (Grad, 1972). The business of the court is not to decide upon environmental matters, but to ensure that the Congressional intent is properly interpreted and that responsible administrative agencies carry out well-established procedures (public hearings, publicly documented environmental impact statements, commissions of inquiry) in a manner that is neither arbitrary nor capricious, so that affected interests have an adequate opportunity to participate and make their views known.

In three years after the passage of NEPA, there were over 200 judicial decisions which clarified the question of standing by private litigants and which established a number of other principles favouring environmental protection (Council on Environmental Quality, 1972, pp. 221–269). In the 'Mineral King' case, the Sierra Club sought action against the U.S. Department of the Interior for lack of compliance with appropriate procedural requirements when granting a road and power line access through a national park and wilderness area known as the Mineral King Valley. (The Department is responsible for the preservation and maintenance of such areas.) The access was necessary to permit construction of a vast recreational complex. The case wound its way through lower courts to the Supreme Court which affirmed a Court of Appeals ruling that the Sierra Club did not have judicial standing since none of its members could demonstrate injury. However, the Court ruled essentially on a technicality. Judicial review of agency procedures must be sought by those who have a direct stake in the outcome. 'That goal would be undermined,' the Court commented, 'were we to construe the Administrative Procedures Act to authorise judicial review at the behest of organisations or individuals who seek to do no more than to vindicate their own value preferences through the judicial process' (Environmental Reporter, Cases, 1972, p. 2044). While the environmentalists have countered this decision by presenting a judicially acceptable plaintiff, Sax (1973) has criticized the Supreme Court decision for its failure to recognize the legal rights of those concerned about possible long-term ecological disruption when major landscape alteration is contemplated.

But the Mineral King opinion is important in that it established the principle that an injury to a non-economic interest such as the 'scenery, natural and historic objects and wildlife of Mineral King' is a significant case for private litigation in the general public interest. 'Aesthetic and environmental well-being, like economic well-being, are important ingredients of the quality of life in our society,' the Court noted, 'and the fact that particular environmental interests are shared by the many rather than the few does not make them less deserving of legal protection through the legal process' (Environmental Reporter, Cases, 1972, p. 2042). In other words, when agency actions are in violation of NEPA and threaten broadly defined environmental interests, appropriately aggrieved citizens are legally entitled to seek judicial redress.

The Court ruling is also significant in that it opened up some interesting aspects of judicial philosophy regarding the extraordinary nature of current environmental questions. Justice Douglas, a staunch ally of the environmentalists, commented on the fact that inanimate objects such as trees, flowers and rivers have no legal rights. He noted: 'contemporary public concern for protecting nature's ecological equilibrium should lead to the conferral of standing upon environmental objects to sue for their own preservation' (Environmental Reporter, Cases, 1972, p. 2044). Justice Douglas was in essence advocating that legal philosophy should incorporate Aldo Leopold's famous articulation of the land ethic (Leopold, 1949). Justice Blackman was most irritated by the ponderousness of the law respecting unprecedented matters. 'Must our law be so rigid and our procedural concepts so inflexible', he wrote, 'that we render ourselves helpless when the existing methods and traditional concepts do not quite fit and do not prove to be entirely adequate for new issues?' (Environmental Reporter, Cases, 1972, p. 2047).

U.S. courts have also supported and reinterpreted the important doctrine of public trust, dating back to Roman times, which places responsibilities upon states to ensure that certain common property resources such as land and lakes are held in trust for free and unimpeded use by the general public (Sax, 1970b). This means that publicly owned resources such as beaches and protected lands cannot be appropriated by private interests without due regard to proper compensation. In such cases, courts have ruled that the onus of proof lies not with the environmentalist to prove harm, but with the proposed developer to show that no injury will ensue, and that specific benefits to the public are being provided to compensate for any losses that the proposed development might create. The defendant must also prove that there is no reasonable alternative to his proposal before proceeding.

An example of the public trust doctrine in operation took place in 1971 when the City of Miami and Dade County, Florida, were stopped from building a modern jet airport which would have upset the critical water regime of the nearby Everglades National Park. While the environmentalists could not block the construction of a small jet training airport in the area, the Everglades dispute highlighted clearly the inability of the modern, complex administrative bureaucracy to assimilate all the relevant environmental information appropriate to a major decision of this kind. The legal injunction was not only a triumph for the environmentalists, it also paved the way for NEPA.

The opening up of citizens' rights to improved information and to the courts has its disadvantages, however. NEPA and the 1967 Freedom of Information Act which forces all government agencies to divulge all relevant but non-confidential information on request has deluged environmental interest groups with a colossal amount of information, much of it highly complex, some of it deliberately so (Wade, 1972; Gillette, 1972a, 1972b; White, 1972). As a result of NEPA the federal government spends at least $65 million and invests thousands of man hours of effort annually on environmental impact statements (Council on Environmental Quality, 1972, pp. 230–247). Approximately 3000 such statements have been produced in the past two years, ranging from a few pages to the nine volume report by the Department of the Interior on the proposed Trans-Alaskan Pipeline (to be discussed in detail later) (Council on Environmental Quality, 1972, p. 247). In addition, almost every major development, particularly in the area of energy production, is delayed by threatened or actual court proceedings (White, 1972). The rise of the legal adversary process has deep-ended the conflicts between environmental protection and the desire for growth which in turn has led to resentment on the part of many politicians and of certain sections of the public who still feel that money in their pockets and jobs on hand are worth 'minor tinkering with the environment'.

As a result, there have been signs of backlash against the legal and administrative advances made by the environmental movement in the United States. Many agencies are avoiding the legal requirements of NEPA, as they are sidestepping the Freedom of Information Act on the grounds that much of the information, regarding say industrial waste disposal, is confidential (Wade, 1972). Alternatively, the agencies are delaying the information, charging excessive fees for documenting all the evidence or simply sprinkling some confidential data among general information so as to classify the whole package as secret. There is also considerable dispute as to the precise role of the CEQ and the environmental impact statements. Many federal agencies regard the statements as a procedural hurdle which should in no way be regarded as a determinant in decision-making (Gillette, 1972a, 1972b). Many statements are delayed, appear late in the decision-making process, or simply are so technical as to be worthless for formal review.

But while there is nothing in the legislation which *forces* an agency to consider the impact statement as part of its evaluative process, a recent court ruling did contend that federal agencies were required to take into *serious consideration all environmental effects* and balance these against any advantages of various alternative proposals. Furthermore, the court ruled that this 'balancing' process be undertaken solely by the agency involved and not by a group of separate agencies acting in concert. This decision, known as the 'Calvert Cliffs' decision, was handed down by the U.S. Court of Appeals in the District of Columbia over a case where environmentalists demanded that the U.S. Atomic Energy Authority review all environmental aspects of a proposed nuclear power generation plant, not simply questions pertaining to possible radiation (Lewis, 1972; Environmental Reporter, Cases, 1972, pp. 1779–1793). This ruling was historic, for it not only clarified the degree of importance that should be attached to environmental impact statements, but it also established the important principle that all federal agencies re-examine their procedures to ensure that all aspects of environmental impact are fully documented and considered.

A subsequent decision by the same court broadened this interpretation even further. In June 1973, the Court ruled that major programmes of research and development must be subject to comprehensive environmental impact statements at an early stage in their progress, and that such statements should review the programme in the broadest possible terms and not simply be confined to specific demonstration plants. As a result the 2-billion-dollar Liquid Metal

Fast Breeder Reactor programme of the Atomic Energy Commission is now subject to a thorough environmental investigation. Further broadening of NEPA regulations may involve environmental analysis of rules and regulations, Congressional legislation and even general policy such as price and income controls (Green, 1972, p. 20).

Though the impact statement is not binding on an agency, it nevertheless is open to public scrutiny and the resulting judicial and political process. While there is no doubt that agencies will be forced to improve the quality of their impact statements over time, there is equally no doubt that the traditional lobbies behind agency operations (oil, minerals, timber, etc.) will seek to ensure that when the agencies 'balance out the equities' their interests are well represented. But at least environmental interests will get a fairer hearing. In passing, it should be noted that NEPA pertains to federal proposals only. In 1972, three states, California, Washington and Massachusetts, passed acts which made environmental impact statements mandatory for all state and private proposals deemed to have a 'significant' impact on the quality of the human environment. It remains to be seen whether the comprehensiveness of this legislation is manageable, for the ability to produce satisfactory environmental impact statements is not widely developed (Ditton and Goodale, 1972).

The role of the CEQ is equally uncertain. It is clear that it has no veto power over proposed actions which it feels may be harmful to the environment, nor is it sufficiently staffed to undertake the necessary independent research to give it teeth. In fact, most of its data and most of its reviews of impact statements are handled by the very agencies and industries it is supposed to regulate. The Council, it appears, is not the independent public watchdog over the environment that many hoped it would be: that would be too politically dangerous since too many well-entrenched vested interests are at stake. Its functions appear to be to facilitate the preparation and standardization of impact statements, rather than to evaluate these or even to comment upon their adequacy. Sax's conclusion, namely that the Council appears to be the arm of the administration rather than the independent advocate of the public interest (Sax, 1970a, p. 92), seems still to hold.

Nevertheless, on balance, the changes in administrative and legislative arrangements regarding environmental protection in the U.S. are most beneficial. Information as to the likely effects of proposed schemes and viable alternatives on both the human and the biophysical environment is thoroughly aired and the techniques of preparing even more detailed impact statements are improving. The decision process is much more open and subject to scrutiny by a variety of articulate and highly competent people advocating public interest. The implementive stage is also much more closely watched, as any administrative failure is subject to legal action. In addition, federal agencies have broadened their definition of 'costs and benefits' and have added to their staff a whole range of interdisciplinary personnel whose job it is to see that comprehensive reviews of all possible effects of a proposed course of action are prepared well in advance of any final decision being made. Furthermore, environmental action groups have received a shot in the arm and are constantly on the watch. All in all, this means that there is a far more comprehensive appraisal of the total environmental planning process from goal setting, anticipatory impact analysis, clarification of the most suitable options (including doing nothing at all or at least delaying until the need is more clearly defined) through decision-making and subsequent implementation and hindsight evaluation.

Recent Administrative and Legislative Responses in Canada

Constitutional Aspects of Environmental Management in Canada

The Canadian constitution as embodied in the British North America (BNA) Act does not grant quite the same freedom of action within the area of environmental management to the federal government vis-à-vis the provinces as is the case in the United States in federal–state relations. In fact the BNA act is considerably hazy with respect to the rights of resource ownership and legislative control (Gibson, 1970). All non-fugative resources, such as land, forests, minerals, power sources, are owned primarily by the provinces who have both legislative and regulatory powers over the allocation and use of such resources. In the case of the two federal territories of the North West Territories and the Yukon, these rights are vested directly with the federal government. Provincial ownership of such resources is not absolute, however, for the federal government does have control over the management and planning of federal property within the provinces such as harbours, roads, canals, etc. and over national parks. In those instances where the resource crosses provincial or national boundaries the federal government has significant regulatory jurisdiction, though the degree of this jurisdiction is by no means clear and is certainly tempered by provincial rights. Fugative resources such as air and water in their natural state are not legally ‘owned’ by either level of government, though both jurisdictions have certain rights to their use. Thus there are very few instances where the federal government may act unilaterally; these are essentially restricted to those special circumstances where the government judges that the national interest is at stake, where it can invoke Section 91 of the BNA which permits federal intervention in provincial matters for the purposes of ‘peace, order and good government’.

Generally speaking, this overlapping jurisdictional arrangement has not caused too many difficulties in the past, as for most resource development issues the areas of responsibility were fairly well defined. But with the rise of public concern over the unintended effects of large-scale resource development, the questions of transprovincial environmental impact and the national public interest have become more real. As a result jurisdictional tensions between the two levels of government have mounted, to the possible detriment of sound environmental protection.

To illustrate: in the area of water and air quality management, the provinces are responsible for establishing environmental quality standards and for issuing and enforcing effluent discharge permits. They are also responsible for ensuring that all activities on nearby lands (such as farm waste practices or logging operations) do not create environmental damage beyond what they regard as reasonable and acceptable to the public interest. But the federal government is responsible for coastal, transprovincial and transnational waters, for anadromous (migratory) fish and for navigation. Environmental quality standards in such areas or cases must be established on a co-operative federal–provincial basis, for neither government has sole jurisdiction. The result is that there are no nationally based and enforceable air or water quality standards in Canada. This situation compares unfavourably to the United States where the basis of federal air and water quality legislation over the past five years has been to establish and to enforce such standards.

The Canadian government plays an important role in monitoring, data collection and the dissemination of information regarding national air and water quality. It may even prescribe desirable national emission standards. In the case of water quality, it has not chosen to do so at present, emphasizing instead the comprehensive river basin management approach in co-operation with the provinces. The 1971 Canada Water Act provides for a comprehensive inventory of natural water quality and quantity, but the 'action' part of the legislation is primarily devoted to the establishment of regional river management authorities in areas with demonstrably critical pollution. To date no such authorities have been established.

Nevertheless, under recent amendments to the 1970 Federal Fisheries Act, the government has attempted to exert its authority over the standardization of specific effluent discharges. In debate over the amendments, the then Minister of the Environment, Mr. Davis, made it clear that through the fisheries legislation the federal government enjoyed total legislative authority over the provinces regarding the setting of standards. In place of the vague terminology as to what constituted 'waste' in the earlier Fisheries Act, the 1970 amendments established clearly defined regulations over the discharge of effluents from pulp and paper mills and chemical plants into streams frequented by fish (*Canada Gazette*, 1971, 1972b). Regulations over the discharge of effluents from other sources will be made later. One of the aims of the regulations, the Minister noted, was to 'avoid the unhappy development of what I might loosely refer to as pollution havens in one part of the country because legislation is not uniform from coast to coast. I am sure changing the Fisheries Act will help to eliminate these so called havens' (Davis, 1970, p. 7).

Nevertheless, the question of enforcement without provincial co-operation is still debatable. Mr. Davis emphasized the need for co-operation rather than policing though he did point out that the Minister may require a company to make changes in plant and/or processing as may be necessary to treat effluent properly. But the history of enforcement is not very favourable to the Ministry. A government legal commentator writes 'It is undoubtedly a valid assumption that there are a number of persons, including existing pulp and paper mills, who are presently discharging deleterious substances in waters frequented by fish contrary to the provisions of the Fisheries Act. Prosecutions for such violations are infrequent' (Morley, 1972a, pp. 16–17).

The 1971 Clean Air Act went one step further, proposing National Air Quality Objectives for five major pollutants: sulphur dioxide, particulate matter, carbon monoxide, photochemical oxidants and hydrocarbons. These were defined at three levels, desirable, acceptable and tolerable, but unless the provinces agree to these objectives, they have no practical significance. Air pollutants in many Canadian cities frequently exceed all three of these levels. Therefore, despite the Minister's boast that these objectives will be truly national in scope to prevent industry from seeking pollution havens (Davis, 1971, p. 2), these standards apply only to federal establishments within the provinces, and to non-federal works only where there is a 'significant' danger to human health. Nevertheless, the Act has established an important principle in that the federal government can intervene in the setting of air quality standards within provincial boundaries for all kinds of emissions, and this, in the Canadian context, is an important step forward in the area of federal–provincial co-operation in environmental matters (Morley, 1972b, pp. 16–20).

As a consequence of constitutionally defined provincial responsibilities, Canadian environmental management policy is far more regionalized in comparison with the U.S. situation, but the federal government is cautiously exerting even more pressure on the provinces to develop more uniform environmental management policies, despite the impediments of the hazy constitutional provisions. This is partly a result of increased public concern for better environmental protection, partly a consequence of widely differing provincial responsiveness to environmental matters, but also partly because the federal government can control provincial matters through its fiscal and grant-aiding programmes, along with its responsibilities regarding fisheries, navigation, commerce and national security. For example, under its powers to regulate commerce the federal government has restricted the lead content of petrol imported into or manufactured in Canada to 2·5 grams per gallon, thereby effectively removing from provincial jurisdiction the question of lead emission controls. The federal government shares in meeting the costs for all kinds of pollution control equipment both for the public sector and the private sector, and it can make its grants conditional on certain standards being met. However, this becomes a matter for delicate provincial–federal negotiation and, in the bargaining that ensues, environmental matters may take second place.

In the administrative sector, in 1970, the Canadian government amalgamated many of its environmental management agencies under a new Department of the Environment. In effect, the Department is responsible for federal activities purely in the renewable resources field and compared to the American EPA its powers are far more restricted, due to the need for provincial co-operation.

However, it has co-ordinated all federal activities relating to pollution in particular and environmental quality generally with respect to the natural (i.e. non-urban and non-human) environment. While the Department was originally designed to act as an environmental counterweight to the established interests of economic development and urban growth, it has not demonstrated its effectiveness in this regard. The emphasis has been more upon reaction to pressures and to public concern than to the careful initiation of comprehensive environmental policies.

Canadian Environmental Law

Canadian law has not developed to the same extent as American law in the area of environmental safeguards. To begin with, Canadian courts have only limited powers to interpret the law: the power of the law is vested primarily with the legislatures which in turn devolve this power to the administration by providing the administrators with very considerable powers of discretion. Consequently, the courts have not faced the question of legal standing as they have done in the United States, with the result that the individual citizen has no recourse to litigation in such cases where he cannot prove special injury. Even class actions, i.e. actions taken by a number of plaintiffs all equally injured on behalf of a class of people similarly affected, are not possible under Canadian law.

A classic example of the weakness of this omission took place in 1970 when about 350 fishermen in Placentia Bay, Newfoundland, were deprived of their livelihood on account of a massive fish kill caused by the discharge of toxic chemicals from a nearby phosphorous production plant. The Supreme Court of Newfoundland ruled that the fishermen did not have a particular proprietary right to the fish in the ocean and so could not prove special and discriminatory harm. The Court also ruled that the plaintiff could not pursue a personal action since the damage he suffered was common to a number of his fellow fishermen and therefore not peculiar to him. Nevertheless, the fish were dead, and many of the fishermen were forced on to welfare for lack of alternative employment opportunities. Deprived of their livelihood by a clear case of environmental mischief, the fishermen could seek no compensation from the offending company (Morley, 1972a, p. 20).

The rule of proprietary interest automatically excludes any individual who has no legal property rights from any nuisance litigation. This means that citizen action taken in the public interest against environmental damage is at present impossible in Canada. Only the Attorney General can take action in such cases: a fact of relatively rare occurrence in environmental matters in Canada.

Canadian litigation differs from the U.S. in a number of other respects. The citizen has no automatic right to information, nor can he insist upon a public hearing in the provinces with regard to any proposed development. The decision to call a public hearing over a proposed effluent discharge application is entirely at the discretion of the director for pollution control in the province concerned and he may set clear limits as to what matters are discussed (Lucas, 1969, pp. 79–85). There is no guarantee of environmental impact legislation nor any other form of anticipatory analysis of potential spillover effects within current provincial legislation, nor does the citizen have to be consulted when standards for air, water or land pollution are being established.

On the face of things, the Canadian citizen has no environmental rights and can only seek redress to environmental wrongs through the ballot box. However, the experience in the United States has not passed unnoticed by Canadian lawyers, administrators and politicians. A group of young Ontario lawyers have established a Canadian Environmental Law Research Fund with the aim of providing low-cost legal assistance to aggrieved citizens. Within the past two years there have been three workshops on the subject of Canadian Environmental Law where federal and provincial officials have met with lawyers and citizen-activitists to design new strategies (Morley, 1973).

Their efforts have not been in vain, for recently there has been some quite clear evidence that the Canadian government is willing slowly to restore some equity to the citizen. But because of the enormous difficulties of province by province co-operation in such matters, the federal government has developed new legislation only in the remaining area where it has sole authority, namely the Canadian Arctic. In 1971, it passed the Northern Inland Waters Act and amended the 1951 Territorial Lands Act to establish water management and lands use management areas respectively under each legislation (Figure 19.1). Once these management areas have been delineated, any proposal other than of a minor nature must be subject to an environmental impact analysis and may be subject to a public hearing should any citizen object who can show that his interests would be affected.

In terms of the current status of environmental legislation in the provinces, these two acts, for all their failings, represent a significant advance in protecting the environmental rights of the citizen. Already the Northern Inland Waters Act has had its effect; a public hearing held to investigate a proposal by the Northern Canada Power Commission, a federal body, showed that an inadequate case had been made for a proposed power scheme at Aishihilc Lake in Yukon Territory and that the impact statements were far from satisfactory. As a result the scheme has been delayed.

Nevertheless, even these two innovative pieces of legislation have serious pitfalls. To begin with, management areas have to be legally established: this decision depends upon the discretion of the local administrator. Unlike the American NEPA the nature of the environmental impact statement is not clearly outlined, therefore the precise quality of this information is also at the discretion of the local administrator. In addition, the environmental impact statement need not be made public: again the amount of publicity is discretionary. Canadians do not enjoy the equivalent of the U.S. Freedom of Information Act. Finally, minor damage, and this is not always clearly

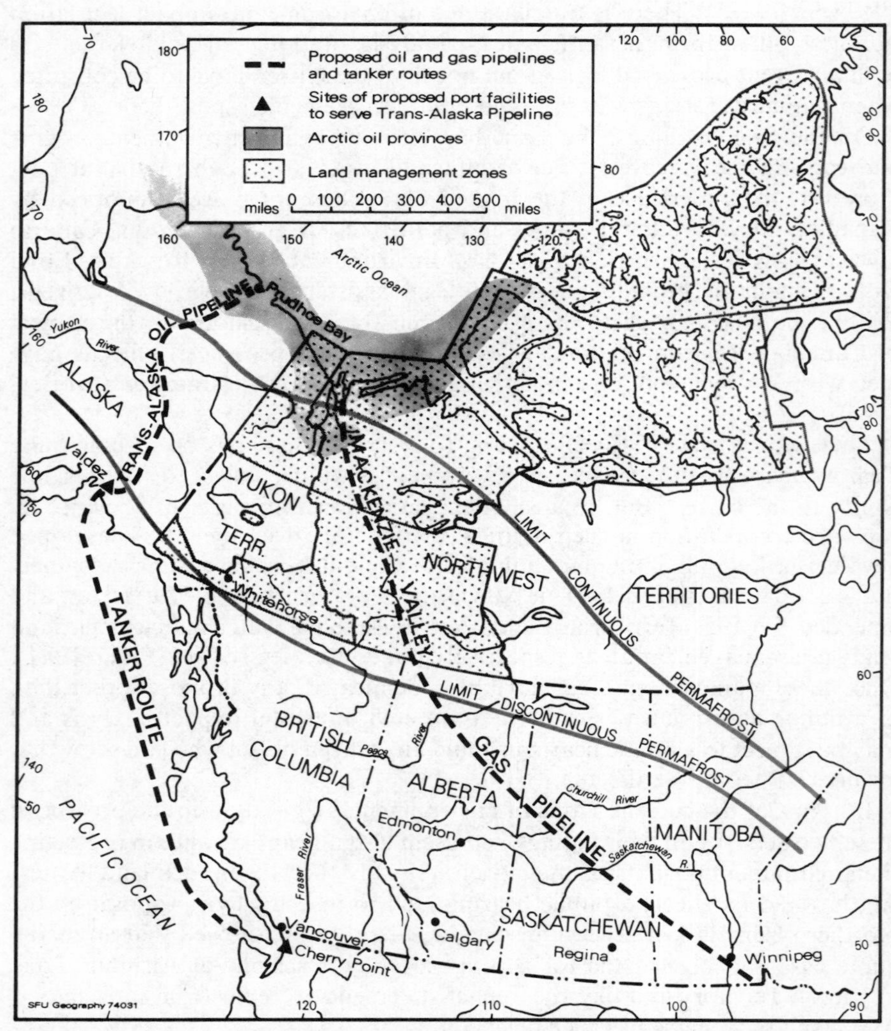

Figure 19.1. Location of the Trans-Alaskan Oil Pipeline, with its connecting tanker route to an oil refinery at Bellingham, Washington, and the proposed McKenzie Valley Gas Pipeline. Land Use Management Areas as designated under the amended Territorial Lands Act are also shown

defined, is not subject either to public hearing or to citizen litigation. In the Canadian south, this may be tolerable, but in the Canadian north, where the biological processes are extremely slow and little understood, environmental impact from even quite minor schemes may prove to be significant over a long period of time.

Canadian Initiatives in International Environmental Legislation

Despite the difficulties of managing environmental affairs within its bounds, Canada has been remarkably active in seeking to safeguard the environmental rights of third-party nations innocently vulnerable to the actions of others. Canada was one of the co-sponsors of the U.N. Conference on the Human Environment, and played a leading role throughout the conference. It was particularly instrumental in establishing two important legal principles as part of the U.N. Declaration on the Human Environment, principles upon which new and much needed international law can be based. These are:

> 'that states have ... the sovereign right to exploit their own resources and the responsibility to ensure that activities within their jurisdiction or control do not cause damage to the environment of other states or of areas beyond the limits of national jurisdiction'

and

> 'states shall cooperate to develop further the international law regarding liability and compensation for the victims of pollution and other environmental damage caused by activities within their jurisdiction or control of such states to areas beyond their jurisdiction'

Canada was also eager to include a principle that states should have access to appropriate information from other nations where they have reason to believe that such information is needed to avoid the risk of significant adverse effects to their national territories. This important principle was blocked by countries who wished to place considerations of national sovereignty above all else (Canadian Stockholm Delegation, 1972).

It is in the area of international maritime law to protect against marine pollution that Canada has been most aggressive. The debate over the Arctic Waters Pollution Prevention Act has already been referred to in detail, for this Act not only represents a landmark in the area of environmental protection but has also been a factor in determining the future shape of Canadian–American relations. Canada also passed an amendment to the Canada Shipping Act in 1971 which provided further safeguards against oil pollution by tankers plying Canadian waters. This Act was spurred on by an incident off the coast of Newfoundland on 4 February 1970, where a Liberian tanker called the 'Arrow' struck a submerged rock. About 16,000 gallons of Bunker C fuel were spilled, resulting in damage to over 160 miles of coastline. The subsequent mopping-up operation (which cost $4 million) was a remarkable achievement in its own way, for ecological damage was minimized and largely short-lived. But a lesson was learnt through this disaster. A Royal Commission commented: 'The difficulty of recovering damages which may very well be suffered in substantial amounts by private citizens and companies from oil spills on our coasts makes it imperative that some better arrangement for their protection should be worked out for the future' ('Arrow' Royal Commission, 1971, p. 109).

The amended Shipping Act attempted to meet these needs. It established rigorous regulations over structural safeguards against oil spill (for example, double hulls, self-contained chambers and automatic valve cut-off mechanisms), and insisted that a Canadian pilot be aboard all such vessels when they entered Canadian waters (*Canada Gazette*, 1971b). Also, of more significance to environmental protection, the Act established a Maritime Pollution Claims Fund which came into effect on 15 February 1972. The Fund imposes a levy of 15 cents per ton on all oil carried into maritime ports. At present, all oil consumed east of Ottawa River and Montreal in Canada is imported. The revenue from this Fund is used to meet the costs of pollution damage where specific liability cannot be proven or obtained. The costs of damage include not only clean-up costs, but also compensation to property owners and fishermen where injury can be proven.

The Maritime Pollution Claims Fund is an important innovation in the area of international environmental protection as it places the burden of use of the sea upon the user, and provides a revenue entirely to meet unsatisfied claims: i.e. damage which cannot be compensated for under existing law. This is a significant improvement on the existing situation where liability for marine pollution is so difficult to ascertain (Goldie, 1965; Jordan, 1970). Nevertheless, before the Fund is used to settle any claim, the claim must be proven in court and all legal remedies must be exhausted. The Fund itself is growing steadily, it is now about $15 million, but the government does not plan to reassess the levy until it has reached $25–30 million, probably in four or five year's time. Naturally, the oil companies were most unhappy about this levy—for a typical modern tanker the levy may be as high as $25,000 per shipment—and two companies, Continental Oil Company of New York and Gulf Oil, even threatened to abandon proposed oil refineries in New Brunswick and Nova Scotia respectively. However, the Canadian government has held firm in its intent, and while the price of oil products has recently risen by as much as 30 cents per gallon in eastern Canada, both the danger of an oil spill and the problems of compensation following an oil spill have been reduced (Fletcher, 1972).

In the area of international fishery regulation, Canada has also played a leading role and in so doing has considerably raised her international stature as a conservationist-minded nation. On the west coast, when there was a noted diminution in the stocks of Pacific salmon, Mr. Davis, then the Minister of the Environment, announced a new restrictive licensing policy for commercial salmon fishermen. Only the larger boats were to be granted licences, the smaller and uneconomic boats which tended to overfish being eliminated and compensated. This was a harsh and politically unpopular act, but it was necessary for Canada to show strength and determination in her negotiations with the International Pacific Salmon Commission to rationalize the overfishing of commercial salmon in international Pacific waters. As a result, the Commission has put pressure on Japanese and Russian fishing vessels to ensure that unconventional fish-catching methods are restricted and that fish catches in excess of certain quotas are reduced. In such circumstances enforcement of these agreements is a difficult problem, but the Canadian policy did set an example of national commitment which is hard for other nations to ignore. To finance an elaborate fish hatchery on the west coast, the federal government established a special licence fee for all non-Canadians bringing boats across the border to fish for salmon for sport. This fee, which is about $25 depending upon the size of the boat, is the first of its kind to be levied on recreationists in Canada, and netted about $300,000 in 1972 (Davis, 1972a, p. 1).

On the east coast Canada took even more drastic action. During the early 1960s the Danes discovered vast schools of Atlantic salmon feeding on herring which in turn was feeding on marine flora associated with the Labradorian ice floes. In the absence of proper international safeguards, the salmon were overfished and the schools reduced by 80% in a matter of five years: 1967–1971 (Davis, 1972b, p. 2). The Canadian government took the view that it had something of a proprietary interest in the salmon since the vast majority spawned in Canadian rivers. As a result, it pressed for international regulation of Atlantic salmon catches, and an interim ban on all salmon fishing in the high seas while it built up more hatcheries. As an act of singular exemplitude, the government made the most difficult decision to ban all fishing of Atlantic salmon by east-coast fishermen for a period of five years, 1971–1976. This was a particularly agonizing decision to make since many fishermen relied upon the Atlantic salmon as a major source of income. In provinces where fishing is a way of life and structural unemployment is widespread, this policy was particularly hard for the maritime fishermen. The government has compensated the fishermen (though how satisfactorily is open to question) by providing a fund of $2 million to buy out fish, boats and gear, taking the view that in the long run a healthy Atlantic fishery will be more profitable for the maritimers. One consequence of this bold act was to place great international pressure on the Danes, the West Germans and the Faroese, who initially were not willing to recognize restrictions on the catching of Atlantic salmon, yet who were members of the International Commission on the North Atlantic Fishery. Eventually, these nations agreed to the ban, under pressure by the Commission. Partly as a result of the Canadian action, Atlantic salmon stocks had recovered substantially by 1973.

Both of these examples point out the enormous difficulties in guaranteeing international co-operation in the management of communally owned resources, since in the absence of mutually enforced regulations the tendency is to overuse these resources, to the ultimate detriment of everybody. It is not easy to provide solutions to this problem, but certainly to set an example by tough responsible action must be regarded as an exemplary first step. In addition, Canada is on safer ground when she wishes to control international exploitation of her own resources. Hence Canadian pressure to amend international maritime law so that if a ship, while outside territorial waters, causes damage to a coastal state, that state should have legal recourse to commandeer the ship or any other ship of the same line as soon as it enters territorial waters if no liability is admitted or compensation paid. Largely due to the Canadian initiative, this fundamental principle was established in the U.N. Law of the Sea Conference held in Caracas, Venezuela, in 1974.

The Canadian experience in responding to the needs for environmental protection differs markedly from American reactions. The Canadian government, tied by constitutional limitations, has made only limited inroads into streamlining its management of internal environmental issues, and toward facilitating the role of the citizen in shaping his own surrounding. However, the Canadian contribution to the handling of international environmental matters has been most impressive, most positive and far more active than the American effort in the same areas. This dichotomy in the Canadian contribution emphasizes all the more clearly the influence of constitutional and other legal constraints in developing creative environmental management policies, but it should be pointed out that such constraints are the product of the collective perceptions of men who were responsible for the founding and early development of the nation.

A different culture, political structure and legal philosophy add up to quite significantly distinct responses to environmental questions within Canada and the U.S., despite the similarity of the problems facing the two nations. But the fascinating feature of all this is that the two responses are steadily merging: towards administrative concentration and co-operation, towards the appraisal of environmental spillover effects as an integrated part of government decision-making in all matters, and towards greater citizen participation and surveillance, both legally and politically. Both countries have much to learn from each other and much to offer the rest of the world.

International Implications of Major Resource Developments

Environmental quality has risen to the status of a primary national goal in both Canada and the United States. The upward shift has been slow and agonizing for both countries, but the significance of this shift comes most clearly to light in particular case studies where the classic dualities of conservation—preservation vs. development, resource exploitation for the many vs. the integrity of a life style for a few, growth vs. stabilization, corporate interest vs. the public interest—come sharply into conflict. In this section three case studies will be analysed in detail, the Trans-Alaskan oil pipeline, the proposed McKenzie Valley natural gas pipeline and the recently completed Great Lakes water quality agreement. Discussion of this last issue will also include a review of the changing role of the International Joint Commission, a rare example of a bilateral institution for resolving international environmental questions.

The Trans-Alaskan Pipeline System (TAPS)

Oil people have long believed that there was oil in the Arctic north. The problem was not so much whether it was there but when the necessary technology would be developed and when prices would rise sufficiently to make northern oil and gas exploitation a feasible proposition. These factors seem to have coalesced in the late '60s. In February 1968, the Atlantic Richfield Company (ARCO) confirmed the existence of vast reserves of oil beneath the north-eastern Alaskan shoreline around Prudhoe Bay. Estimates of these reserves vary enormously, but some 10 billion barrels of recoverable oil have already been proved, while up to 150 billion barrels of oil and over 26 trillion cubic feet of natural gas may be present (Norgaad, 1972, p. 85). This is by far the largest reserve of oil and gas remaining in the United States, but its attractiveness was further enhanced by an estimated drilling price of 20–50 cents per barrel (Norgaad, 1972, p. 86). (The world market price of oil, now highly unstable, is currently $11·95 per barrel.) These remarkable oil and gas finds were duplicated on the Canadian side of the Alaska–Yukon border and throughout the Arctic archipelago. Here, estimated crude oil reserves lie currently at 43 billion barrels and potentially recoverable natural gas reserves at 260 trillion cubic feet (National Energy Board, 1969, p. 52).

Of course, the problem facing the oil interests was how to transport the oil to the appropriate markets. Needless to say, at no time was there any question that this would be an impossible task. The attitudes toward frontier resource exploitation have hardly changed over the intervening century. The oil is there, therefore it must be developed: there is no question of leaving it alone. Mr. Milton Bradshaw, the President of ARCO, in testimony before the Senate Committee on Interior and Insular Affairs, typifies this viewpoint. 'The basic question at hand, I believe, is how seriously we need Alaskan oil. Do we need it badly enough to press for the earliest possible delivery at the lowest cost, both environmental and economic? Or can we afford to leave it where it is another two, five or ten years, until our needs are more critical?' (Bradshaw, 1973, pp. 2–3). The faith in the triumph of technology remains undiminished: the oil companies have invested over $100 million in research funds into assessing the most appropriate method of oil transportation, but while there is no proof that all the unintended effects have been minimized, there seems to be an implicit belief that following such vast expenditures of money surely nothing can really go wrong. Environmental impact was recognized but never regarded as a serious constraint. In many respects, then, the perceptions of oil companies' officials and consultants viewing the untapped reserves of the Arctic north must have been very reminiscent of the feelings of the early gold diggers along the northern Californian river valleys, 100 years earlier.

Three alternatives for oil transportation were considered. One was a tanker route through the North-west Passage to east-coast ports, but this was abandoned due to technical and climatic problems, and the very real possibility of Canadian intervention following the passage of the Arctic Waters Pollution Prevention Act. Canadian feelings on this matter were made clear during the trial voyage of the Humble Oil tanker 'Manhattan' in July 1969. The second alternative was a pipeline along the McKenzie Valley to the pipeheads in Alberta and thence to eastern markets. This route was favoured by the Canadian government, who were considering a 'corridor' of pipelines from the north-west Arctic to northern Alberta, carrying both oil and natural gas along with an access road and communications facilities. Indeed, Canada was so eager to benefit from joint American financing of this project (the gas pipeline alone is estimated to cost at least $5 billion) that the Canadian Minister of Energy, Mines and Resources made it clear that Canada was willing to supply additional quantities of oil to the U.S. for at least two years to compensate for the extra time needed to construct such a route (*Toronto Globe and Mail*, 29 August, 1972). Later, in June 1973, he dramatically reversed his views by announcing that in view of the current energy shortage all non-contracted Canadian exports of oil and gas to the U.S. would cease for a 'temporary' period (later explained as 'not less than two years'). Subsequently, the Canadian government has announced its intention to phase out oil exports altogether.

The oil companies, however, favoured what they thought would be the cheapest and fastest alternative, namely an 800-mile pipeline from Prudhoe Bay to the ice-free part of Valdez on the south Alaskan shore (Figure 19·1). From Valdez, large tankers were supposed to ship the oil to west-coast U.S. markets, though ARCO officials admitted that 'up to 25 per cent' of the oil might be exported to Japan or that some of the oil might be shipped to the Virgin Islands, where it would be refined and transported to the lucrative east-coast markets (Corrigan, 1971; Cichetti, 1972). (The reasons for these proposals are fully developed by Cichetti (1973), and basically involved the manipulation of existing U.S. shipping and tax laws to maximize oil corporation profits.) In preparation for the pipeline, ARCO constructed a $100 million refinery at Cherry Point in north-west Washington State, about 10 miles south of the Canadian border. About 10% of Alaskan oil was to be shipped through the Juan da Fuca straits lying between Vancouver Island and the Olympic peninsula and up the Puget Sound to Cherry Point. This area is highly prized for its recreational value.

It is difficult to isolate all the arguments in favour of the TAPS alternative. Certainly it would be cheaper than the Canadian route—$3 billion is the current estimate though this is probably seriously conservative—but the terrain would pose more difficulties as the Trans-Alaskan route involves the crossing of three mountain ranges and three major rivers, compared with only one mountain barrier and no major rivers for the Alaskan–Canadian route. In addition, it would involve oil loading and massive transportation by tanker through dangerous waters. Cichetti (1972) has completed a rigorous economic analysis and has found that the Canadian route would be superior economically as well as environmentally, adding a present value of $1 billion of oil tax revenue to Alaska and $5 billion in net profit to the oil companies before tax.

Probably the most significant reason favouring TAPS was that the oil companies were dealing with American governments and American states where the rules of conduct are already well established. ARCO has a 27·5% interest in the Prudhoe Bay discovery, and it is interesting to note that its president, Robert Anderson, is a National Republican Committee man from New Mexico whose affairs and those of ARCO were handled by Richard Nixon's law firm, who is reported to have contributed $44,000 to the 1968 Republican campaign and whose company sponsored the 1972 visit of President Nixon to Mainland China (Weisberg, 1971, p. 134; Ablett, 1972, p.4).

The cynic is also reminded of the fact that the Secretary of the Interior responsible for granting the oil leases in 1969 was Walter J. Hickel, a former governor of Alaska and an associate of the oil interests in the state. Alaska stands to gain much from TAPS. Revenue from land leases in the oil-producing areas is estimated to lie between $2·4 and $4·8 billion over the next 30 years: this is the equivalent of $10–15,000 per Alaskan resident in 1970 (Norgaard, 1972, p. 91). In addition, state royalties on the projected 2 million barrels of crude oil passing through the pipeline daily will amount to $1 million (Norgaard, 1972, p. 90). Clearly, Alaska sides with the oil companies in promoting TAPS and the west-coast tanker route. Another factor may be the Jones Act, a piece of U.S. legislation which states that goods moving between American ports must be shipped in American boats. The building of the 41 supertankers proposed for the TAPS route would be a lucrative prize for the currently unemployed workers in depressed U.S. shipbuilding yards. Finally, because oil corporations enjoy enormous tax benefits (the average corporate income tax in this sector is only 10% compared with an average U.S. corporate income tax of 50%), the quicker the pipeline was to be built the better their cash flow. Delay is much more expensive to oil companies than to society generally.

In September 1969, ARCO, British Petroleum and Humble Oil combined with other companies to form the Alyeska consortium. B.P. has since merged with Standard Oil of Ohio on the condition that it produces 600,000 barrels of oil per year by 1977; presumably it feels that Alaskan oil is the speediest source to meet this commitment (Cichetti, 1973, p. 11). The consortium paid $900 million to the State of Alaska for leases in the Prudhoe Bay area, and in early 1970 applied to the Department of the Interior, which is responsible for the federally owned Alaskan interior, for permission to grant the necessary leases to construct a road to parallel the proposed pipeline.

The Department of the Interior made a cursory 8-page analysis of the environmental impact of a proposed access routeway to the pipeline, but before it could issue its special land-use permits to allow construction of the access route, the whole pipeline proposal was stopped on 23 April 1970, by a preliminary injunction granted by Judge George Hart, Jr. of the U.S. District Court for the District of Columbia. Justice Hart found the Department of the Interior in apparent violation of Section 102 of NEPA (the environmental impact statement Section) and Section 28 of the Mineral Leasing Act of 1920 that permits only a 54-foot right of way on public land for pipelines of this type. (The Alyeska group were asking for 200 feet (Environmental Reporter, Cases, 1970, p. 1335).) Therefore the Department of the Interior was forced by the court order to embark on the monumental task of providing a comprehensive environmental impact statement for TAPS.

During this period the environmental arguments against TAPS became more publicly known. Alaska is vulnerable to severe earthquakes which no pipeline could withstand. The 'new' port of Valdez where a huge superport would be built is 4 miles north-west from its old site, which was demolished by the Alaska earthquake of 1964. The tanker route between Alaska and Washington is treacherous for navigation, due to the frequency of severe Pacific storms and the dangers of partially charted, rocky shoreline. The temperature of the oil moving through the pipeline will be 145° F, but there is no known insulating material which can keep the outside of the pipe below freezing temperatures. A warm pipe cannot be laid underground as is the current practice since it would melt the permafrost and create stresses which would eventually shatter the pipe. To place the pipeline above ground does not substantially lessen the danger of oil spills on account of the earthquake danger, and it is stil not clear whether the migratory Alaskan caribou will cross beneath the 800-mile barrier. Oil spills are considered likely. Carter (1969, p. 96) reports that in 1968 there were nearly 500 oil-pipeline leaks in the U.S., mostly involving pipes of small diameters. Already there are numerous examples of spillages from existing oil workings in Alaska, with drastic effects on the sensitive northern flora and fauna. Oil concentrations as low as 3 mg per litre are reportedly toxic to fish. Arctic plant cover is particularly vulnerable to oil spills, and when this is removed the permafrost layer is exposed to warm summer temperatures with resulting degradation and erosion of the soil (Department of the Interior, 1972; Legget and MacFarlane, 1972).

To be fair to the Alyeska consortium, the court injunction gave them time to embark on extensive ecological research on their own. Much of this required them to reconsider pipeline routeways and pipeline design. This is a significant point. One hundred years ago, the idea of extensive environmental surveys associated with 'frontier' resource development was unthinkable: in 1969 the oil companies did not regard the environmental question as particularly significant as there was so much money and political indebtedness at stake. By 1970 the consortium had invested over $100 million in sophisticated ecological investigation in order to provide satisfactory evidence for NEPA's requirements. The wild Alaskan interior was not to be dismissed as an 'unimportant wasteland', for though millions of Americans will never see this area for themselves, they have expressed a desire to safeguard its unusual ecology.

Canadians were also most unhappy with TAPS, but the Canadian government refused to censure the scheme publicly. In British Columbia there was far more alarm over the proposed tanker route, and the B.C. legislature went on record as opposing the whole scheme. In May 1973, B.C. Premier Barrett went to Washington with a much publicized proposal for a rail link from Alaska through B.C. to Cherry Point, but nothing came of this. Pressure was brought to bear on Ottawa by the B.C. government but to no avail. Nevertheless, David Anderson, a Liberal Member from Esquimault-Saanich, a coastal riding near Victoria, and chairman of a Parliamentary Sub-Committee on Environmental Pollution, embarked on a personal crusade against the tanker route. Anderson did all he could to embarrass the ruling Liberals. He demanded public hearings by his Committee on the west coast but was refused. Along with the Canadian Wildlife Federation and its affiliated organizations, including the B.C. Wildlife Federation, he took suit against the U.S. Department of the Interior and the major oil companies involved (Environmental Reporter, Cases, 1972, p. 1101).

This is a dramatic example of the way in which the law has shifted to help the environmentalists. Anderson and the B.C. Wildlife Federation claim a 'special interest' in protecting the wildlife off the B.C. shoreline and seek an injunction against tankers on behalf of all Canadians who share an equal interest. As B.C. Wildlife Federation members own property on the west coast, they have a proprietary interest in the issue. However, Anderson, a Canadian, is taking legal action against U.S. governmental officials and a U.S. oil company under Section 101(c) of NEPA which establishes the recognition by Congress that every person (interpreted as regardless of nationality) should enjoy a healthful environment, and that everyone has a responsibility to contribute to the preservation and enhancement of their environments. That legal action is an unprecedented example of international civil legal action by an individual. The case has not yet been adjudicated, for it will certainly go to the U.S. Supreme Court. It appears doubtful that the injunction will be upheld, but the very fact that such a case has risen through the U.S. courts is an indication of the revolution that is taking place in jurisprudence with respect of environmental matters.

The Canadian government responded to the growing concern from the west coast by commissioning a socio-economic impact study of the effects of oil spills off the B.C. coastline. The study, which was issued in May 1972, confirmed U.S. reports that oil spills would be 'inevitable', that Canada should provide emergency clean-up facilities immediately (costing about $5 million) and 'should pursue with the utmost vigour every possible avenue to ensure international protection of the environment from oil spills' (Paish, 1972). It also recommended a west-coast version of the Maritime Pollution Claims Fund with a suggested levy of 17 cents per ton established on an international basis, and concluded that identifiable losses from a tanker spill could be of the order of $5 million, though clean-up costs could rise as high as $24 million.

Only a month later, ironically during the very week when the U.N. Conference in Stockholm was dealing with resolutions about national responsibility for international pollution problems (4–11 June), an ARCO tanker unloading Venezuelan oil into its Cherry Point refinery accidentally discharged some oil. ARCO officials were quick to announce that only 10–20 barrels escaped, but

Figure 19.2. Protective booms surround a tanker carrying Venezuelan oil at the Cherry Beach refinery near Bellingham, Washington. Leakage from a connecting valve caused an oil spill of some 300 barrels. (Photograph courtesy of the *Vancouver Sun*)

the U.S. Coast Guard estimated the spillage at 300 barrels (*Vancouver Sun*, 6 June 1972, p. 1). This was enough to create a 14-mile-long oil slick that soiled 25 miles of resort beaches in north-western Washington and about 3 miles of southern B.C. beaches. Thus, while the Canadian Minister of the Environment was pressing for international safeguards for coastal states, 2000 young volunteers were mopping up oil, spilled from operations in the U.S., on a Canadian beach (Figures 19.2 and 19.3). Diplomatic pressure on the U.S. and in turn upon ARCO was intense, and in August 1972, admitting liability out of court, ARCO paid $40,000 to the Municipality of Surrey, B.C. to meet the costs of clean-up. The volunteers were never compensated. However, the damage was done: Anderson made the most of this occasion. 'I hate to say I told you so,' he commented to the press, 'but those idiots (ARCO) keep telling us what a great system they have and that there is no human error. Well, this shows what their assurances are worth' (*Vancouver Sun*, 6 June 1972, p. 6). Once again, technology is proven vulnerable to the fickle hands which operate it.

Nevertheless, the lesson was learnt. On 14 July, Mr. Davis met with Russell

Figure 19.3. Part of the oil spill from Cherry Beach was carried northwards on to the British Columbia coastline where many young volunteers spent days cleaning up the beach. (Photograph courtesy of the *Vancouver Sun*)

Train, the Chairman of CEQ, and the two men produced a communique which reaffirmed their support for the U.N. Declaration on the Human Environment, especially Principle 21, which says that states have the responsibility to ensure that activities within their jurisdiction do not cause damage to third parties, and Principle 22, which expects nations to co-operate in such situations. As a result, continuing discussions are to be held between the two countries regarding bilateral management of the coastal waters. Contingency plans for combating oil spills have now been signed for both the Atlantic and Pacific Coasts and the Great Lakes. The effectiveness of such plans remains to be demonstrated, but important matters of principle have been established.

On 15 August, 1972, Justice Hart of U.S. District Court ruled that the U.S. Department of the Interior had complied with all the regulations under NEPA and that he was not willing to be responsible for any further costly delays (Environmental Reporter, Cases, 1972, pp. 1467–1468). In addition, he was confident that the final decision on the matter rested with the Supreme Court. The Justice's decision reflects the legal interpretation of Section 102 of NEPA. The TAPS impact statement was nine volumes and over 4000 pages, and though it did not favour either the Alaskan or the Canadian routes, its findings were not binding on Secretary of the Interior Rogers Morton. Justice Hart was satisfied that a 'reasonable' impact statement had been prepared.

But the story is not over. On 15 August 1972, the U.S. District Court of Appeals argued that the pipeline could not be constructed as long as Section 28 of the Mineral Leasing Act (limiting construction right of way to 54 feet) was not rescinded by an act of Congress (Environmental Reporter, Cases, 1972, pp. 1977–2039). The court, however, did not rule on the appeal over the adequacy of the environmental impact statement, contested by environmentalist groups because of its insufficient analysis of the advantages of a Canadian alternative routeway.

By March 1973 there were four bills before Congress, two seeking to amend the offending sections of the Mineral Leasing Act and two requesting that Congress authorize the pipeline specifically, regardless of contrary provisions in other legislation (such as NEPA). By late 1973, a number of coalescing events—the instability in the Middle East, the Arab oil embargo, supply shortages caused by oil companies ill-equipped to adjust their huge inventories to a changing demand situation, wildly fluctuating oil prices and the threat of economic recession–all of this made Congress view the matter of U.S. energy independence with the highest priority. The Alyeska Consortium capitalized on this mood, stressing that Canadians could no longer be trusted to supply the needed amounts of oil and gas.

In a series of historic decisions both Houses of Congress acquiesced to the powerful oil lobby. During July the Senate passed a bill which not only amended the right of way restriction but also authorized the construction of the pipeline without the need for any further court actions under NEPA. The vote for this latter amendment was 49 to 49: (the tie was broken by Vice-President Agnew). The House also accepted this provision though it did not accept the full Senate bill. On 16 November 1973 the compromise bill was finally signed into law by President Nixon, who commented that he was sorry that environmental standards might have to be relaxed to increase production of domestic energy supplies. The overriding of NEPA is potentially a most significant move and it was followed by strenuous attempts to undermine its effectiveness further (Chan, 1973). But under questioning from a congressional committee Russell Peterson, the newly appointed chairman of the CEQ, noted: 'I do not think the delay [in the Alaska pipeline] merits the scuttling of the NEPA provision. It would be a major step backward to our country if we de-emphasise the importance of the accomplishments of this legislation.'

The Alaskan pipeline case highlights many of the conflicts associated with large-scale resource decisions outlined in this chapter. The issue was politically important yet an environmental test case: it brought to the fore the difficulty of making sound environmental judgements when traditional values of national security and economic growth are at stake. In accepting the pipeline, Congress extracted some important concessions from the President and the oil companies. First there is to be a $100-million liability fund for any offshore oil spill and a $50-million liability fund for any land oil spill. Second, an effort must be made to discuss the feasibility of alternative routes. Third, the Federal Trade Commission is endowed with far greater powers to investigate and to prosecute the anti-trust practices of large corporations. These important concessions may be of little consolation to Canadians who undoubtedly lost favour in the eyes of the Nixon Administration due to their own increasingly independent energy posture, but they serve to highlight that environmental decisions are essentially political decisions, which, in a democracy, reflect the public mood.

The Canadian Gas Pipeline Decision

The National Energy Board (NEB) of Canada is a regulatory body established in 1951 to oversee all aspects of Canadian energy development, transmission and pricing between provinces and between Canada and the U.S.A. In 1969, it issued a report dealing with national energy supply and demand in Canada in which it recognized that, unless there was massive exploration in the Canadian north, Canadian natural gas supplies would not be sufficient to meet expected Canadian needs and U.S. import demands (National Energy Board, 1969). Attention was therefore to be turned to the rich Canadian north, where a year previously some large deposits of oil and natural gas had been discovered. The Canadian Petroleum Association estimated the ultimate recoverable natural gas reserves at 725 trillion cubic feet, but warned that to meet all the predicted demands in 1990, at least one-third of these would have to be developed and piped to southern markets, including the United States (National Energy Board, 1969, p. 64).

In 1967, the Canadian government joined with a number of oil firms including British Petroleum, Humble and Standard Oil (Ohio) to form a consortium named Panarctic. The Canadian government holds 45% of the shares in the consortium, and Canadian interest controls the company. Within six months, Panarctic had made some remarkable finds, having made sufficient test drills to prove that large reserves of oil and gas lay under the Yukon–Alaska Border, and that extensive reserves of gas existed under Baffin Island and Ellsmere Island. Within another six months, over 564 million acres of oil rights had been leased by the Canadian government, of which Panarctic held rights to 55 million acres.

The Canadian government was placed in a dilemma. It was committed to opposing the Trans-Alaskan pipeline on environmental grounds, yet it was eager to promote a Canadian pipeline 'corridor' routeway consisting of oil, gas, roads, transmission lines and communications facilities, even though it had little knowledge of the possible environmental impact of such a scheme. At the same time, it was condemning the U.S. for failing to review the wider consequences of TAPS, and it was under considerable pressure in the House of Commons for promoting a continental energy policy without fully evaluating Canadian energy needs. The proposed gas pipeline, which would carry Alaskan as well as Canadian natural gas, would be 2400 miles long and either 48 or 56 inches in diameter. If 48 inches, the pipe would carry 3·6 billion cubic feet of gas to the U.S. daily (present Canadian demand is 3 billion cubic feet daily), but a 56 inch diameter pipe would have a capacity of 6 billion cubic feet (Wilson, 1972, p. 18): Figure 19.1. In view of the very serious current shortage of gas in the U.S., and with such a throughput, gross revenue from the natural gas would be $7 million daily, or a little over $80 per second, so it seems clear that the lucrative exports of natural gas to the gas-poor eastern American markets is a primary reason for the construction of the pipeline at this time. In addition, financing of the pipeline cannot be made by Canadians alone, for the total cost of the scheme is estimated at $5 billion, an investment that would seriously damage the Canadian balance of payments and banking stability should it be made unilaterally (*Toronto Globe and Mail*, 10 March 1973).

But the Canadian Minister then responsible for northern development, Mr. Cretien, was clearly in favour of the pipeline, even at the price of massive American investment. 'The Canadian north can be regarded as one of the underdeveloped regions of the world,' he told a symposium on petroleum economics and evaluation in Dallas in March 1971 (Cretien, 1971, p. 4), 'but to develop the great potential of the North, to overcome the great technical challenge of exploration, production and transportation, we are going to need help. we are going to need skills, we are going to need capital. At the present time,' he continued, 'some Canadians are asking themselves if there are ways of measuring their own investments in their own economy, or if the present controls are adequately protecting Canadian interests. The point I want to emphasise to you today is that foreign capital need not fear such questioning, for we will remain an open country seeking positive, not negative, answers to these questions' (Cretien, 1971, p. 8).

Mr. Cretien displays some of the attitudes of the early exploiters of resources in virgin territories. Oil and gas development is the key to economic take-off in the north and the 60,000 people of the area will benefit accordingly. He cannot conceive of leaving the Arctic alone. 'Those who say "freeze the Arctic",' he commented, 'are in effect advocating a welfare economy for northerners or forced migration to the south. They would be pushing northerners into a new Dark Age' (Cretien, 1971, p. 15). He recognizes that there may be an impact on the native people of the area but regards this as a challenge, not an impediment. 'These people are going through the difficult process of adapting themselves to the modern technologies and ways of life being introduced from southern Canada, while at the same time attempting to maintain their own identity and their own culture. This period of transition . . . is in many ways unavoidable' (Cretien, 1971, p. 3).

The main question for the government, it seems, is to ensure that northern peoples have access to the jobs that are available. However, even the most optimistic estimates of the labour requirements for the construction phase of the pipeline put manpower needs at only 6000, most of whom will be labourers, an occupation not highly regarded by native populations in the north (Indian Affairs and Northern Development, 1971). For the operation and maintenance phase, current estimates require only a mere 150 men, mainly administrative, so what will happen to the local population, some of whom will be flushed with high southern wages for a short period, is still largely unanswered.

The impact of southern values on the Indian and Eskimo cultures has already resulted in considerable social distress and tension, manifested in alcoholism, drug trafficking, disrupted families, poor housing and social services and low standards of health and education. The Indians have suffered most, particularly since the majority of Eskimos still live in areas as yet relatively untouched by southern development. The National Indian Brotherhood of Canada is gathering political strength to protect their interests and ensure their rights. They have been spurred on by the successful claim by the Alyeska Indians in Alaska who proved in court that they enjoy usufructuary rights to the Alaskan territories which were never settled when the U.S. bought this area from Russia in 1888. As a result, the U.S. government paid $500 million to the Alyeska Indians to settle the dispute, and to buy out their claims on the rich land leases over the Alaskan oil-fields. The Canadian Indians are pressing for a similar claim, but it is less certain that this will be successful, as Indian rights in Northern territories were never officially recognized via any formal treaties. Currently, the legal status of Indian lands is not clear as there have been no test cases applicable to the present situation (Bird, 1972, pp. 24–30). However, the Brotherhood was successful in March 1973 in obtaining a temporary injunction against any construction of the pipeline until this issue had been settled. Despite a federal government attempt to invoke a writ of prohibition on the restraining order, the Appelate Court of the North-west Territories upheld the injunction.

The case will probably be decided politically rather than legally. The National Brotherhood is growing in political stature and its case has been significantly helped by a November 1973 ruling by the Quebec Superior Court that Quebec Indians have a right to a particular way of life and to the land upon which their livelihood depends. This latter decision, made in connection with an enormous hydroelectric development in James Bay in northern Quebec, in effect recognized the usufructuary rights of the native Canadian Indian to an unspoiled environment. But the National Brotherhood is equally interested in self-determination and the right to administer its own affairs through more autonomous Territorial Councils, not through the paternalistic administration still determined by Ottawa.

The case for the Eskimo is in many ways even more significant, as in the very far north the Eskimo is still largely untouched by southern influence. The development of oil and gas leases in these far northern areas is already having its effects on these people, but the impact of pipeline construction will be far more dramatic. The Eskimo is not as politically articulate as the Indian and may suffer even more than the Indian, as few southerners realize how vastly different the two northern cultures are and tend to see 'the native problem' as embracing Indians and Eskimos alike. The Eskimo fears over the northern pipeline are most poignantly expressed in their own words:

> 'Eskimos were once known as the "laughing people" and it was true. We were the happiest people in the world. But this is no longer true. The people in the more developed, or should I say more civilized, communities such as Frobisher Bay, Cape Dorset and Resolute Bay in the Eastern Arctic, are no longer what they used to be. The Eskimo no longer laugh, they no longer smile. They are a confused people. And you can hardly blame them.
>
> In the earlier days, they didn't have such problems as alcoholism and drugs, having to worry about a monthly rent, pollution, politics, new diseases.
>
> In the more isolated communities such as Grise Fiord, Pond Inlet, Arctic Bay, Broughton Island, you will still find the last of the laughing people. But as progress and the so called civilized world reaches them, they too, I fear, will also disappear. I am not saying that we are going to try and stop development in the north, because we couldn't anyway, but we would like thing to slow down a little bit. Give the Eskimo a chance to prepare himself for things to come. That is all we ask' (Man and Resources Workshop, 1972).

The Canadian government is genuinely concerned over the possible environmental and social impact of the northern oil and gas finds and the proposed McKenzie Valley gas pipeline. Because all of this development takes place on federal lands, the government has taken over total control of the whole operation. In addition to its investment in Panarctic, it has issued a set of guidelines for companies to follow when preparing an application for the National Energy Board. While the guidelines (which cover such matters as environmental production, employment and training facilities for northerners, Canadian investment and a corridor concept) are not legally binding, they will be taken into consideration by the NEB when assessing the merits of any pipeline proposal. The government has also insisted upon comprehensive environmental impact statements for all proposals under the terms of the Northern Inland Act and the amended Territorial Lands Act. These statements are to show (a) the suitability of the route chosen compared with other routes, (b) an assessment of both environmental and social impacts on nearby settlements and nearby or existing transportation systems and (c) a comparison of the environmental and social impacts of the proposed routes with alternative routes. These statements are very reminiscent of the U.S. NEPA statements, but it will be recalled that neither the contents nor the implications of the statement need be made public.

In May 1972, the pipeline and oil companies joined into one huge consortium to pool their resources for meeting the enormous costs of research and preparation. Prior to this time there were two large consortia, The Gas Arctic System Study Group consisting of 14 companies, and The North West Project Study Group of 6 companies. By midsummer 1972, the Gas Arctic Group had spent over $14 million on its Prudhoe Bay test pipeline while the North West Group had committed more than $21 million on its Inuvik experimental pipeline (Leggett and MacFarlane, 1972, pp. 199–207). This research covered a variety of environmental factors, including the impact on wildlife, Arctic flora, permafrost studies and the development of special kinds of metals, insulating materials and pumping mechanisms. All of this research is co-ordinated by a quasi-governmental Canadian Arctic Resources Committee, and by an interagency Task Force on Northern Oil Development.

With over $30 million of private money and $15 of public money invested in the gas pipeline it would seem unlikely that the project will be stopped. Nevertheless the pipeline is not yet a certainty. On 30 November 1973 the *Toronto Globe and Mail* published a report by a respected brokeıage firm cautioning against further investment in the McKenzie Valley gas pipeline, since following the approval of the Alaskan oil pipeline American gas interests were promoting the construction of a parallel gas pipeline to transport American natural gas to Valdez and thence to American markets. But, undeterred, the Canadian government is holding extensive public hearings on the McKenzie valley gas pipeline throughout 1975, following the publication of an environmental impact statement by the Gas Arctic consortium and quasi-judicial public hearings on this statement conducted by the Canadian Arctic Resources Committee.

These public hearings which must be held throughout the nation will doubtless bring into focus the classical questions of this issue: political representation and political autonomy of the north; the preservation of native cultures; the public interest and the corporate interest; legal redress by Canadians over matters affecting their natural surroundings; the power of the administrative state vis-à-vis judicial and political forums; finally the whole issue of growth, energy needs and demands, Canadian control over its resources and the international implications of continental energy policies. The pipeline debate is only beginning: as it unfolds public awareness of what the environmental movement is all about in North America today will widen.

The International Joint Commission

Modern issues of environmental concern are testing a corpus of international law that is not fully suited to deal with the peculiar nature of such issues. Where two or more states share a common resource, such as a water body, there is incomplete agreement as to the apportionment of rights between the nations both as to the appropriate use of the resource and as to ensuring its aesthetic quality. Generally the equitable utilization doctrine is invoked (Jordan, 1970; Utton, 1973). This doctrine restricts territorial sovereignty to the extent that each basin state is entitled to a reasonable and equitable share within its own territory of the beneficial uses of the resource. The interpretation of 'reasonable and equitable' is a matter for international arbitration, but prior to the establishment of the International Court of Justice in The Hague in 1945, there were few formal international adjudicatory boards which were capable of handing down such judgments.

Canada and the United States have long recognized this dilemma, and sought to establish an international organization to harmonize the allocation of water flowing across the extensive boundary between the two nations. In 1909, the two countries passed the Boundary Waters Treaty which established the International Joint Commission, a six-man adjudicatory body which was responsible for investigating transboundary water problems and advising their respective governments as to the appropriate course of action. The Treaty was designed primarily to protect the levels and navigability of the Great Lakes and other boundary waters against unilateral diversion or obstruction, but in its preamble it states clearly that its purpose is to 'prevent disputes regarding the use of all boundary waters' and to make provision for the settlement of all questions involving the rights, obligations or interests of either nation along their common frontier. From its inception therefore, the International Joint Commission was created to consider any question of mutual interest in a reasonable and amicable manner. In defending the IJC in Congress in 1913 Mr. Elihu Root, former U.S. Secretary of State, who signed the treaty, noted: 'I do not think we shall ever see the time when this commission will not be needed to dispose of controversies along the boundary line in their inception, furnishing a machinery ready at hand for people to get relief and redress without going into the long process of diplomatic correspondence. I think it will have to continue as long as the ordinary courts of the country continue' (*Congressional Record*, 1913, p. 4172).

Two articles of the Treaty are of particular significance. Article 4 states that 'boundary waters, or waters flowing across the boundary, shall not be polluted on either side to the injury of health or property on the other.' This article reflects the doctrine of equitable apportionment, and recognizes that basin states can pollute as long as this pollution is not unreasonable and not injurious to the other party. Article 9 of the Treaty invests the Commission with broader powers of investigation into any matters of mutual concern to the two governments on the proviso that either one or both governments refer specifically to the IJC to investigate the matter. It should be noted that the IJC has no independent powers of inquiry: it was conceived as a reactive body which must await government-initiated references before it can work. This stipulation links to the thorny question of territorial sovereignty, for neither government wishes to see an independent international body publicly investigating faults for which each feels they may be ultimately responsible.

The Commission is involved in two kinds of work. The first kind deals with applications made to it by individuals or companies who wish to alter the quantity or quality of transboundary water in any way. Power schemes, navigable water levels, permits to abstract or discharge wastes all fall into this category. The burden is upon the applicant to furnish all the necessary information and data. Interested persons, who can show that they may be adversely affected by the effects of the proposed scheme, may intervene. This is followed by public hearings, held by the Commission, on both sides of the border, after which the Commission hands down its 'order' concerning the proposed scheme; this order is binding on both governments.

In the other case, where a problem of mutual concern to the two governments arises, the Commission is given a 'reference' to investigate the matter. The reference is usually quite specific in its circumscriptions and intentions, though it need not be restricted to matters concerning boundary waters or even closely related problems. The Commission may appoint one or a number of international technical boards to investigate various aspects of the problems. These boards prepare a preliminary report to the Commission which is published and followed by a series of public hearings. On the basis of all the evidence provided, the Commission publishes its final report and recommendations which, though not binding on the governments, are naturally considered with great care. Should the governments agree that action be taken, the Commission can establish one or more International Advisory Boards which oversee subsequent implementation and which may call further public hearings. As of December 1972, the Commission had received a total of 95 dockets, 58 applications and 37 references. The Commission is at present responsible for 13 boards of control, 7 boards of investigation and 8 surveillance boards (Canada–United States University Seminar, 1973, p. 29).

The Great Lakes Agreement

The problem of water pollution in the Great Lakes has always been of interest to the IJC. In August 1912, almost immediately after it was established, the IJC was sent a reference to examine the extent to which the waters of the Great Lakes had been polluted 'so as to be injurious to the public health and unfit for domestic or other uses' and to seek remedies. The Commission reported in 1918, using strong language. It found 'a situation along the frontier which is generally chaotic, everywhere perilous and in some cases disgraceful' (International Joint Commission, 1918, p. 31). Untreated sewage from municipalities, industries and ships was the major problem, a matter which was easily remedied by the construction of treatment plants and strict control over waste discharges. The Commission was willing to take the responsibility of executing this clean-up programme should the governments wish to widen its areas of jurisdiction, but the two governments hedged on this issue, requesting instead that the IJC furnish a reciprocal legislation or treaty for the two nations themselves to carry out the recommendations. The Commission submitted a draft treaty two years later, but both governments were reluctant to conclude an agreement, owing to overlapping state and provincial jurisdictional authority in the area of public health control. Another reason for the hesitation was that the Commission sought in the draft treaty to expand its area of control, particularly the setting of standards, the investigation of violations and the degree of adjudication (Bilder, 1972, pp. 479–483). The Commission was most firm on this point since it was convinced that the only way to halt transboundary pollution was to control it at its source: only if the Commission was clothed with such powers would this be possible. However, neither government wished to grant the Commission such powers of authority at that time, for neither wished to see fact-finding and the identification of injury taken entirely out of their hands.

The result of all this was that while minor remedies were sought, pollution in the Great Lakes worsened. The increasing population and industrial growth around the Great Lakes basin led to widespread discharge of municipal and industrial wastes, while the rapid increase of shipping as a result of the opening up of the St. Lawrence Seaway meant harbour dredging and the dumping of sludge in the lakes, and the discharge of wastes (particularly oil) from large ocean-going ships. A pollutant of particular significance is the modern synthetic household detergent which contains large amounts of phosphate salts as cleansing agents. Phosphates also enter the lakes via industrial wastes and drainage from heavily fertilized agricultural lands, though 50–70% of all phosphate in Lakes Erie and Ontario can be attributed directly to municipal effluent (International Joint Commission, 1971, p. 22). The inorganic phosphates are not broken down by conventional waste treatment processes and therefore they build up in the slow-moving lakes in vast quantities (in 1967, 62 million lb in Lake Erie, 21 million lb in Lake Ontario) (International Joint Commission, 1971, p. 49). Phosphates provide a major nutrient source for primary freshwater biota, especially algae and water weeds (*cladophora*). The result was that the lakes, especially Lakes Erie and Ontario, became prematurely eutrophic, the algae and *cladophora* decayed on the lake bottoms, reducing the dissolved oxygen content: this in turn led to the elimination of major commercial fish species such as blue pike, white fish and cisco (International Joint Commission, 1971, p. 56). As once pleasant and popular beaches became littered with decaying vegetation, dead fish and oil discharged from passing ships, so the general public joined the commercial fishermen in the chorus of protest.

These protests forced the governments to act, which they did finally in 1964 when the IJC was furnished with one of its most significant and broad-ranging pollution references to embark on one of the most extensive water quality investigations ever undertaken. Two advisory boards were established, one for Lake Erie and the other for Lake Ontario and the International Section of the St. Lawrence; twelve federal agencies and as many state and provincial agencies were directly involved, and ten semi-annual reports and three interim reports were prepared before the final reports of the two advisory boards appeared in 1969. Following this, the IJC held public hearings in eight U.S. and Canadian cities.

The final report of the Commission was published in January 1971. In this the Commission recognized that transboundary pollution of a serious nature was occurring, but wisely decided not to apportion the blame. Instead it recommended advanced but expensive combined industrial and municipal waste treatment for all communities around the two lakes, suggested guidelines for the control of navigation, discharges from shipping and the dumping of waste, and produced a list of appropriate and compatible water quality standards for specific parameters and pollutants. The total cost of the proposed clean-up was estimated to be $211 million for the Canadians and $1373 million for the U.S. side. The Commission's report was not limited to technical evaluations and recommendations. It dealt also with appropriate means of monitoring, implementing and enforcing its regulations and recomendations. It also suggested that comprehensive investigations be made of the two upper lakes, Huron and Superior, including their tributary land areas. Finally, in a further attempt to broaden its powers, it requested that the two governments authorize the Commission to establish a permanent regional bilateral board to assist it in carrying out these duties and to delegate to this board with necessary powers of implementation and enforcement.

These recommendations go as far, and in terms of implementation even further, than did those the Commission made in its submission for a draft treaty in 1920. But times and events have changed; on 10 June 1971, in a communique signed by Russell Train, Chairman of the Council on Environmental Quality, and Jack Davis, Canadian Minister of the Environment, the two governments accepted all these recommendations with the exception of those relating to implementation and enforcement. The communique noted that 'in designing the agreement, it was accepted that programs and other measures established to meet urgent problems would in no way affect the rights of each country to the use of its Great Lakes waters' (paragraph 17). National sovereignty was in no way to be undermined.

When President Nixon visited Prime Minister Trudeau in April 1972, the two leaders signed the historic Great Lakes Water Quality Control Agreement. Both leaders stressed the importance of the Agreement, though Mr. Trudeau underscored its ecological rather than its economic perspective. 'The Agreement,' he noted, 'does little for the respective economic problems, international security issues or common social ills which affect both nations. Nevertheless,' he added, 'it marks our recognition of the fragility of our planet and the delicacy of the biosphere on which all life is dependent. This agreement deals with the most vital of all issues–the process of life itself' (Trudeau, 1973, p. 1). Common water quality objectives for the lower lakes are now established: programmes and measures to obtain these objectives are agreed upon, and the Commission is charged, through its regional Great Lakes Water Quality Board, with the monitoring of the whole agreement. The Commission lost its battle for ultimate authority, however, for the vital question of enforcement of the agreement is left to the two national governments and the respective states and province. But the Commission, through its Great Lakes Board, can carry out independent investigations, which is a step beyond its reactive role, and it may publish the results of these investigations through the medium of public hearings.

The Canadian government was quick to implement the findings of the Commission, restricting the amount of phosphates in detergents to 20% in 1971, 5% in 1973, with an eventual ban on phosphates by 1974 (Press Release, Environment Canada, 15 April 1972; 5 October 1972). It also provided very low interest loans to all lakeside municipalities for advanced sewage treatment plants. In addition, the Ontario Water Resources Commission has acted with diligence in seeing that the appropriate water quality standards are met for the appropriate classes of effluent.

The thrust of the concern over the Great Lakes had always stemmed from Canada. Most of Canada's industry and much of her population is concentrated around the Great Lakes, but while Canada is entitled to share equally with the U.S. in the use of the lakes, Canadians rightly feel that most of the pollution stems from the American side of the border. Thus Canada has always regarded this question seriously (as it has with all matters of international pollution) and in acting quickly and decisively expects the U.S. to play her part and embark on stringent measures for pollution control. Indeed, the Canadians are so concerned over possible U.S. delay in implementing the Agreement that they have already embarked on an ambitious programme of advanced waste treatment, and have urged the IJC to follow up its comprehensive study with a specific statement of needs and priorities facing both countries in meeting the water quality objectives (Mutch, 1973).

The U.S., on the other hand, has never regarded the IJC very seriously. Virtually no attention has been paid to it by Congress, and only on one occasion was a Presidential nomination to the Commission ever questioned (Canada–United States University Seminar, 1973, pp. 27–47). Related to this, the U.S. has always considered the Great Lakes pollution problem as somewhat of a minor regional issue, and is reluctant to burden its lakeside municipalities with the heavy taxes required to furnish the necessary treatment plants, nor has it banned phosphates in detergents. Nevertheless, as one of its last acts, the 92nd Congress passed a Water Pollution Control Act which authorizes the President to spend up to $24 billion over the next 15 years to clean-up water pollution. A substantial part of this money was destined for the Great Lakes communities. The President, however, vetoed the Act on the grounds that 'its laudable intent is outweighed by its unconscionable price tag. I am prepared for the possibility that my action on this bill will be overriden,' he declared, 'but even if the Congress defaults its obligation to the taxpayers, I shall not default mine' (Strout, 1972). Congress unanimously rejected the President's veto but even so the President authorized only half the approved funds for the fiscal years 1973 and 1974. The Democrat-controlled Congress reacted swiftly. Senator Hubert Humphrey wrote to Mr. Nixon: 'Your proposed expansion of impoundment threatens the constitutionally mandated relationship between the executive and legislative branches of government' (Strout, 1972). Nevertheless, the American communities never received their full allocation of funds, although clean-up operations are now progressing. This particular battle incorporates much wider issues, such as the respective roles of the legislature and the administration, but it serves to show that for President Nixon, at least, economic stability and tax control took precedence over environmental improvement. These differences of reaction and initiative between the two countries are fascinating: not only do they touch on questions of national priorities, but they reflect more fundamental differences in the nature of the two political systems.

The Trail Smelter Case

The Commission's work has not been restricted to water problems alone. Under Article 9 of the Boundary Water Treaty it may investigate upon request 'any question or matter of difference between the two countries involving the rights, obligations or interests of either.' In 1928, the Commission was asked to report on an incidence of damage to property in the State of Washington caused by a smelter in Trail, B.C. In its report to the two governments, the Commision found that such damage had occurred, that compensation amounting to $350,000 should be made to the State of Washington for distribution to the property owners affected, and established clearly the principle 'that no state has the right to use or permit the use of its territory in such a manner as to cause injury . . . to the territory of another or the property or persons therein, when the case is of serious consequence and the injury is established by clear and convincing evidence' (International Joint Commission, 1941). Canada paid up. This ruling is widely regarded as a landmark in the quest for international liability for transboundary pollution, and points up the authority and impartiality of the Commission which could easily have voted along national lines in such a case.

The Skagit Valley Issue

Another and very recent reference of great significance for the Commission was the Skagit Valley Reference of 1971. The Skagit River runs south from B.C. into Washington through tremendously attractive country. It is widely regarded as a fine fly-fishing stream, but equally prized for its power potential. In 1920, Seattle City Light Power Company applied for a permit from the U.S. Federal Power Commission to dam the Skagit about eight miles south of the border. The first stage of the proposed dam was expected to back up water only a few hundred yards into B.C., and the B.C. government was unconcerned about the issue so long as property rights in the area were not violated. Seattle City

Figure 19.4. The Skagit Valley at the border of the U.S. and Canada. Back-up water from Ross Dam to the South (left of picture) runs into the Canadian portion of the valley, causing unsightly damage and environmental deterioration. Seattle City Light and Power would like to raise the existing Ross Dam by an additional 200 feet thereby resulting in more environmental damage to British Columbia. (Photograph courtesy of the *Vancouver Sun*)

Light bought up the only property affected and the dam was completed in 1940. The following year Seattle City Light made an application to the IJC to raise the height of its Ross Dam by an additional 130 feet: the effect of this would be to back up water over some 5000 acres of the lower Skagit Valley (Figure 19.4).

Since it was war time, the IJC held one rather perfunctory hearing in Seattle, with no hearings at all held in Canada. There was virtually no discussion of the environmental consequences of the flooding and few people even heard about the proposal. Environmental questions were simply not an issue at that time. According to one local newspaper 'experts testified that the land is of little value either for crops or timber. The Games Commissioner for British Columbia expressed the hope that the project will not affect fly fishing in the upper Skagit' (*Vancouver Sun*, 14 September 1941). The IJC granted Seattle City Light its application and recommended that it arrange the necessary terms of compensation with the B.C. government. In mid-1952, representatives of Seattle and B.C. had tentatively agreed on terms of compensation, a lump sum payment to B.C. of $255,508. While the City of Seattle officially authorized this in May 1953, the Province was not satisfied with the terms and demanded a renegotiation of the agreement. The IJC intervened in the dispute and in June 1954 an Interim Agreement was signed by both parties every year until 1966. In 1967, the B.C. government concluded a final agreement whereby the City agreed to pay an annual sum of $34,566·21 (U.S.) as compensation for flooding. The B.C. government then announced that it was establishing a provincial park around the affected area to capitalize on the existence of the man-made lake.

However, a number of conservation groups in B.C. did not accept this matter as final, since they regarded the untouched lower Skagit as an irreplaceable recreational asset. Pressure mounted when a member of the B.C. legislature, Mr. Brousson, turned the Skagit issue into a personal crusade.

Mr. Brousson and a local citizens' action group, Run Out Skagit Spoilers (ROSS), combined with a number of fish and game clubs in the Vancouver area to produce a vociferous protest. The issue soon lost its focus, for it became associated on the Canadian side with a growing dislike of American domination of Canadian interests. A popular song at the time put it more dramatically:

> 'Skagit Valley, Skagit Valley
> They would turn you to a mud pond
> To run the Coca Cola coolers in Seattle, U.S.A.'

Furthermore, the Canadian government was put under some pressure by its Vancouver M.P.s for failing to pursue its obligations under the Navigable Waters Act, which permits the federal government to intervene over the provinces where navigable waters are affected by obstructions or other impediments.

In May 1971, the two governments sent a reference to the IJC 'to investigate the environmental and ecological consequences in Canada of the raising of the Ross Lake . . ., taking into account relevant information about environmental and ecological consequences elsewhere on the Skagit River' (International Joint Commission, 1972, p. 10). There was much concern in Canada over these terms of reference as they appeared not to allow the Commission to investigate whether the dam should be raised at all. However, these suspicions were allayed by the Chairman of the U.S. Section of the Commission, who stated publicly that the Commission was free to interpret the terms of reference as it saw fit and that, indeed, it could recommend that the loss of environmental values outweighed the advantages of short-run power (*Vancouver Sun*, 5 June 1971, p. 1). After carrying out its own investigation and holding two public meetings the Commission issued its final report in May 1972. In effect, this report constituted an international environmental impact statement in which social values and ecological effects were considered equally with dollar and cents issues of power generation and recreation.

The Commission concluded that the Skagit Valley was 'an uncommon and non-restorable area and has important social values' (International Joint Commission, 1972, p. 30). It also felt that it had been given insufficient time to complete the necessary ecological studies to assess fully the consequences of the new water levels, but it did conclude that the effects would be 'far reaching and permanent.' As a result of hearing many submissions from both Americans and Canadians who were concerned about the proposal, it noted that 'there was a real willingness to sacrifice either money, electric power or general economic growth for the preservation of the present state of the Skagit. These submissions reflect both a social preference for retaining natural areas and a desire to keep social options open' (International Joint Commission, 1972, p. 27). These options it defined as:

—'the right of each citizen to have access to an uncommon natural environment of the quality of the Skagit Valley;' and

—the duty of our present generation to avoid making irrevocable decisions which would unnecessarily diminish our children's natural heritage' (International Joint Commission, 1972, p. 25).

These statements reflect clearly the depth of understanding held by the IJC over the Skagit issue. The question is not simply a question of international legal obligation, as many legal advisers to the Commission noted, but one pertaining to the preservation of a unique natural setting for the pleasure and benefit of both Canadian and U.S. citizens. The delay caused by the protest has largely undermined the argument put forward by Seattle City Light in seeking to raise the dam in 1942, that the provision of short-term peaking power was the primary objective.

The final decision on the Ross Dam still awaits a ruling by the Federal Power Commission. The Commission is not bound by the IJC findings but clearly must take these carefully into account. In November 1973 the Canadian government unanimously passed a resolution that the Skagit Valley be saved from further flooding and the EPC effectively killed the whole proposal by demanding a further two years of ecological investigations before it could reach a decision.

The International Joint Commission is a remarkable institution. It is the only international authority of its particular kind in the world and has been studied as much by students of international law as has the Tennessee Valley Authority by water resource researchers. Its success lies in its impartiality and benign arbitration. Neither the Canadian nor the U.S. commissioners are chosen for their political affiliations and one of its former chairmen complimented it for its lack of marked national bias (Heeny, 1967, pp. 7–8). It has earned a well-deserved reputation for credibility and expert deliberation. It has also been praised for its flexibility, its wide range of investigations and its sound judgment. It provides a unique forum for the amicable settlement of international environmental issues (though non-environmental issues may be discussed), and establishes a channel through its public hearings for the concerned citizen to be impartially and fairly heard. As a result, the IJC has, until recently, never been a source of political debate.

However, with changing priorities, the IJC is attempting to redefine its role. It would like to play a more active part in investigation and enforcement than it does at present: it would like to be freed of the limitations of a tight budget which is dependent upon the number of applications and references sent to it: it is eager to see an extension of its authority to plan programmes of research and investigation as it sees fit and to ensure that its recommendations are being fully carried out (Bilder, 1972, pp. 540–550). Naturally the state, provincial and national governments are loath to grant these wishes, for the IJC could well become a sovereign power in its own right. But the peculiar problems that environmental issues bring with them, third-party effects, arbitration of incommensurables, allocation of conflicting but legitimate demands, require new institutional arrangements. The fact that the IJC can 'interpret' its own terms of reference, and the fact that it did initiate further investigations and that it has established its own permanent monitoring agency as a result of the Great Lakes Agreement, plus the fact that its technical advisory committees are becoming increasingly staffed with personnel who are responsible for their agency operations and who are simply talking as interested individuals: all these facts imply that the IJC is moving steadily towards new and wider powers while still maintaining its advantages of adaptability, fairness, impartiality and amicable adjudication.

Appraisal

This chapter has deliberately focused upon the present and the future, since policy-making involving environmental matters makes innovations almost daily. The two most important conclusions we can reach about the impact of environmental issues on Canadian–American relations are: first, the *pervasiveness* which these issues now have on all aspects of national and international decision-making and second, the *rapidity* with which institutions have changed to accommodate the environmental question generally. As a result of NEPA and its state counterparts, all major policy issues and all significant proposals are subject to environmental scrutiny in the U.S., and it is only a matter of time before similar legislation is passed in Canada. Already the IJC is undertaking *de facto* environmental impact analysis. Environmentalism is not simply the protection of tundra or rare butterflies; it links to deep philosophical questions of national growth and progress, to political democracy and the integrity of national and regional sub-cultures, to economic and cultural independence and to the balancing of measurable and unmeasurable values in the short and long terms. The debates and negotiations which are continuing permit North American society to evaluate more fully its preferences for the kind of landscape, forged by a unique but ever-changing fusion of individual wishes, social preferences and ecological constraints, which constitutes the 'American Environment.'

References

Ablett, D., 1972. His Pipeline's a Bit Clogged. *Vancouver Sun*, 25 March 1972, p. 4.

Anderson, F. R., 1973. *NEPA in the Courts: A Legal Analysis*, Washington: Resources for the Future.

'Arrow' Royal Commission, 1971. Report of the Royal Commission on the Pollution of Canadian Waters by Oil and Formal Investigation into the Grounding of the Steam Tanker 'Arrow'. Ottawa: Information Canada.

Bradshaw, M. F., 1973. Testimony before the U.S. Senate Committee on Interior and Insular Affairs, 27 March 1973. Washington, D. C.: Senate Committee on Interior and Insular Affairs.

Bilder, R. B., 1972. Controlling Great Lakes Pollution: A Study in U.S.–Canadian Environmental Cooperation. *Michigan Law Review*, **70,** 469–556.

Bird, M. J., 1972. *An Analysis of Federal Interests Affected by the Proposed James Bay Hydro-Development*. Ottawa: Environment Canada.

Cahn, R., 1973. 'Energy crisis' vs. environment. *The Christian Science Monitor*, 21 November 1973, p. 21.

Canada Gazette, 1971a. Pulp and Paper Effluent Regulations, *Canada Gazette*, **105,** 1886–1892.

Canada Gazette, 1971b. Canada Shipping Act: Oil Pollution Prevention Regulations. *Canada Gazette*, **105,** Part II, 1723–1734.

Canada Gazette, 1972a. Regulations Under the Arctic Waters Pollution Prevention Act, 1970. *Canada Gazette*, **106,** Part II, 1033–1037.

Canada Gazette, 1972b. Chlor-Alkali Mercury Regulations. *Canada Gazette*, **106,** Part II, 436–440.

Canada–United States University Seminar, 1973. *A Proposal for Improving the Management of the Great Lakes of the United States and Canada*. Ithaca, N. Y.: Cornell University Water Resources and Marine Sciences Centre.

Canadian Stockholm Delegation, 1972. Report by the Canadian Delegation to the U.N. Conference on the Human Environment held in Stockholm, Sweden, 4–11 June 1972. Ottawa: Environment Canada.

Carter, L. J., 1969. North Slope Oil Rush. *Science*, **166,** 85–92.

Cicchetti, C. J., 1972. *Alaskan Oil: Alternative Routes and Markets*. Baltimore: Johns Hopkins Press.

Cicchetti, C. J., 1973. The Wrong Route. *Environment* **15,** 4–12.

Corrigan, R., 1971. Resources Report: Japan may get some Alaskan Oil; Foreign Flag Shipping of Exports is Likely. *The National Journal*, 31 July 1971.

Council on Environmental Quality, 1972. *Environmental Quality: The Third Annual Report of the President's Council on Environmental Quality*. Washington, D.C.: Government Printing Office.

Crampton, R. C. and Boyer, B. B., 1972. Citizen Suits in the Environmental Field: Peril or Promise? *Ecology Law Quarterly*, **2,** 407–436.

Cretien, J., 1971. Northern Development and Northern Pipelines. Ottawa: Press Release from Mr. Cretien's Office, 9 March 1971.

Daly, H. E., 1973. The Steady State Economy: Toward a Political Economy of Biophysical Equilibrium and Moral Growth. In *Toward a Steady State Economy* (Ed. Daly, H. E.), San Francisco: Freeman, pp. 149–174.

Davis, J., 1970. Speech of Hon. Jack Davis to the House of Commons on Amending the Fisheries Act, 20 April 1970. Ottawa: Press Release from Mr. Davis' Office.

Davis, J., 1971. Speech by Hon. Jack Davis on the Clean Air Bill, 19 February 1971. Ottawa: Press Release from Mr. Davis' Office.

Davis, J., 1972a. Our West Coast Fishery: A Progress Report. Ottawa: Press Release from Mr. Davis' Office, 17 March 1972.

Davis, J., 1972b. Saving Our Atlantic Salmon. Notes for a Meeting with Commercial Salmon Fishermen, Chatham, N. B., 27 April, 1972. Ottawa: Press Release from Mr. Davis' Office.

Department of the Interior, 1972. *Environmental Impact Statement, Trans Alaska Pipeline, Alaska*, Washington, D.C.: Government Printing Office, 9 vols.

Ditton, R. B. and Goodale, T. I., 1972. *Environmental Impact Analysis: Philosophy and Methods*, Madison: University of Wisconsin Sea Grant Programme.

Environmental Reporter, Cases, 1972. *An Annual Compilation of Environmental Cases*, Washington, D.C.: Bureau of National Affairs.

Fletcher, T., 1972. Nobody Knows if Oilspill Fund Will Pay. *Vancouver Sun*, 18 March 1971, p. 15.

Gibson, D., 1970. Constitutional Aspects of Environmental Management in Canada. A Report prepared for the Privy Council of Canada, Ottawa: Information Canada.

Gillette, R., 1971. Environmental Protection Agency: Chaos or 'Administrative Tension?' *Science*, **173,** 703–707.

Gillette, R., 1972a. National Environmental Policy Act: Signs of Backlash are Evident. *Science*, **176,** 30–33.

Gillette, R., 1972b. National Environmental Policy Act: How Well is it Working? *Science*, **176,** 146–150.

Goldie, L. F. E., 1965. Liability for Damage and the Progressive Development of International Law. *International and Comparative Law Quarterly*, **4,** 1189–1258.

Goldie, L. F. E., 1970. International Principles of Responsibility for Pollution. *Columbia Journal of Transnational Law*, **9,** 283–330.

Grad, F. P., 1972. Review of Sax's Book 'Defending the Environment'. *Natural Resources Journal*, **12,** 125–131.

Green, H. P., 1972. *The National Environmental Policy Act and the Courts*, Washington, D.C.: The Conservation Foundation.

Haefele, E. T., 1971. Environmental Quality as a Problem of Social Choice. In *Managing Environmental Quality* (Eds. Kneese, A. V. and Bower B. T.), Baltimore: Johns Hopkins Press, pp. 281–332.

Hall, G., 1972. *Ecology: Can we Survive under Capitalism?*, New York: International Publishing Co.

Hawkes, N., 1972. Human Environment Conference: Search for a Modus Vivendi. *Science*, **175,** 736–738.

Head, I., 1972. The Foreign Policy of New Canada. *Foreign Affairs*, **50,** 237–257.

Heeny, A. D. P., 1967. Along the Common Frontier: The International Joint Commission. Canadian Institute of International Affairs, 26.

Indian Affairs and Northern Development, 1971. Manpower Requirements for the Construction Phase of the Proposed Mackenzie Valley Natural Gas Transmission Line. Ottawa: Press Release from Mr. Cretien's Office, 27 May 1971.

International Joint Commission, 1918. *Final Report on the Pollution of Boundary Waters Reference*, Ottawa: International Joint Commission.

International Joint Commission, 1941. Opinion of the Ad Hoc Tribunal on the Trail Smelter Reference. *American Journal of International Law*, **35,** 684–716.

International Joint Commission, 1971. *Pollution of Lake Eire, Lake Ontario and the International Section of the St. Lawrence River*, Ottawa: Information Canada.

International Joint Commission, 1972. *Environmental and Ecological Consequences in Canada of Raising Ross Lake in the Skagit Valley to Elevation 1725*, Ottawa: International Joint Commission.

Jordan, F. J. E., 1970. Recent Developments in International Environmental Control. *McGill Law Journal*, **15,** 279–291.

Laxer, J., 1971. *The Energy Poker Game*, Toronto: New Press.

Legget, R. F. and MacFarlane, I. C., 1972. *Proceedings of the Canadian Northern Pipeline Research Conference*, Ottawa: National Research Council.

Leopold, A., 1949. *A Sand County Almanac*, New York: Oxford University Press.

Lewis, R., 1972. *The Nuclear Power Rebellion*, New York: Viking.

Lucas, A. R., 1969. Water Pollution Control Law in British Columbia. *U.B.C. Law Review*, **4,** 56–86.

Lucas, A. R., 1971. Legal Techniques for Pollution Control: The Role of the Public. *U.B.C. Law Review*, **6,** 167–191.

MacNeill, J. W., 1971. *Environmental Management: A Report to the Privy Council on Environmental Management in Canada*, Ottawa: Information Canada.

Man and Resources Workshop, 1972. Statement by Eskimo Delegation to the Man and Resources Workshop, Montebello, Quebec, 3 November 1972. Montreal: Canadian Council of Resource and Environmental Ministers.

Morley, C. G., 1972a. *Legal Developments in Canadian Water Management*, Ottawa: Environment Canada.

Morley, C. G., 1972b. *A Cooperative Approach to Pollution Problems in Canada*, Ottawa: Environment Canada.

Morley, C. G. (Ed.), 1973. *Canada's Environment: The Law on Trial*, Winnipeg, Monitoba: Agassiz Centre for Water Studies.

Mutch, D., 1973. Canadians Eye Lake Cleanup: U.S. Lags in Pact for Water Quality. *The Christian Science Monitor*, 29 May, 1973, p. 11.

National Energy Board, 1969. *Energy Supply and Demand in Canada and Export Demand for Canadian Energy*, Ottawa: National Energy Board and Information Canada.

Norgaard, R. B., 1972. Petroleum Development in Alaska: Prospects and Conflicts. *Natural Resources Journal*, **12,** 83–107.

Paish, H. and Associates, 1972. *West Coast Oil Threat in Perspective*, Ottawa: Environment Canada.

Sax, J., 1970a. *Defending the Environment: A Strategy for Citizen Action*, New York: Knopf.

Sax, J., 1970b. The Public Trust Doctrine in Natural Resource Law. *Michigan Law Review*, **68,** 471–492.

Sax, J., 1973. Standing to Sue: A Critical Review of the Mineral King Decision. *Natural Resources Journal*, **13,** 76–88.

Sax, J., and Connor, T., 1972. Michigan's Environmental Protection Act of 1970: A Progress Report. *Michigan Law Review*, **70,** 1003–1021.

Senate Committee on Interior and Insular Affairs, 1973. *Congress and the Nation's Environ-*

ment. U.S. Senate Committee on Interior and Insular Affairs, Washington, D.C.: Government Printing Office.

Strout, R. L., 1972. Epic Clash Shaping over Clean Water. *The Christian Science Monitor*, 2 December 1972, pp. 1, 3.

Trudeau, P. E., 1973. Notes for Remarks by the Prime Minister following Signing of Great Lakes Water Quality Control Agreement. Ottawa: Press Release, Office of the Prime Minister.

Utton, A. E., 1973. International Water Quality Laws. *Natural Resources Journal*, **13,** 382–314.

Wade, N., 1972. Freedom of Information: Officials Thwart Public Right to Know. *Science*, **175,** 498–502.

Waterton, D., 1970. *The Continental Waterboy: The Columbia River Controversy*, Toronto: Clark Irwin.

Weisberg, B., 1971. *Beyond Repair: The Ecology of Capitalism*, Boston: Beacon Press.

White, G. F., 1972. Environmental Impact Statements. *The Professional Geographer*, **24**, 302–309.

Index

All references to authors and their published works will be found in the Bibliography to each chapter.